IET ENERGY ENGINEERING SERIES 178

Integrated Motor Drives

Other volumes in this series:

Volume 1 **Power Circuit Breaker Theory and Design** C.H. Flurscheim (Editor)
Volume 4 **Industrial Microwave Heating** A.C. Metaxas and R.J. Meredith
Volume 7 **Insulators for High Voltages** J.S.T. Looms
Volume 8 **Variable Frequency AC Motor Drive Systems** D. Finney
Volume 10 **SF_6 Switchgear** H.M. Ryan and G.R. Jones
Volume 11 **Conduction and Induction Heating** E.J. Davies
Volume 13 **Statistical Techniques for High Voltage Engineering** W. Hauschild and W. Mosch
Volume 14 **Uninterruptible Power Supplies** J. Platts and J.D. St Aubyn (Editors)
Volume 15 **Digital Protection for Power Systems** A.T. Johns and S.K. Salman
Volume 16 **Electricity Economics and Planning** T.W. Berrie
Volume 18 **Vacuum Switchgear** A. Greenwood
Volume 19 **Electrical Safety: a guide to causes and prevention of hazards** J. Maxwell Adams
Volume 21 **Electricity Distribution Network Design, 2nd Edition** E. Lakervi and E.J. Holmes
Volume 22 **Artificial Intelligence Techniques in Power Systems** K. Warwick, A.O. Ekwue and R. Aggarwal (Editors)
Volume 24 **Power System Commissioning and Maintenance Practice** K. Harker
Volume 25 **Engineers' Handbook of Industrial Microwave Heating** R.J. Meredith
Volume 26 **Small Electric Motors** H. Moczala *et al.*
Volume 27 **AC-DC Power System Analysis** J. Arrillaga and B.C. Smith
Volume 29 **High Voltage Direct Current Transmission, 2nd Edition** J. Arrillaga
Volume 30 **Flexible AC Transmission Systems (FACTS)** Y-H. Song (Editor)
Volume 31 **Embedded generation** N. Jenkins *et al.*
Volume 32 **High Voltage Engineering and Testing, 2nd Edition** H.M. Ryan (Editor)
Volume 33 **Overvoltage Protection of Low-Voltage Systems, Revised Edition** P. Hasse
Volume 36 **Voltage Quality in Electrical Power Systems** J. Schlabbach *et al.*
Volume 37 **Electrical Steels for Rotating Machines** P. Beckley
Volume 38 **The Electric Car: Development and future of battery, hybrid and fuel-cell cars** M. Westbrook
Volume 39 **Power Systems Electromagnetic Transients Simulation** J. Arrillaga and N. Watson
Volume 40 **Advances in High Voltage Engineering** M. Haddad and D. Warne
Volume 41 **Electrical Operation of Electrostatic Precipitators** K. Parker
Volume 43 **Thermal Power Plant Simulation and Control** D. Flynn
Volume 44 **Economic Evaluation of Projects in the Electricity Supply Industry** H. Khatib
Volume 45 **Propulsion Systems for Hybrid Vehicles** J. Miller
Volume 46 **Distribution Switchgear** S. Stewart
Volume 47 **Protection of Electricity Distribution Networks, 2nd Edition** J. Gers and E. Holmes
Volume 48 **Wood Pole Overhead Lines** B. Wareing
Volume 49 **Electric Fuses, 3rd Edition** A. Wright and G. Newbery
Volume 50 **Wind Power Integration: Connection and system operational aspects** B. Fox *et al.*
Volume 51 **Short Circuit Currents** J. Schlabbach
Volume 52 **Nuclear Power** J. Wood
Volume 53 **Condition Assessment of High Voltage Insulation in Power System Equipment** R.E. James and Q. Su
Volume 55 **Local Energy: Distributed generation of heat and power** J. Wood
Volume 56 **Condition Monitoring of Rotating Electrical Machines** P. Tavner, L. Ran, J. Penman and H. Sedding
Volume 57 **The Control Techniques Drives and Controls Handbook, 2nd Edition** B. Drury
Volume 58 **Lightning Protection** V. Cooray (Editor)
Volume 59 **Ultracapacitor Applications** J.M. Miller
Volume 62 **Lightning Electromagnetics** V. Cooray

Volume 63 **Energy Storage for Power Systems, 2nd Edition** A. Ter-Gazarian
Volume 65 **Protection of Electricity Distribution Networks, 3rd Edition** J. Gers
Volume 66 **High Voltage Engineering Testing, 3rd Edition** H. Ryan (Editor)
Volume 67 **Multicore Simulation of Power System Transients** F.M. Uriate
Volume 68 **Distribution System Analysis and Automation** J. Gers
Volume 69 **The Lightening Flash, 2nd Edition** V. Cooray (Editor)
Volume 70 **Economic Evaluation of Projects in the Electricity Supply Industry, 3rd Edition** H. Khatib
Volume 72 **Control Circuits in Power Electronics: Practical issues in design and implementation** M. Castilla (Editor)
Volume 73 **Wide Area Monitoring, Protection and Control Systems: The enabler for Smarter Grids** A. Vaccaro and A. Zobaa (Editors)
Volume 74 **Power Electronic Converters and Systems: Frontiers and applications** A.M. Trzynadlowski (Editor)
Volume 75 **Power Distribution Automation** B. Das (Editor)
Volume 76 **Power System Stability: Modelling, analysis and control** A.A. Sallam and B. Om P. Malik
Volume 78 **Numerical Analysis of Power System Transients and Dynamics** A. Ametani (Editor)
Volume 79 **Vehicle-to-Grid: Linking electric vehicles to the smart grid** J. Lu and J. Hossain (Editors)
Volume 81 **Cyber-Physical-Social Systems and Constructs in Electric Power Engineering** S. Suryanarayanan, R. Roche and T.M. Hansen (Editors)
Volume 82 **Periodic Control of Power Electronic Converters** F. Blaabjerg, K. Zhou, D. Wang and Y. Yang
Volume 86 **Advances in Power System Modelling, Control and Stability Analysis** F. Milano (Editor)
Volume 87 **Cogeneration: Technologies, Optimisation and Implentation** C.A. Frangopoulos (Editor)
Volume 88 **Smarter Energy: from Smart Metering to the Smart Grid** H. Sun, N. Hatziargyriou, H.V. Poor, L. Carpanini and M.A. Sánchez Fornié (Editors)
Volume 89 **Hydrogen Production, Separation and Purification for Energy** A. Basile, F. Dalena, J. Tong and T.N. Veziroğlu (Editors)
Volume 90 **Clean Energy Microgrids** S. Obara and J. Morel (Editors)
Volume 91 **Fuzzy Logic Control in Energy Systems with Design Applications in Matlab/Simulink®** İ.H. Altaş
Volume 92 **Power Quality in Future Electrical Power Systems** A.F. Zobaa and S.H.E.A. Aleem (Editors)
Volume 93 **Cogeneration and District Energy Systems: Modelling, Analysis and Optimization** M.A. Rosen and S. Koohi-Fayegh
Volume 94 **Introduction to the Smart Grid: Concepts, technologies and evolution** S.K. Salman
Volume 95 **Communication, Control and Security Challenges for the Smart Grid** S.M. Muyeen and S. Rahman (Editors)
Volume 96 **Industrial Power Systems with Distributed and Embedded Generation** R. Belu
Volume 97 **Synchronized Phasor Measurements for Smart Grids** M.J.B. Reddy and D.K. Mohanta (Editors)
Volume 98 **Large Scale Grid Integration of Renewable Energy Sources A. Moreno-Munoz (Editor)**
Volume 100 **Modeling and Dynamic Behaviour of Hydropower Plants** N. Kishor and J. Fraile-Ardanuy (Editors)
Volume 101 **Methane and Hydrogen for Energy Storage** R. Carriveau and D.S-K. Ting
Volume 104 **Power Transformer Condition Monitoring and Diagnosis** A. Abu-Siada (Editor)
Volume 106 **Surface Passivation of Industrial Crystalline Silicon Solar Cells** J. John (Editor)
Volume 107 **Bifacial Photovoltaics: Technology, applications and economics** J. Libal and R. Kopecek (Editors)

Volume 108 **Fault Diagnosis of Induction Motors** J. Faiz, V. Ghorbanian and G. Joksimović
Volume 109 **Cooling of Rotating Electrical Machines:** Fundamentals, modelling, testing and design D. Staton, E. Chong, S. Pickering and A. Boglietti
Volume 110 **High Voltage Power Network Construction** K. Harker
Volume 111 **Energy Storage at Different Voltage Levels: Technology, integration, and market aspects** A.F. Zobaa, P.F. Ribeiro, S.H.A. Aleem and S.N. Afifi (Editors)
Volume 112 **Wireless Power Transfer: Theory, Technology and Application** N.Shinohara
Volume 114 **Lightning-Induced Effects in Electrical and Telecommunication Systems** Y. Baba and V.A. Rakov
Volume 115 **DC Distribution Systems and Microgrids** T. Dragičević, F. Blaabjerg and P. Wheeler
Volume 116 **Modelling and Simulation of HVDC Transmission** M. Han (Editor)
Volume 117 **Structural Control and Fault Detection of Wind Turbine Systems** H.R. Karimi
Volume 118 **Modelling and Simulation of Complex Power Systems** A. Monti and A. Benigni
Volume 119 **Thermal Power Plant Control and Instrumentation: The control of boilers and HRSGs, 2nd Edition** D. Lindsley, J. Grist and D. Parker
Volume 120 **Fault Diagnosis for Robust Inverter Power Drives** A. Ginart (Editor)
Volume 121 **Monitoring and Control using Synchrophasors in Power Systems with Renewables** I. Kamwa and C. Lu (Editors)
Volume 123 **Power Systems Electromagnetic Transients Simulation, 2nd Edition** N. Watson and J. Arrillaga
Volume 124 **Power Market Transformation** B. Murray
Volume 125 **Wind Energy Modeling and Simulation Volume 1: Atmosphere and plant** P.Veers (Editor)
Volume 126 **Diagnosis and Fault Tolerance of Electrical Machines, Power Electronics and Drives** A.J.M. Cardoso
Volume 128 **Characterization of Wide Bandgap Power Semiconductor Devices** F. Wang, Z. Zhang and E.A. Jones
Volume 129 **Renewable Energy from the Oceans: From wave, tidal and gradient systems to offshore wind and solar** D. Coiro and T. Sant (Editors)
Volume 130 **Wind and Solar Based Energy Systems for Communities** R. Carriveau and D.S-K. Ting (Editors)
Volume 131 **Metaheuristic Optimization in Power Engineering** J. Radosavljević
Volume 132 **Power Line Communication Systems for Smart Grids** I.R.S. Casella and A. Anpalagan
Volume 134 **Hydrogen Passivation and Laser Doping for Silicon Solar Cells** B. Hallam and C. Chan (Editors)
Volume 139 **Variability, Scalability and Stability of Microgrids** S.M. Muyeen, S. M. Islam and F. Blaabjerg (Editors)
Volume 142 **Wind Turbine System Design: Volume 1: Nacelles, drive trains and verification** J. Wenske (Editor)
Volume 143 **Medium Voltage DC System Architectures** B. Grainger and R.D. Doncker (Editors)
Volume 145 **Condition Monitoring of Rotating Electrical Machines** P. Tavner, L. Ran, C. Crabtree
Volume 146 **Energy Storage for Power Systems, 3rd Edition** A.G. Ter-Gazarian
Volume 147 **Distribution Systems Analysis and Automation 2nd Edition** J. Gers
Volume 151 **SiC Power Module Design: Performance, robustness and reliability** A. Castellazzi and A. Irace (Editors)
Volume 152 **Power Electronic Devices: Applications, failure mechanisms and reliability** F. Iannuzzo (Editor)
Volume 153 **Signal Processing for Fault Detection and Diagnosis in Electric Machines and Systems** M. Benbouzid (Editor)
Volume 155 **Energy Generation and Efficiency Technologies for Green Residential Buildings** D. Ting and R. Carriveau (Editors)
Volume 156 **Lithium-ion Batteries Enabled by Silicon Anodes** C. Ban and K. Xu (Editors)

Volume 157 **Electrical Steels, 2 Volumes** A. Moses, K. Jenkins, Philip Anderson and H. Stanbury
Volume 158 **Advanced Dielectric Materials for Electrostatic Capacitors** Q. Li (Editor)
Volume 159 **Transforming the Grid Towards Fully Renewable Energy** O. Probst, S. Castellanos and R. Palacios (Editors)
Volume 160 **Microgrids for Rural Areas: Research and case studies** R.K. Chauhan, K. Chauhan and S.N. Singh (Editors)
Volume 161 **Artificial Intelligence for Smarter Power Systems: Fuzzy Logic and Neural Networks** M.G. Simoes
Volume 165 **Digital Protection for Power Systems 2nd Edition** Salman K. Salman
Volume 166 **Advanced Characterization of Thin Film Solar Cells** N. Haegel and M. Al-Jassim (Editors)
Volume 167 **Power Grids with Renewable Energy** Storage, integration and digitalization A.A. Sallam and B. OM P. Malik
Volume 169 **Small Wind and Hydrokinetic Turbines** P. Clausen, J. Whale and D. Wood (Editors)
Volume 170 **Reliability of Power Electronics Converters for Solar Photovoltaic Applications** F. Blaabjerg, A.l Haque, H. Wang, Z. Abdin Jaffery and Y. Yang (Editors)
Volume 171 **Utility-scale Wind Turbines and Wind Farms** A. Vasel-Be-Hagh and D.S.-K. Ting
Volume 172 **Lighting interaction with Power Systems, 2 volumes** A. Piantini (Editor)
Volume 174 **Silicon Solar Cell Metallization and Module Technology** T. Dullweber (Editor)
Volume 175 **n-Type Crystalline Silicon Photovoltaics: Technology, applications and economics** D. Munoz and R. Kopecek (Editors)
Volume 180 **Protection of Electricity Distribution Networks, 4th Edition** J. Gers and E. Holmes
Volume 182 **Surge Protection for Low Voltage Systems** A. Rousseau (Editor)
Volume 184 **Compressed Air Energy Storage: Types, systems and applications** D. Ting and J. Stagner
Volume 186 **Synchronous Reluctance Machines: Analysis, optimization and applications** N. Bianchi, C. Babetto and G. Bacco
Volume 191 **Electric Fuses: Fundamentals and new applications 4th Edition** N. Nurse, A. Wright and P.G. Newbery
Volume 193 **Overhead Electric Power Lines: Theory and practice** S. Chattopadhyay and A. Das
Volume 194 **Offshore Wind Power Reliability, availability and maintenance, 2nd edition** P. Tavner
Volume 196 **Cyber Security for Microgrids** S. Sahoo, F. Blaajberg and T. Dragicevic
Volume 198 **Battery Management Systems and Inductive Balancing** A. Van den Bossche and A. Farzan Moghaddam
Volume 199 **Model Predictive Control for Microgrids: From power electronic converters to energy management** J. Hu, J.M. Guerrero and S. Islam
Volume 204 **Electromagnetic Transients in Large HV Cable Networks: Modeling and calculations** Ametani, Xue, Ohno and Khalilnezhad
Volume 208 **Nanogrids and Picogrids and their Integration with Electric Vehicles** S. Chattopadhyay
Volume 210 **Superconducting Magnetic Energy Storage in Power Grids** M.H. Ali
Volume 211 **Blockchain Technology for Smart Grids: Implementation, management and security** H.L. Gururaj, K.V. Ravi, F. Flammini, H. Lin, B. Goutham, K.B.R. Sunil and C. Sivapragash
Volume 212 **Battery State Estimation: Methods and Models** S. Wang
Volume 215 **Industrial Demand Response: Methods, best practices, case studies, and applications** H.H. Alhelou, A. Moreno-Muñoz and P. Siano (Editors)
Volume 213 **Wide Area Monitoring of Interconnected Power Systems 2nd Edition** A.R. Messina
Volume 217 **Advances in Power System Modelling, Control and Stability Analysis 2nd Edition** F. Milano (Editor)

Volume 225 **Fusion-Fission Hybrid Nuclear Reactors: For enhanced nuclear fuel utilization and radioactive waste reduction** W.M. Stacey

Volume 238 **AI for Status Monitoring of Utility Scale Batteries** Shunli Wang, Kailong Liu, Yujie Wang, Daniel-Ioan Stroe, Carlos Fernandez and Josep M. Guerrero

Volume 905 **Power system protection, 4 volumes**

Integrated Motor Drives

Edited by
Xu Deng and Barrie Mecrow

The Institution of Engineering and Technology

Published by The Institution of Engineering and Technology, London, United Kingdom

The Institution of Engineering and Technology is registered as a Charity in England & Wales (no. 211014) and Scotland (no. SC038698).

First published 2022

The Institution of Engineering and Technology
Futures Place
Kings Way, Stevenage
Hertfordshire SG1 2UA, United Kingdom

www.theiet.org

British Library Cataloguing in Publication Data
A catalogue record for this product is available from the British Library

ISBN 978-1-83953-186-6 (hardback)
ISBN 978-1-83953-187-3 (PDF)

Typeset in India by MPS Limited

Cover image photographed by the editors

Contents

About the editors

Xu Deng is a Newcastle University academic track (NUAcT) fellow in electrical power with the Electrical Power Research Group at Newcastle University, UK. Her main research interests are integrated motor drives (IMDs) and advanced control techniques for electric motor drive systems.

Barrie Mecrow is a professor with the Electrical Power Research Group at Newcastle University, UK. Prior, he was a turbo-generator design engineer at NEI Parsons, UK. His research interests include fault-tolerant drives, high-performance permanent magnet machines, and novel switched reluctance drives. He has numerous industrial links, including being the advisor to Rolls-Royce on electrical machines and technical advisors to both Dyson and Protean Electric.

Chapter 1

Introduction

Xu Deng[1], Barrie Mecrow[1], Tom Cox[2] and Simon Lambert[1]

1.1 Concept of integrated motor drives

The integration of electrical components is receiving increased attention in the field of power electronics and electrical drives research. The integrated motor drive (IMD) emerged as a structural integration of an electric motor with a motor drive as a single unit, with at least 20% less volume and 30–40% lower overall system costs of installation and manufacturing. The elimination of expensive components such as shielded connection cables, a separate housing for the power converters, a centralised controller cabinet, and high-voltage and current bus bars is the primary driving force in increasing the power density and lowering the cost. It also leads to an improved electromagnetic interference (EMI) behaviour due to the direct connection of the motor to the drive without additional cables.

Depending on the innovative integration concepts, integrated systems are commonly smaller, have greater structural flexibility and enhanced thermal management, while ensuring that system integration and modularity are not mutually exclusive. When IMDs are employed as a direct replacement of less efficient, fixed speed, direct on-line motors, they offer significant advantages, including lower cost of installation and enhanced system efficiency when compared to discrete motor drive systems.

There are varying degrees of integration found under the overall field of integrated machines, beginning with very mild integration – whereby the IMD is essentially two identifiably separate entities (i.e., the motor and the adjustable speed drive) with minimal integration beyond mechanical fixing – through to IMDs with very intricate integrations whereby the multi-physics interactions are deeply interlinked and highly complex. The philosophies and impacts of various degrees of integration are ongoing debates in the field.

[1]School of Engineering, Newcastle University, UK
[2]Department of Electrical and Electronic Engineering, Power Electronics and Machine Control Group, University of Nottingham, UK

1.2 Types of IMDs

The first level, standard frame mounting, covers the integration of a distinct drive unit onto the frame of a machine, as shown in Figure 1.1. In this integration form, both motor and drive are nominally an integrated unit; however, there remains a clear delineation between the motor and the drive. Although interconnected electrically and physically, the two may have distinct cooling paths and mechanisms, can have a reasonable degree of mechanical isolation in terms of anti-vibration mounting and can be robustly EMI shielded from one another due to the presence of the motor casing between them.

The second level of integration, surface and endplate-mounted integration involves the drive being more fully integrated into the machine, as shown in Figure 1.2. Endplate-mounted integration involves the integration of the drive into the endcap of the machine. This increases the level of integration, leading to more challenges, particularly regarding thermal isolation. This method of integration typically allows the drive and machine to share a cooling path via forced air cooling from machine rotation. This level carries a closer burden of mechanical integration. However, the drive and the machine may still be kept thermally, and to a limited extent mechanically isolated from one another, with limited EMI concerns given the distinct and separate regions still occupied by the machine and drive.

The third and final level of integration concerns the integration of the drive within the casing of the machine, as shown in Figure 1.3. While there have been various approaches to this, this commonly consists of individual drive modules associated with segments of the winding that are closely integrated into the machine. In some cases, these may share a direct mechanical mounting to the active

Figure 1.1 First-level integration: standard frame mounting

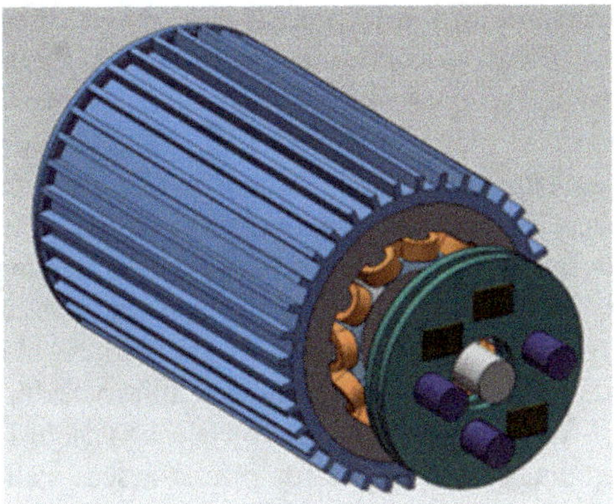

Figure 1.2 Second-level integration: endplate mounting

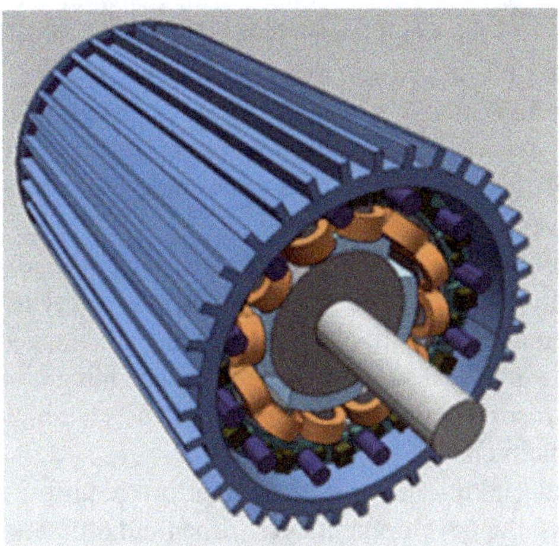

Figure 1.3 Third-level integration: in casing mounting

material, while in others, these will be closely integrated but have their own distinct cooling systems and paths. This is the closest level of integration, and while this carries large potential benefits in terms of modularisation, minimisation of volume and design for performance, this is also the area where the challenges of integration are most extreme in terms of mechanical, thermal and EMI issues.

The three levels of integration will be explored and discussed through the book, looking at the key benefits and challenges.

1.2.1 Standard frame mounting

Whilst the advantages of using Adjustable Speed Drives (ASDs) with electric machines are numerous and well understood, direct on-line (DOL) start motors account for the vast majority of electric machines worldwide and, for low-power cases (< 2.2 kW), these account for as much as 97% of the worldwide installations [1]. This market domination is due to the prevalence of applications which do not require complex control functionality that ASDs offer such as fans, pumps, compressors, and HVAC systems.

To facilitate interchangeability and mechanical and electrical commonalities – which aid some aspects of installation and system design – industry has adopted a number of standards for machine frame sizes [2]. This has led to the result that whilst there are many hundreds of electrical machine manufacturers worldwide, for any given supply voltage and power level, the range of available standard DOL machines tends to have a very similar appearance and footprint.

A similar, albeit less pronounced convergence of design has occurred in the ASD market with products tending towards plastic-cased, wall-mounted modules commonly with a display and interface.

With standard frame mounting, the machine and the drive are integrated together insofar as the drive is mounted onto the machine frame, either directly or via a mounting plate. Even in its simplest form, many of the more obvious and financial advantages of integrated drives are largely achievable simply through the collocation of motor and drive (e.g. removal of cable lengths, single installation, cable emissions, etc.). Whilst a simple mechanical coupling may not yield optimised integration – and without careful thought can result in lifetime deterioration through vibration of the electronics, etc. – the underlying subsystems, the motor and the drive need very little reengineering.

Considering less specialised applications where the extra investment of including or retrofitting an ASD as a separate system along with a DOL installation may not be commercially attractive, the functional advantages of ASDs may be unlocked through the marginal savings made in a basic integrated drive installation. This is particularly true in economies where labour costs are high thus the reduction of installation and maintenance time is important.

Indeed, since the additional engineering effort required in basic integrated drives is less arduous, the development cycle time (and ultimately cost) is far lower and can even be achieved by third parties outside the original engineering manufacturer of the drive or the motor. This may mean that even for small production runs or specialised projects, the IMD is a viable solution in achieving the functionality of the ASD. It also has clear implications for larger volume manufacturers where the smaller development costs may increase profitability.

A particular point of interest in surface frame-mounted integrated machines is that this supports modularity and end user customisation. As the standard DOL machine will fit a frame size, machines can be employed with or without ASD or even retrofitted to improve system, response, performance or efficiency. With a frame mount IMD, there is the potential to separate the DOL machine and ASD as products. This will allow a configurable solution to be marketed that is capable of including a surface mount ASD, may or may not include one, depending on user and application requirements.

An example of this can be seen in the commercial literature for NIDEC's commander ID300 system [3]. As shown in Figure 1.4, it is marketed as an

Figure 1.4 NIDEC commander 3 drive mounted to IMfinity induction motor. Courtesy of Leroy-Somer.

'Integrated drive solution', where the IMfinity® induction motors are available as standalone DOL machines, an integrated ASD and motor and the integrated ASD as a retrofit – effectively broadening the product range of the company with little or no significant additional engineering investment. Here it can be argued that the concept is that, to the system designer, the ASD is an option for consideration which provides additional functionality much as a gearbox or brake would be. All these options can be integrated to the base motor and hence could be considered IMDs but the underlying motor in the integrated package is no different to the DOL option.

Some manufacturers have taken this flexible concept even further, for example, Danfoss have a product range which are marketed as 'Decentral Drives' with one example, the VLT® DriveMotor FCP 106, effectively being a motor-mounted drive (it can also be wall mounted), where the motor is not included as part of the product, as shown in Figure 1.5. Instead, a motor adaptor plate (supplied in three sizes depending on the drive power) is available.

The aim here is to allow for retrofitting of an ASD onto a range of motors based on frame size and compatibility, and while unlikely to be completely universal this approach potentially allows for a much greater range of motor choice and adaptability. This also has significant potential benefits in retrofitting, where the addition to rather than replacement of an existing DOL machine would be a significant cost factor.

This concept of third-party or after-market drives which offer a variety of connectivity interfaces (e.g. Modbus, etc.) and can be integrated to mass-market machines either at original point of supply or latterly as a retrofit, further

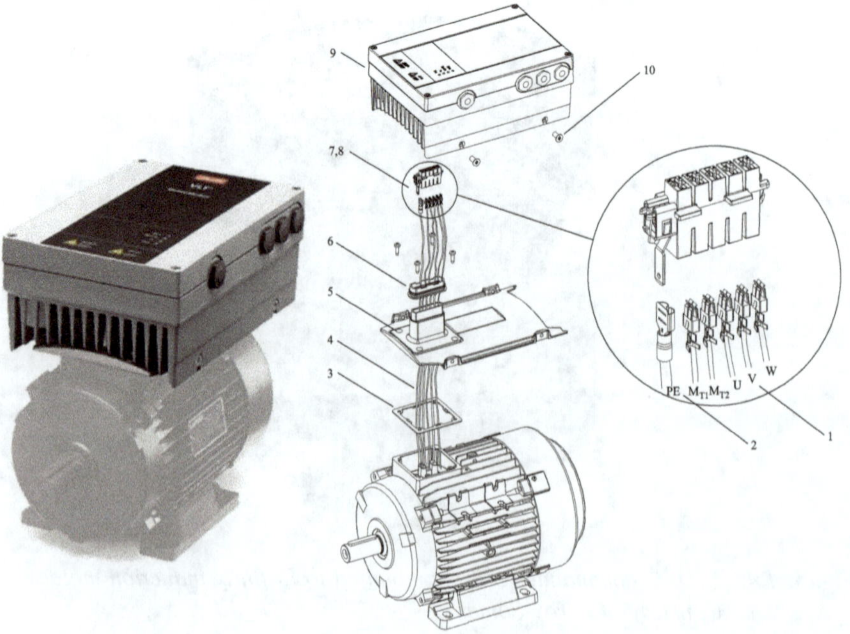

Figure 1.5 Danfoss decentral drive concept with motor adaptor plate [4]. Figure, courtesy of Danfoss.

removes the challenge of locking into one particular supplier. Given that one of the key concepts of standard frame size industrial machines is the ability to use a wide range of suppliers for an application, this again may be a key benefit in integration.

1.2.2 Endplate mounted integration

At this level of integration, endplate mounting, the power electronics passives, etc. are integrated within the end cap of the machine. This level of integration takes the distinct step that it is not possible to employ an off the shelf commercial drive as may be used with surface mounting. The geometry of endplate integration requires that the drive is designed around the available space envelope, namely that it will fit into a cylindrical area of similar diameter to the motor while maintaining a physical clearance for the motor shaft to pass through. A good example can be seen in Figure 1.6, showing the individual elements of the end cap integrated drive as well as their integration into a full system.

A potential benefit of this integration type is that it offers minimal changes to the drive space envelope compared to a DOL motor only configuration. While the end cap length will likely be increased to accommodate the drive, the machine frame size and radial space profile can typically remain unchanged, allowing for a significant degree of retrofitting. This may also be of benefit in applications such as electric vehicles where the space profile is typically quite constrained.

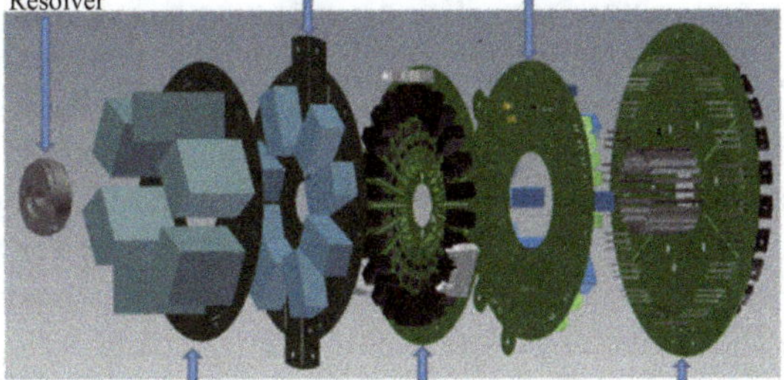

(a)

(b)

Figure 1.6 A high-speed end cap-integrated machine. (a) Drive sub layers and (b) fully assembled drive [5]

This type of integration offers some degree of mechanical isolation. Although the drive electronics are more closely integrated than with surface mounting, the drive till has a distinct area separated from the stator and so some degree of mechanical separation can be maintained.

The thermal challenges of this type of integration are larger. The end cap in machines is commonly fitted with a shaft-mounted fan that provides a reasonable degree of forced air cooling when the machine is operating at speed. In classical machines, this fan has a clear cooling path, driving air from the casing end across the machine end windings and through the airgap and any core vents in the stator and rotor to provide simple but effective cooling of all components. When introducing power electronics to the end cap, there is an obvious challenge that the cooling path gets more complex, as this direct cooling path is obstructed. This means that a greater degree of consideration must be given to the cooling design, to ensure that the motor cooling is not negatively impacted.

End cap integration does also provide an opportunity in thermal terms, as the power electronics are included in the forced air-cooling path of the machine; this behaviour can also be used to cool the power electronics. Further, as the flow of ambient air is over the power electronics followed by through the motor, this also works well with the typical thermal limitations of power electronics packaging and passives which are much lower than those of the motor itself. This means that power electronics and motors can share a cooling path, but with the flow of air, both the power electronics and motor can be operated at their respective temperature limits without the risk of thermal problems, if correctly designed. As with all direct fan-cooled machines, care should be taken due to the air flow rate being linked to rotation speed. This can be a challenge at low-speed operation for both power electronics and the motor itself; however, the power electronics are undoubtedly more thermally sensitive with lower thermal capacity than the motor itself and so care should be taken with thermal design margins.

In EMI terms, there is also an increased degree of challenge. Motor winding to drive interconnections can now be both shortened (when compared to non-integrated drives) and confined within the motor body and so external EMI can potentially be significantly reduced and more effectively screened. There is an internal challenge however, as the end cap converter is mounted in reasonable proximity to the end windings of the machine. This creates an internal EMI design challenge, where care must be taken at the design stage to ensure that the close proximity of these motor windings does not adversely affect the power electronics operation.

1.2.3 Mounting within the machine casing

In this topology of integration, either the power electronics, the drive passives or both are integrated directly into the casing of the machine. This may include cases where the drive is integrated fully onto the active material or supporting structure of the machine, directly sharing cooling flows and paths, or it may cover machines that exhibit close mechanical integration but aims to maintain distinct cooling paths for different subcomponents (motor, power electronics, passives, etc.). This topology can be considered at the extreme edge of integration, both in terms of the potential benefits this brings and in terms of the technical challenges this introduces. A fine example of this level of integration can be seen in [6] and Figure 1.7, with an individual integrated module including power electronics and phase conductor which may be assembled into a full machine.

The benefits of integration directly onto the active material of the machine are that connection lengths are minimised with inherent benefits for reducing losses and susceptibility to EMI, machine volume is reduced, power density improved, and thermal management systems can be consolidated. This level of close integration also lends itself well to modularity with benefits to both manufacturability and fault tolerance.

This level of integration also brings huge challenges in terms of the systems to be integrated. These challenges may be categorised into the following areas:

- *Thermal*: Both the power electronics and the stator of the electrical machine share a cooling path, with limited ability to differentiate the thermal paths of the two, and so the thermal conditions of one subsystem are at least partially dictated by the other. This necessitates a reasonably unified thermal tolerance for both machine winding and power electronics, as well as a high level of awareness around the design of the cooling path.
- *Mechanical*: With close integration, any mechanical stresses on the stator can easily be imparted onto the power electronics, which historically are much less

(a)

(b)

Figure 1.7 A high-speed end cap-integrated machine. (a) drive-phase module, (b) 6-phase demonstrator unit mounted on dynamometer with test machine. © 2023. Reproduced, with permission from, IEEE [6].

robust to mechanical stress and vibration. Increases in mechanical stress could easily lead to lifetime deterioration of the power electronics, which may be a major issue considering that the lifetime of a typical drive is generally less than that of a typical stator.

- *EMI*: In classical machines, the magnetic fields are largely contained within the core of the machine, with leakage fields still generally well confined within the casing. When closely integrating power electronic devices onto the active material, it can be expected that significant fields may be encountered, which cannot be allowed to interfere with the operation of these devices. Although shorter lead and connector lengths from integration may mitigate this, it is a concern that must be carefully considered at the design stage.

Yet, this level of full integration has only been applied in limited commercial cases. An example is integration in-wheel drive system from Protean seen

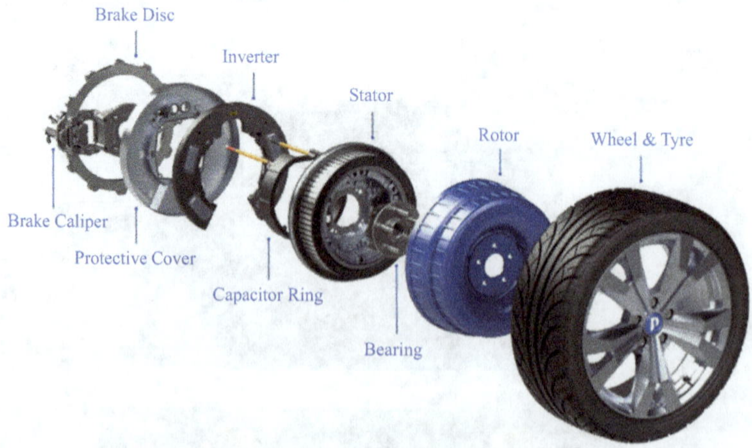

Figure 1.8 An integrated in-wheel drive system from Protean

below, which aims for a fully integrated electric drive and braking system (Figure 1.8).

In terms of academic works, however, this level of integration has had significant study, and many examples may be found in the literature, ranging from studies of individual components and subsystems to fully realised and prototyped integrated machines.

For the integration of passive components into the active material of the machine, existing passives as commonly used in drive exhibit severe limitations. While the winding structure itself is generally as thermally robust as the stator winding, the widely used ferrite core type of inductor is often thermally limited, with typical operating temperatures suitable for air cooling. For operation at high temperatures, discrete powder cored inductors are available that can operate at temperatures up to 200°C.

For closer integration of the drive inductors, several authors have solutions. The integration of inductors into the active parts of the machine has been proposed in works by the Universities of Illinois [7], Newcastle University [8] and the University of Nottingham [9], though with different implementations.

In terms of the challenges highlighted for passive device integration, inductors are arguably one of the simpler components to integrate, when compared to capacitors and power electronics.

Regarding the outlined challenges, in thermal terms, the structure of inductors is similar to that of the main stator, using wound, insulated copper windings with similar thermal tolerances to those of the primary stator windings. Further, integrated inductors commonly employ parts of the core structure for their magnetic path, meaning that again their core material construction and hence tolerances are the same as the stator core.

In terms of mechanical vibration and tolerances, this is another area where all the benefits of robustness, encapsulation and mechanical support enjoyed by the stator windings can also be enjoyed by the integrated inductors.

EMI is a prime area of concern for these inductors. With multiturn coils in close proximity to the stator core, there is a large potential for coupling and interference between the stator windings and wound inductors and much care must be taken to minimise coupling at the design stage.

The integration of inductors into the thermal path and even active area of the motor has been accomplished as described and relies in large part on the similar thermal tolerances of inductor and machine winding materials. The other key passive component, the capacitor, poses a separate but no less distinct challenge for close integration. The DC link capacitor is a primary factor and limiter in terms of capability to develop highly integrated drives. In [10,11], where work is focussed on the reduction of DC link capacitor sizing, it is reported that the DC link capacitor alone is responsible for up to 30% of the volume of modern power electronics.

There are multiple fundamental technologies available for DC link capacitors, including multi-layer ceramic capacitors, electrolytic capacitors, film capacitors and elector-chemical double layer (or "supercapacitors") [12]. Of these technologies, ceramic capacitors distinguish themselves for IMD applications based on their high current rating, ability to withstand temperatures up to 200°C and small size. All of these features make ceramic capacitors apt for use in close integration, where thermal and volume challenges are key. One major disadvantage is in mechanical robustness, as ceramic capacitors are susceptible to mechanical shock and vibration, which is a key challenge when closely integrated with the active material. Of the other types, electrolytic, film and electro-chemical double layer capacitors are all thermally limited to temperatures lower than those typical of an active machine, which makes their use in close integration challenging, necessitating careful design and management of thermal losses and cooling paths.

Where a closely integrated machine was developed in [13], a key focus of the work was to minimise the size of the DC link capacitance in order to better facilitate machine integration in terms of overall volume. The work discounted the use of electrolytic capacitors, based on their previously highlighted thermal limitations. Through design optimisation, a solution was developed using a combination of film and ceramic capacitors. This work supports a key challenge of machine integration that of unified multi-disciplinary design where the DC link sizing is fundamentally coupled to the machine design in terms of both electromagnetic and thermal aspects but also in terms of the control of the machine.

Figure 1.9 shows detail of dc link capacitances on an integrated module [6]. In [14], the concept of holistic design is taken further. This work sets out a comprehensive design method for DC link architecture, aiming to account for end application, supply constraints and hardware limitations. The work again demonstrates a strong multidisciplinary link between the application and the design, with a holistic approach necessary to develop an optimised integrated solution. Not only is the optimum sizing linked to the physical and electrical system characteristics, but the control strategy has a significant bearing on this sizing, particularly if the integrated machine has been extensively modularised, giving increased options for cooperation between drive modules.

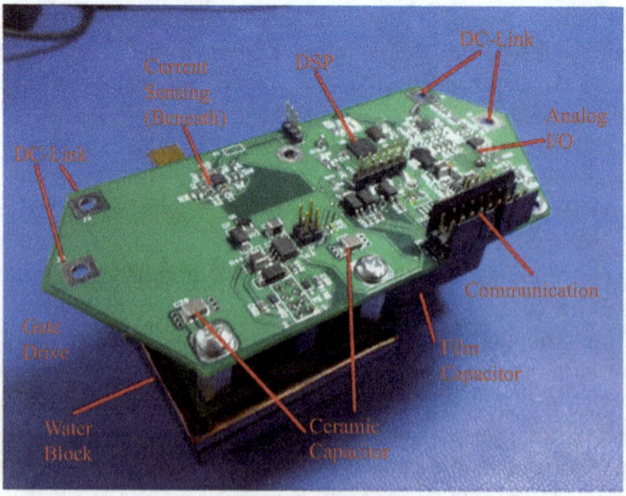

Figure 1.9 DC link capacitance on integrated module. © 2023. Reproduced, with permission from, IEEE [6].

As well as passives, the integration of full drives necessitates the inclusion of the power electronics themselves into the thermal path of the machine, and even onto the active area itself. In order for this to be successful when the cooling path is shared, the power electronics generally need to have the same levels of thermal tolerance as the machine (typically limited by the machine winding insulation, a typical value would be NEMA Class H at 180°). With the advent of commercially available wide band gap (WBG) devices, options have opened up for the application of power electronics at equivalent temperatures to motor insulation, where traditional silicon-based power electronics would prove unreliable.

A key issue remains in the form of packaging. While WBG devices have shown themselves capable of effective operation at machine peak temperatures, the device cannot be considered in isolation. Where a pre-packaged module is used, the associated package thermal limits are often similar to those found in more conventional Si devices, making them unsuitable for application. Even when going to the device level, thermal challenges remain for the integration of WBG devices at high temperature, both in the electrical connections to the device and the physical and thermal connection to the supporting substrate. High temperatures and thermal cycling can lead to mechanical stress and failure in both wire bonds and die attachment. In [15], various approaches are detailed to combat this, ranging from the use of differing attachment materials more suited to high temperature, to the fundamental redesign of connection methods developed specifically to be more forgiving of the thermal challenges of high temperature and cycling.

A further concern is the control and sensing sections of the integrated drive. While not significant thermal sources in themselves, close integration of microcontrollers potentially brings them into contact with the thermal loads from the machine and drive that they are not innately designed to deal with. Although in some cases, the control is external to the integrated machine, the ideal of an integrated drive would include the motor, drive and control elements, so that the machine connections are limited to the provision of power, and the overall performance demand input (along with any performance feedback from the machine that may be required by the application.

To realise the full in-casing integration of machines, designers must bring together the disparate strands of passives, power electronics and control, while as has been explained benefits of integrating, some of these elements fully within the case can also be achieved.

1.3 Contents of the book

The electrical machine typically comprises the largest and heaviest component of an IMD and hence needs to have a chapter devoted to it. This book will not attempt to give a comprehensive review of all electrical machines – a topic which could occupy several books in its own right. Instead, it will focus on aspects of electrical machines which are particularly relevant to integrated drives. Chapter 2 covers two main topics. It first deals with construction methods and modularity. With close integration, it is possible to split the system into different modules and the case to stay with three phase becomes weaker. This creates a large range of options and topologies. The second topic concerns materials: advances in materials have had a major impact, particularly with regard to insulation systems, winding technology and magnetic materials. Sections have hence been dedicated to each of these in turn.

Power electronics, sensors, controllers and the advanced control techniques will be introduced in Chapter 3. Converter topologies, including two-level, multi-level and multi-phase topologies, will be discussed first. Integration of magnetic passive components and converter modularity will also be introduced as approaches for power electronics integration. When it comes to choosing sensors for IMDs, the size and the robustness of the sensors are the key features and need careful consideration. Current, position and speed sensing technologies will then be reviewed and summarised with suggested IMD application. The control of an integrated drive is very much the same as that of a conventional machine of similar dynamic performance. Steady-state efficiency may be slightly improved in IMDs through the use of more than three phases, which also offers improved fault tolerance, but performance is dominated by the thermal design. The last section of this chapter will use a brief overview of conventional control schemes as a basis for identifying the specific features arising from the integration of the power electronics with the machine, which can enhance performance, or need additional design considerations.

In Chapter 4, we will explore the potential impact of WBG power switches on the future development of IMD technology. This examination will begin with a concise discussion of the advantages offered by the key material properties of WBG semiconductor materials compared to silicon for IMD applications. This section will be followed by a brief discussion of the major types of WBG power semiconductors that are currently available as candidates for use in IMDs. The next section of this chapter will focus on a balanced discussion of the advantages and special challenges associated with the introduction of WBG power switches into IMD systems using voltage-source inverter (VSI) technology that has dominated adjustable-speed motor drive applications since the 1980s when IGBTs and MOSFETs became the leading choices for new motor drive applications. Candidate techniques for addressing the identified challenges will be discussed. The final section of the chapter will be devoted to exploring the opportunities opened by WBG switch technology for their application in IMD systems using current-source inverter (CSI) technology in the place of conventional VSIs. Opportunities will be discussed for taking advantage of the CSI approach to overcome the significant challenges imposed by inserting WBG power switches into conventional VSI topologies as drop-in replacements for IGBTs and MOSFETs.

Both system integration and thermal management have been identified as essential to development of the next-generation IMDs. Chapter 5 will provide a detailed technology overview of thermal management of IMDs across a variety of applications, starting with a generic classification of alternative thermal management techniques and system integration topologies. The technology overview then follows with several commercial and research examples of IMDs illustrating the key concepts for an integrated heat removal. The chapter continues with alternative technology avenues for advanced thermal management of integrated motors and drives. The authors' observations and forecasts of the further evolution to thermal management of IMDs are discussed in the concluding section of the chapter.

The role of passive components in a motor drive system is often overlooked at the early stages of the design and in many academic research papers. The size, mass, cost and often temperature rating and lifetime expectation of the drive system may be dominated by the passive components rather than the semiconductor components and control topology. In Chapter 6, the role of passive components in battery powered and mains powered applications will be discussed. The latest technology in passive components will also be presented in this chapter with a discussion on how these can be integrated within an integrated drive and motor system including sharing inductor magnetic circuits with that of the machine, through minimisation and customisation of capacitive filters.

Modern automotive transmissions require high-power density, low weight, high efficiency, low noise and reduced manufacturing costs. This can only be achieved with the application of precise and effective gearbox design, manufacture, high-quality materials and heat treatment and effective quality control. Gear technology is challenging and often poorly understood, and the objective of Chapter 7 is to direct engineers to appropriate gear-related resources and to improve their understanding in general. Throughout this chapter, we will use International Standards Organization (ISO) terms and symbols which are in effect the international language of gears. ISO standards and their supporting technical reports provide a useful source of reference material for those involved with gears and gearboxes and will be referenced throughout.

In Chapter 8, we will review several cutting-edge IMD research prototypes and commercial products. With all the essential parts of an IMD being separately discussed in previous chapters, we will be demonstrating how academics and engineers have used the enablers and dealt with all the above-mentioned challenges in real life. The development of a high-speed, permanent magnet, SiC-based drive with integrated input filters will be introduced first. Its focus on magnetic integration and power electronics integration will be explained in detail, along with all the advanced manufacturing techniques that have been employed. A wide bandgap based current source inverter (CSI) IMD will then be discussed. To overcome the drawbacks of the conventional VSI IMD, WBG-based current-source power converters have been selected which do not require any bulky dc-link capacitors nor complicated distributed dc-link design. Special attention is devoted to the power density and thermal performance of the WBG-based CSI developed for this IMD. To further integrate power electronics with an electric motor, there are challenges in providing solutions to both domains of thermal and electrical integration of a drive system. The third prototype in this chapter will show you a novel approach to thermal and electrical integration of high-power density drives in automotive applications. Industry has also shown interests in this IMD concept. The ProteanDrive by Protean Electric, which is an in-wheel electric motor system for passenger cars and light commercial vehicles, is a good example of the adoption of IMDs in commercial drives. In the last section of this chapter, we will show the motivation of integration, the integration approach being used, the operational experience and also future integration possibilities of this commercial IMD.

At end of the book, Chapter 9, the main challenges in the design, manufacturing and operation of IMD are recalled and discussed. Some of these challenges pose limitations in the application of IMD to higher power level or to production process scalability. A representative selection of emerging concepts is then considered, maintaining the focus on the challenge that they address. In this way, it is possible to include in the selection, concepts that have not yet been applied to the IMD world but are eligible as technology enablers. These emerging concepts are presented following a sensible classification: design concepts and assembly technologies.

References

[1] H. Dai, T. M. Jahns, R. A. Torres, M. Liu, B. Sarlioglu, and S. Chang, "Development of high-frequency WBG power modules with reverse-voltage-blocking capability for an integrated motor drive using a current-source inverter," in *2018 IEEE Energy Conversion Congress and Exposition (ECCE)*, 2018. New York, NY: IEEE, pp. 1808–1815.

[2] *IEC 60072-1:2022*, IEC, 2022.

[3] Nidec. Commander ID300. Available at: https://www.leroy-somer.com/documentation_pdf/5595_en.pdf

[4] Available at: https://files.danfoss.com/download/Drives/MI03O102.pdf

[5] X. Deng, M. A. S. Mohamed, S. Lambert, and B. Mecrow, "Development of a high-speed, permanent magnet, SiC-based drive with integrated input filters," *IEEE Transactions on Energy Conversion*, vol. 35, no. 2, pp. 863–874, 2020, doi:10.1109/TEC.2020.2973449.

[6] A. Shea and T. M. Jahns, "Hardware integration for an integrated modular motor drive including distributed control," in *2014 IEEE Energy Conversion Congress and Exposition (ECCE)*, 14–18 September 2014, pp. 4881–4887, doi: 10.1109/ECCE.2014.6954070.

[7] B. M. Nee and P. L. Chapman, "Integrated filter elements in electric drives," in *2007 IEEE Vehicle Power and Propulsion Conference*, 9–12 September 2007, pp. 148–153, doi:10.1109/VPPC.2007.4544115.

[8] M. A. S. Mohamed, S. M. Lambert, B. C. Mecrow, X. Deng, S. Ullah, and A. C. Smith, "Integrating the magnetics of an LCL filter into a high speed machine with pre-compressed coils," in *2017 IEEE International Electric Machines and Drives Conference (IEMDC)*, 21–24 May 2017, pp. 1–7, doi:10.1109/IEMDC.2017.8002292.

[9] M. R. Khowja, C. Gerada, G. Vakil, C. Patel, and P. Wheeler, "Design optimization of integrated rotational inductor for high-speed AC drive applications," in *2017 IEEE International Electric Machines and Drives Conference (IEMDC)*, 21–24 May 2017, pp. 1–8, doi:10.1109/IEMDC.2017.8002145.

[10] J. V. Damme, L. Verkroost, H. Vansompel, F. D. Belie, and P. Sergeant, "A holistic DC link architecture design method for multiphase integrated modular motor drives," in *2019 IEEE International Electric Machines & Drives Conference (IEMDC)*, 12–15 May 2019, pp. 1593–1598, doi: 10.1109/IEMDC.2019.8785391.

[11] L. Verkroost, J. V. Damme, H. Vansompel, F. D. Belie, and P. Sergeant, "Module connection topologies and interleaving strategies for integrated modular motor drives," in *2019 IEEE International Electric Machines & Drives Conference (IEMDC)*, 12–15 May 2019, pp. 559–564, doi: 10.1109/IEMDC.2019.8785248.

[12] L. Verkroost, J. V. Damme, D. V. Bozalakov, F. D. Belie, P. Sergeant, and H. Vansompel, "Simultaneous DC-link and stator current ripple reduction with interleaved carriers in multiphase controlled integrated modular motor drives," *IEEE Transactions on Industrial Electronics*, vol. 68, no. 7, pp. 5616–5625, 2021, doi:10.1109/TIE.2020.2992965.

[13] R. U. A. Shaikh, A. Saeed, and R. Kumar, "Review on present and future integration techniques for capacitors in motor drives," in *2018 International Conference on Computing, Mathematics and Engineering Technologies (iCoMET)*, 2018. New York, NY: IEEE, pp. 1–8.

[14] J. Van Damme, L. Verkroost, H. Vansompel, F. De Belie, and P. Sergeant, "A holistic DC link architecture design method for multiphase integrated modular motor drives," in *2019 IEEE International Electric Machines & Drives Conference (IEMDC)*, 2019. New York, NY: IEEE, pp. 1593–1598.

[15] T. M. Jahns and H. Dai, "The past, present, and future of power electronics integration technology in motor drives," *CPSS Transactions on Power Electronics and Applications*, vol. 2, no. 3, pp. 197–216, 2017, doi:10.24295/CPSSTPEA.2017.00019.

Chapter 2
Electrical machines for an integrated drive

A.C. Smith[1], T.M. Jahns[2] and B.C. Mecrow[3]

2.1 Machine issues for integrated drives

The electrical machine typically comprises the largest and heaviest component of an integrated drive and hence deserves to have a chapter devoted to it. This chapter will not attempt to give a comprehensive review of all electrical machines – a topic which could occupy several books in its own right. Instead, it will focus on aspects of electrical machines which are particularly relevant to integrated drives.

The chapter covers two main topics. It first deals with construction methods and modularity. With close integration, it is possible to split the system into different modules and the case to stay with three phases becomes weaker. This creates a large range of options and topologies, as discussed in Section 2.2. The second topic concerns materials: advances in materials have had a major impact, particularly with regard to insulation systems, winding technology and magnetic materials. Sections have hence been dedicated to each of these in turn.

2.2 Motor modularisation

2.2.1 Rationale for modular motor drives

Modular motor drives (MMDs) comprise a special class of motor drives in which the electric motor and its power electronic converter are broken into modular sub-units which all operate in a coordinated manner to convert electrical power to mechanical power or vice versa. This class of modular motor drives can, in turn, be further divided into a variety of architectural sub-classes that can be defined sufficiently broadly to include multiple identical stator–rotor assemblies, each with its own dedicated motor drive unit, all sharing the same shaft (Figure 2.1(a)) [1]. However, attention in this section will be limited to the combination of a single rotor with a single-stator assembly that has multiple modular stator windings, each excited by its dedicated modular power converter (Figure 2.1(b)). Combining the modular power converters

[1]University of Manchester, UK
[2]University of Wisconsin – Madison, USA
[3]Newcastle University, UK

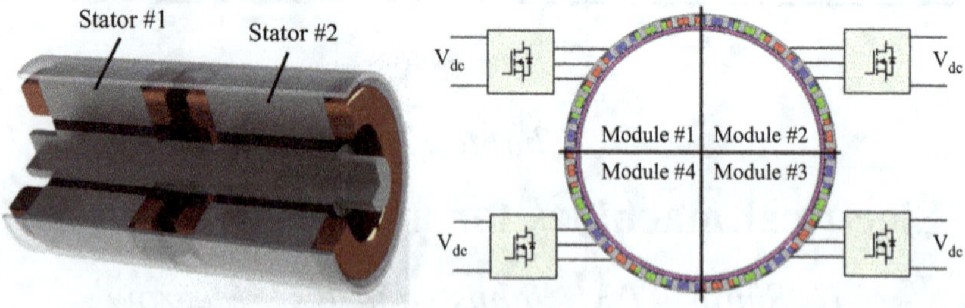

Figure 2.1 *(a) Modular motor drive with two motors sharing same shaft with two inverters . © 2023 Reproduced, with permission, from IEEE [1]; (b) Modular motor with single rotor and single stator with multiple modular stator windings, each excited by a dedicated modular power converter.*

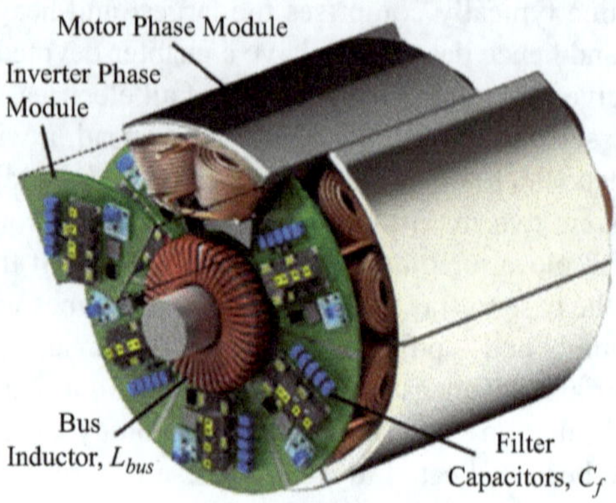

Figure 2.2 *Integrated Modular Motor Drive (IMMD) concept highlighting segmented stator phase modules, each accompanied by a dedicated inverter phase module implemented using a current-source inverters (CSI).*

into the same housing as the modular motor leads to the concept of an integrated modular motor drive (IMMD) which is of particular interest for this discussion [2].

There are a number of reasons that modular motor drives are appealing for different classes of motor drive applications, and three of them that are attracting growing attention are highlighted here:

- *Manufacturing improvements:* Modularity offers a powerful means of achieving faster and lower cost production of manufactured goods when the product can be assembled from several identical modules that are manufactured in large volumes by specialized manufacturing equipment that is optimized for rapid, high-volume production. An example is the IMMD concept shown in Figure 2.2. The IMMD shown in this figure is composed of six identical phase-drive units, each of which incorporates a laminated segmented stator tooth-pair. Each tooth in the tooth-pair is encircled by its own concentrated winding.

These two windings are connected in series, comprising a phase winding that is excited by a dedicated single-phase inverter mounted at the end of the tooth assembly. Fitting six identical wedge-shaped phase-drive units together to form an annular ring as shown in the figure completes the core of the IMMD's machine stator-plus-drive assembly. This complete assembly fits into a single stator housing, eliminating the need for a separate drive enclosure.

- *Fault-tolerant motor drives*: Demanding applications such as aircraft electric propulsion motor drives require extremely high levels of reliability with a mean time between failure (MTBF) that can be as high as 10^{10} hours (1.14 million years) [3], exceeding the MTBF values of the highest-quality industrial drives by more than 10,000 times. One of the best-known avenues to bridge this huge reliability deficit is to adopt fault-tolerant modular motor drive (FT-MMD) architectures in which individual modules can fail without causing a failure of the complete drive because the remaining healthy modules can continue delivering power until the faulty module is repaired. The IMMD architecture in Figure 2.2 is a serious candidate for this type of application if each of the individual phase-drive units is specially designed to continue operating when any of the other phase-drive-units fail.
- *Machine winding voltage stress reduction:* In a conventional voltage-source motor drive, the peak line-to-line voltage experienced by the machine windings is the dc bus voltage V_{bus}. As the motor drive rating increases above ~250 kW, the value of the dc bus voltage gradually increases into the medium voltage range (i.e., $V_{bus} > 1$ kV), requiring more complicated machine windings (e.g., form-wound windings) to avoid the risk of dangerous partial discharges due to the high electric fields associated with the rising winding voltages. One approach that uses modular motor drive concepts to avoid these high winding voltages is

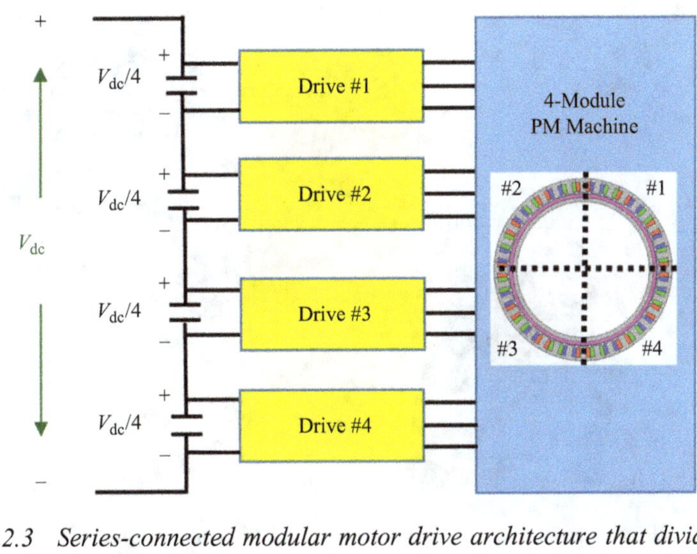

Figure 2.3 Series-connected modular motor drive architecture that divides the total dc link voltage into lower values that excite the modular drives connected in series.

shown in Figure 2.3. According to this approach, the input dc buses of the modular inverters that excite the multiple sets of 3-phase windings in the Figure 2.1(b) MMD architecture are connected in series. Consequently, the dc bus voltage of each of the modular drives in Figure 2.3 is reduced to $V_{bus} = V_{input}/n$, where n is the number of series-connected modular drives. As a result, each of the n sets of 3-phase voltage windings is exposed to a maximum line-to-line voltage which is a factor of n smaller than the voltage associated with a single inverter, regardless of whether the inverter is implemented as a two-level or multi-level inverter. This advantage takes on special importance for aerospace motor drives operating at high altitudes where partial discharge (PD) problems can arise at peak winding terminal voltage levels as low as 300 V due to the degrading impact of reduced air pressure on the PD inception voltage.

2.2.2 Machine compatibility with modular stator windings

Different types of machines have varying levels of compatibility with modular stator windings which has a significant impact on their compatibility with the major modular motor drive architectures. Since this is a rich topic that cannot be thoroughly addressed in this book, attention will be focused on highlighting the most important distinguishing features affecting their compatibility characteristics. For the same reason, the major types of machines addressed in this discussion will be limited to permanent magnet synchronous, wound-field synchronous, synchronous reluctance, induction, and switched reluctance machines (Figure 2.4).

One of the most important distinguishing features that will be discussed is the compatibility of different machine types with concentrated windings. Concentrated

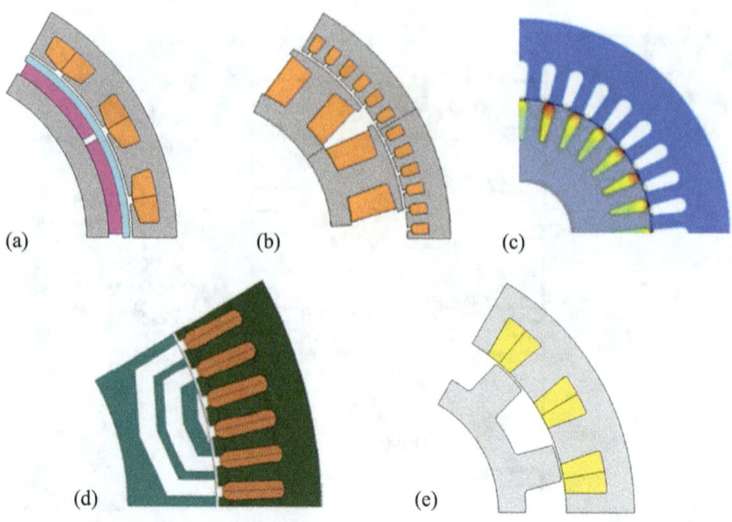

Figure 2.4 Five considered machine types: (a) PM synchronous; (b) wound-field synchronous; (c) induction; (d) synchronous reluctance; (e) switched reluctance

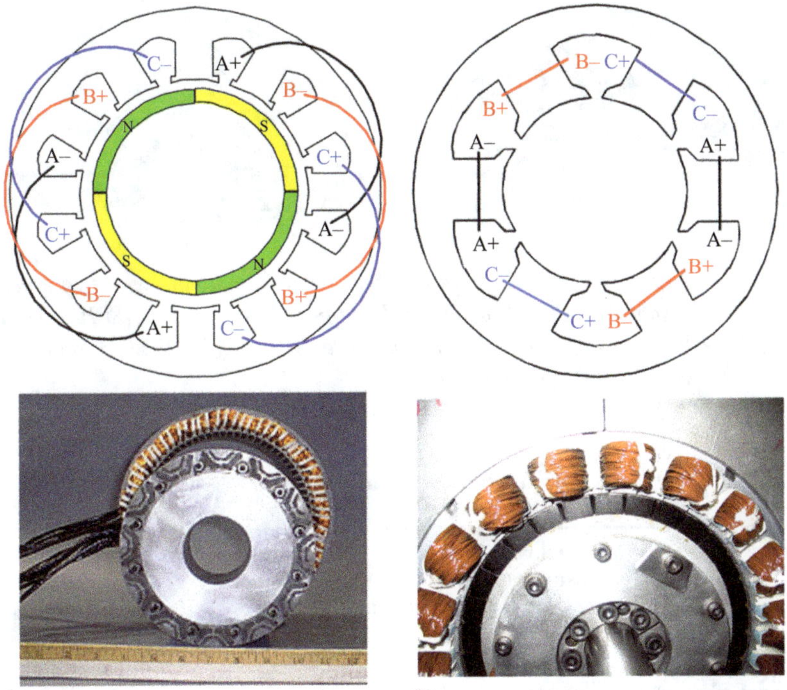

(a) (b)

Figure 2.5 (a) Distributed windings and (b) concentrated windings

windings are stator windings that encircle single stator teeth, while distributed windings are installed in stator slots and span two or more stator teeth, as shown in Figure 2.5. A second winding feature that plays a significant role in this discussion is whether the value of a winding's slot-per-phase-per-pole (SPP) metric has an integral (e.g., 1, 2, etc.) or fractional value (e.g., 2/5, 2/7, etc.), as illustrated in Figure 2.6 [4]. A stator winding that combines the features of concentrated windings and a fractional SPP value is referred to as a fractional-slot concentrated winding (FSCW) configuration that merits special attention during winding discussions for modular motor drives.

The first three rows of Table 2.1 summarize the compatibility of the five machine types with the different types of windings and stator constructions discussed above. Of the five machine types shown in Figure 2.4, only the switched reluctance machine is not compatible with distributed windings. Some researchers continue to work on developing new hybrid machine topologies that seek to combine distributed windings with SR machine rotors, but they will not be discussed here [5].

In contrast, only PM synchronous and switched reluctance machines are highly compatible with concentrated windings (including segmented stators). The switched reluctance machine is particularly well suited to concentrated windings that,

(a) (b)

Figure 2.6 Cross-section of 10-pole PM synchronous machine with two types of stator windings: (a) integral slot-per-phase-per-pole (SPP = 2); and (b) fractional slot-per-phase-per-pole (SPP = 2/5) . © 2023 Reproduced, with permission, from IEEE [4].

Table 2.1 Comparison of Machine Compatibility with Modularisation Features

Machine Type	**PM Synchronous (Surface PM & Interior PM)**	**Wound-Field Synchronous**	**Synchronous Reluctance**	**Induction**	**Switched Reluctance**
Feature					
Compatible with Distributed Windings	X	X	X	X	
Compatible with Concentrated Windings and Segmented Stators	X				X
Compatible with Fractional-Slot Concentrated Windings (FSCW) Topologies incl. Sinusoidal Excitation	X				
Compatible with Single-Phase Lane Inverter Configurations	X	X	X	X	X
Compatible with Multi-Phase Lane Inverter Configurations	X	X	X	X	X

when excited, create highly magnetized salient stator poles. Each excited stator pole exerts a strong attractive force on the nearest salient rotor pole, producing reluctance torque as the result. Of these two machine types, only the PM synchronous machine is an excellent candidate for fractional-slot concentrated winding topologies with sinusoidal excitation. However, the windings in switched reluctance machines can be classified as a special member of the FSCW family.

Although a comprehensive explanation will not be provided here, the other three machine types are not good candidates for concentrated windings (including FSCW topologies) because their rotors do not interact well with the rich combination of spatial harmonic rotating magnetic fields generated by concentrated windings excited by balanced 3-phase (or, more generally, n-phase) sinusoidal voltage/current excitation [6]. For example, these spatial harmonic fields magnetically couple with the rotor cage conductors in induction machines to generate high losses with almost no average torque production. FSCW configurations are not generally regarded as a good match with synchronous reluctance machines because the periodic reluctance variations that lie at the heart of its torque production are spatially filtered and suppressed by the winding configuration. As a result, all of the asynchronous flux harmonics combine to create a large leakage inductance which reduces the power factor of the machine [7,8], but can also be beneficial in some applications by limiting the short-circuit current amplitude during faults.

2.2.3 Manufacturing issues for modular motors with concentrated windings

Consistent with the discussion in the two preceding sections, PM machines designed with FSCW stator topologies are excellent candidates for segmented stators. In addition to the manufacturability advantages that segmented stators can provide that were noted previously, segmented stators with concentrated windings can offer other valuable advantages that include windings with higher slot fill factors (i.e., ratio of copper area to total slot area). That is, segmented stators make it straightforward to wind tightly-packed coils around each stator tooth before the segmented teeth with their windings are assembled into an annular ring. In contrast, stators fabricated using conventional single-piece stator laminations require specialized and expensive winding machines to install each winding through the narrow slot openings between adjacent teeth, bearing some resemblance to the complexity of sewing machines. Higher slot fill factor values typically result in higher torque/power ratings for the segmented-stator machines because of the higher current that can be carried through the slots for the same total winding loss as random-wound windings installed through the slot openings with lower slot fill factors.

However, there are special mechanical design and manufacturing processes that have been developed to address the unique challenges and opportunities associated with the implementation of concentrated windings with segmented stators.

When segmenting a stator core, the manufacturer needs to decide both where the joints between segments are to be placed and how the interlocking between segments should be created. The choice needs to take account of:

- the ability to increase slot fill factor;
- the reduction of scrap resulting from the punching process;
- the number of additional components adding cost to the assembly process;
- the required mechanical rigidity of the assembly and the acoustic noise which results;
- any increase in iron loss due to the increased length of lamination cut edges;
- the accuracy of the joints required to prevent unbalanced flux and increased cogging.

Informal feedback from those creating and running motor production lines is that summation of tolerances on multiple components can make it challenging to maintain a consistent product. However, there are millions of PM synchronous machines manufactured each year around the world using segmented stators and concentrated windings, so these manufacturing challenges can be successfully addressed.

2.2.4 *Motor characteristics for fault-tolerant modular motor drives*

2.2.4.1 Introduction to stator winding configuration subclasses and isolation types

The purpose of this section is to identify which motor characteristics are most important for achieving superior fault tolerance in modular motor drives and which of the five machine types are best suited for providing these characteristics. As noted in Section 2.2.1, attention in this discussion is focused on modular motors consisting of a single stator with multiple modular stator windings, each excited by a dedicated power converter as shown in Figure 2.1(b). This class of modular motor drives will be broken down further into two subclasses illustrated in Figure 2.7. The first of these shown in Figure 2.7(a) is referred to as a single-phase lane modular motor drive because each phase winding (with two winding leads) is excited by a dedicated single-phase inverter which typically takes the form of an H-bridge inverter. The combination of a single-phase winding that is electrically isolated from all of the other windings and its dedicated single-phase inverter is referred to as a *lane*, consistent with terminology defined in [9,10]. Each lane inverter can be connected to an independent dc power source for maximum fault tolerance, or all lane inverter inputs can be supplied from the same dc power source if independent dc sources are not available.

The second MMD subclass shown in Figure 2.7(b) is referred to as a multiphase lane MMD modular motor drive because it replaces the single-phase windings in Figure 2.7(a) with balanced n-phase winding sets, each of which is electrically isolated from all of the other n-phase winding sets and excited by its dedicated n-phase inverter as drawn in the figure. While 3-phase windings are a

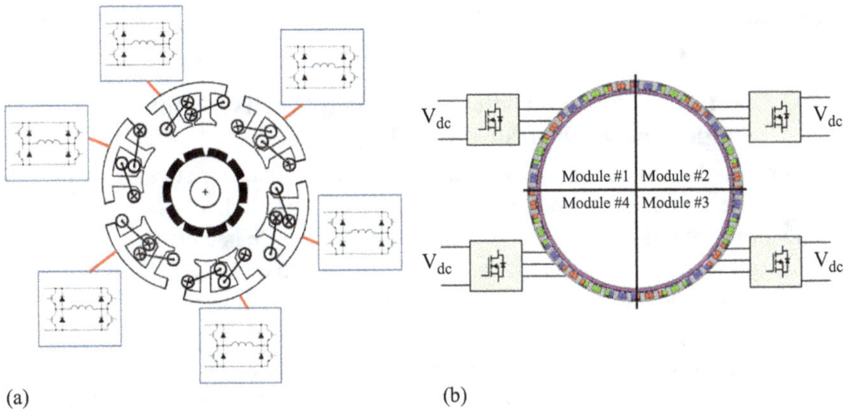

Figure 2.7 (a) Single-phase lane MMD and (b) multi-phase lane MMD configurations

popular choice for this architecture, the number of phase windings in the balanced n-phase set can be any integer number n that is greater than or equal to 2. It is assumed that the n-phase windings in each balanced winding set are connected in either a wye or delta configuration so that only *n* machine leads must be excited for each *n*-phase winding set. This means that an *n*-phase full-bridge voltage-source inverter consisting of *n* identical phase-legs connected to the same dc bus is a serious candidate for exciting the *n* phases in the winding set.

Table 2.1 shows that all five of the machine types are compatible with both the single-phase lane MMD and multi-phase lane MMD configurations for fault tolerant applications. However, this does not mean that all 10 of these MMD configurations are equally good for practical fault-tolerant MMD implementations. This important observation will be discussed in more detail after presenting the module isolation characteristics of the five machine types.

The fault tolerance of any modular systems depends on the ability of each of the modules to isolate itself from the effects of a fault in one of the other modules. For modular machines, this isolation must be considered in four dimensions [10]:

- *Electromagnetic isolation:* Are the machine modules electromagnetically decoupled from all of the other machine modules?
- *Electrical isolation:* Are the machine modules galvanically isolated from all of the other machine modules?
- *Physical isolation:* Can the machine modules be conveniently protected from physical contact with all of the other machine modules?
- *Thermal isolation:* Can the machine modules be installed in a manner that suppresses the thermal conductivity between them so that a fault in one module will not overheat any of the other motor modules?

For the single-phase lane MMD configuration in Figure 2.7(a), a motor module consists of a single two-lead phase winding, while a motor module for the multiphase lane MMD configuration in Figure 2.7(b) consists of an n-phase balanced winding set.

2.2.4.2 Isolation characteristics of five machine types for single- and multi-phase windings

Having made this important distinction, the isolation characteristics of each of the five machine types must be evaluated separately for each of these two motor module designations. Table 2.2 present the isolation characteristics of the five machine types for all 10 combinations. An S or M entry in any cell indicates that the associated machine type can be designed to provide the associated isolation feature using a single-phase or n-phase winding set, respectively.

These motor isolation ratings reveal some interesting trends. They show that all five machine types can be designed to provide any combination of the four types of isolation described above for multi-phase winding sets (marked by M in Table 2.2). An example of a PM synchronous machine using a 3-phase distributed winding is provided in Figure 2.8 that conveniently illustrates two of these isolation types. This machine has 30 slots and 10 poles, yielding a slot-per-phase-per-pole (SPP) value of 1. Each two-pole arc spanning a fifth of the stator circumference represents a module with a separate three-phase winding, as shown. This figure shows that the windings are configured so that the five 3-phase winding set make no physical contact with each other in either the slots or the end-winding regions. If these windings are excited by

Table 2.2 Comparison of Machine Isolation Achievability with Single- and Multi-Phase Drive Units

Machine Type	PM Synchronous (Surface PM & Interior PM)	Wound-Field Synchronous	Synchronous Reluctance	Induction	Switched Reluctance
Feature					
Electromagnetic Isolation Achievability	S,M	M	M	M	S,M
Electrical Isolation Achievability	S,M	S,M	S,M	S,M	S,M
Physical Isolation Achievability	S,M	M	M	M	S,M
Thermal Isolation Achievability	S,M	M	M	M	S,M
Legend	Single-Phase Unit		S		
	Multi-Phase Unit		M		

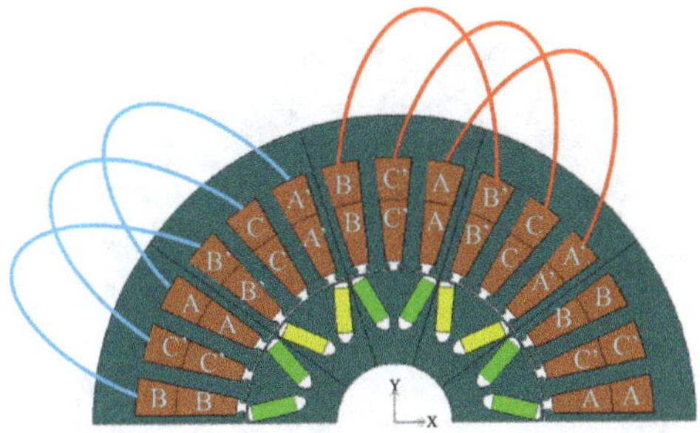

Figure 2.8 Cross-sectional view of a 30-slot, 10-pole ac machine stator with four 3-phase distributed winding sets (SPP=1) that illustrates the basis for achieving both physical and electrical isolation of the five 3-phase winding sets . © 2023 Reproduced, with permission, from IEEE [4].

adopting the multi-phase lane MMD configuration in Figure 2.7(b), the five winding sets are electrically isolated as well. However, this should *not* be interpreted to mean that these isolation features are naturally provided by conventional designs of these machines, with some exceptions that will be discussed in more detail later in this section. In fact, a continuing major challenge for motor designers is how to design fault-tolerant machines that can provide these isolation features with minimum negative impact on machine mass, volume, and cost.

Table 2.2 shows that, in contrast to the multi-phase winding isolation characteristics, all four of the isolation types can be provided for single-phase windings by only two of the five machine types, namely, PM synchronous and switched reluctance machines. The reason is that the achievability of single-phase winding isolation is closely tied to the machine's compatibility with concentrated windings, since a concentrated winding around a single tooth makes it feasible to build barriers to electromagnetic, electrical, physical, and thermal coupling with any of the other phases windings that are similarly concentrated. In contrast, distributed windings use phase windings that are inherently coupled to each other in multiple dimensions, making it much more difficult to achieve any of the four isolation types for the individual phase windings.

Figure 2.9 provides an example that illustrates the isolation opportunities attainable by concentrated windings, showing a cross-section view of a FSCW PM synchronous machine with 12 stator teeth and 10 poles. It is notable that only 6 of the 12 stator teeth are encircled by concentrated windings. The remaining 6 stator teeth that alternate with the wound teeth around the circumference of the machine provide physical barriers between all of the adjacent coils, significantly enhancing their galvanic isolation. Since each coil side fills a stator slot (a defining characteristic of a single-layer winding), there is no opportunity for the coil to make physical contact with any of the other phase windings inside the stator slots.

Finite element analysis of this machine structure reveals that the magnetic flux lines (visible in Figure 2.9) resulting from excitation of the two phase A windings form closed loops that pass through the two wound stator teeth and the stator yoke

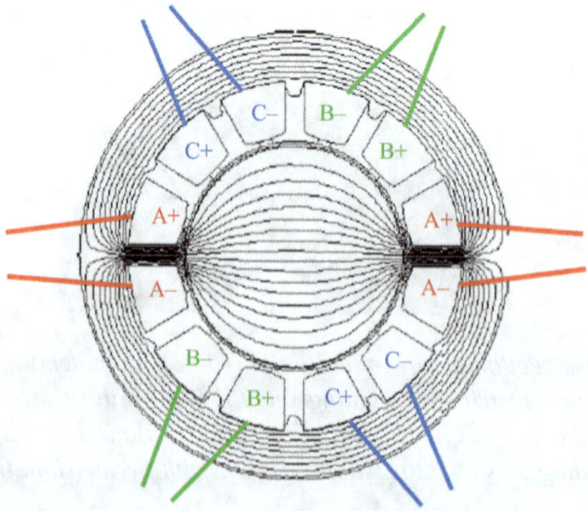

Figure 2.9 *Cross-section view of a 12-slot, 10-pole PM synchronous machine stator with single-layer fractional-slot concentrated windings (SPP=2/5) that illustrates the basis for implementing all four isolation features into each single-phase winding*

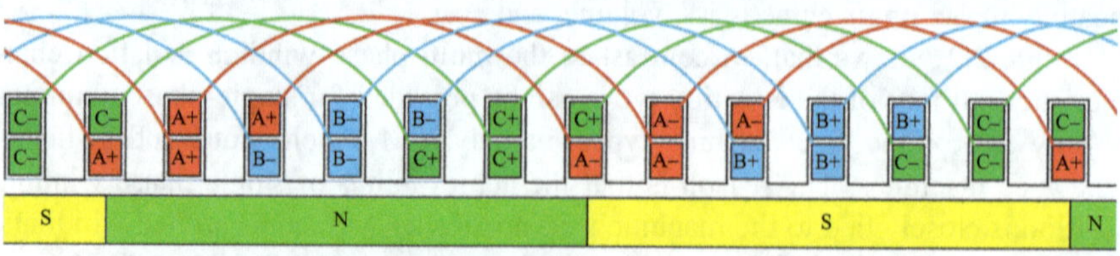

Figure 2.10 *Developed cross section view of two poles of a 24-slot, 4-pole PM synchronous machine stator with short-pitched distributed windings (SPP=2, coil pitch = 5/6) that illustrates the difficulty of achieving electromagnetic, electrical, physical, and thermal isolation for each single-phase winding*

channels between them, exhibiting near-zero electromagnetic coupling with any of the other concentrated phase windings. The presence of the intervening stator tooth between all adjacent windings provides a valuable means to thermally isolate each concentrated winding if it develops a serious (e.g., short-circuit) fault. Finally, electrical isolation can be provided for this machine by adopting the single-phase lane modular motor drive configuration in Figure 2.7(a), eliminating all galvanic electrical connections between the stator windings.

In contrast, Figure 2.10 shows a developed (i.e., unrolled) cross-section view of a two-pole span (approx. half of the stator) for a three phase PM synchronous machine with distributed windings using 24 stator slots and 4 poles, yielding a slot-per-phase-per-pole (SPP) value of 2. Each coil is short pitched by one slot (i.e., coil pitch = 5/6) to help suppress low-order spatial harmonic magnetic fields. Each slot contains two coil sides, and half of the slots for this stator winding contain coil-sides from two different phases because of the short-pitched coils. Each phase is tightly coupled electromagnetically with the other two phases. Physical and thermal isolation is very difficult to achieve because of the slots with the shared coil sides, as well as the overlapping of coils from different phases in the end winding region shown in the top section of Figure 2.10.

It should be noted that it is possible to adopt the single-phase lane modular motor drive configuration in Figure 2.7(a) for these distributed windings, requiring inverter access to both leads of each phase winding in order to achieve electrical isolation of the three phase windings. However, the tight electromagnetic coupling of these windings places a special burden on coordination of the inverter controls for the three phases to insure that the excitation waveforms applied to the three windings do not contain any zero-sequence voltage components that will cause large zero-sequence currents to flow in the three windings.

2.2.4.3 Open- and short-circuit fault tolerance of different machine types

The last topic regarding modular motor fault tolerance that is addressed in this section is the achievability of fault tolerance in the presence of open-circuit and short-circuit faults for the five types of machines. In order to keep this discussion as concise as possible, attention will be limited to fault tolerance in the context of multi-phase winding groups and the multi-phase lane modular motor drive configuration in Figure 2.7(b). It is assumed that the objective is to neutralize the impact of the fault in the multi-phase winding group that experiences the fault by driving all of the phase currents in that group to zero or some safe value and accepting the loss of average torque from the faulted winding group.

Because of significant differences in the threats posed by short-circuit faults appearing at the terminals of multi-phase winding sets and internal short-circuit faults that are developed inside or between coils of multi-phase windings, these two categories will be given individual attention. Although it may seem counter-intuitive to readers who are new to this subject, turn-to-turn faults between adjacent turns in any of the winding coils are generally acknowledged to be the most dangerous member of the family of internal short-circuit faults. The high threat posed by turn-to-turn faults is attributable to the extremely low internal impedance of shorted turns that can lead to damaging levels of fault currents even when the driving voltages are very low. No similar subclassification is required for open-circuit faults.

The fault tolerance achievability values for considered faults are presented in Table 2.3 with rankings at one of three levels. "Natural" (marked by N) indicates that the machine being rated exhibits a natural fault tolerance for the associated fault type with no or minimal extra effort required during the machine design process. "Intermediate" (marked I) indicates that the machine type requires some level of machine design modifications and/or drive control adjustments to

Table 2.3 Comparison of Machine Fault Tolerance Achievability for Major Fault Types

Machine Type	PM Synchronous (Surface PM & Interior PM)	Wound-Field Synchronous	Synchronous Reluctance	Induction	Switched Reluctance
Feature					
Open-Circuit Fault Tolerance Achievability	I	N	N	N	N
Winding Terminal Short-Circuit Tolerance Achievability	I	I	I	I	N
Internal Winding Short-Circuit Tolerance Achievability	C	C	I	I	I
Legend	Natural Intermediate Challenging		N I C		

neutralize the fault in that machine type. "Challenging" (marked C) indicates that even after modifying the machine design to reduce the fault impact level, aggressive/fast control action is required to recognize and respond to the fault to insure its safe neutralization.

Of the two major fault types, open-circuit faults are nearly always more benign than short-circuit faults. The entries in Table 2.3 for open-circuit faults is consistent with this observation, showing that four out of the five machine types exhibit a natural tolerance of this fault type. The PM synchronous machine is ranked as "intermediate" for open-circuit faults because of the special case of PM machines with wide ranges of constant-power (flux weakening) operation that can generate dangerously high back-emf voltages at high speeds if the inverter is suddenly shut down by removing all of the gate-drive signals, resulting in an uncontrolled generator fault [11]. Fortunately, there are well-established fault response techniques that include machine design modifications to limit the machine's characteristic current to safe levels less than or equal to the rated current, combined, in some cases, with special voltage clamping circuits to protect the power electronics from the risk of over-voltage failures.

Short-circuit faults appearing at the terminals of a multi-phase winding group pose the next highest level of average risk to the five machine types. For this fault type, only the switched reluctance exhibits a natural fault tolerance that is attributable to its high level of electromagnetic isolation between winding groups and the absence of any back-emf in the faulted phases when no current is flowing in the short-circuited machine phases. The induction machine might at first seem likely to be safe from post-fault currents because of the passive nature of its squirrel-cage rotor. Unfortunately, any rotor current takes time to decay so that the spinning rotor can deliver a back-emf to the shorted phase-winding group because of the rotor current induced by the healthy phase-winding groups that are still delivering torque. In brief, all four of the remaining machine types are ranked as "intermediate" because they can be custom designed to sufficiently suppress the amplitude of the short-circuit current that flows in the faulted multi-phase group to achieve fault tolerance.

As noted above, short-circuit faults appearing inside the winding coils of a multi-phase winding group pose the biggest risk, particularly for turn-to-turn faults, because of the dangerously high fault currents that can be induced in the shorted turns with very low amplitudes of back-emf excitation. The fact that the magnets in a PM synchronous machine cannot be turned off even if a fault is detected presents a serious challenge, but the prospects for fault tolerance are little better for the wound-field synchronous machine because the remaining healthy multi-phase winding sets will not contribute sufficient useful torque if the field current is dropped to zero. Aggravating the situation, turn-to-turn faults can be notoriously difficult to confidently detect, despite the dangers they present after the faults are fully developed. Although researchers are actively pursuing new solutions, internal winding faults still must be ranked as "challenging" (C) for both of these machine types.

The situation is somewhat better for the synchronous reluctance and induction machines because neither of these machines has active field excitation in its rotor. However, this advantage only becomes meaningful if the internal turn faults can be rapidly detected, resulting in rapid excitation removal from the faulted module windings. As a result, these two machines are ranked as "intermediate" for internal winding short circuits.

The switched reluctance machine holds the highest promise for fault-tolerant response to internal winding faults because of the machine's natural advantages summarized above. As a result, a phase winding with an internal short-circuit fault will not suffer from any current in the short-circuited winding turns, provided that the electromagnetic coupling between the phase windings is designed to be as low as possible and excitation is removed from the faulted phase winding. However, this means that a robust fault detection scheme is required to quickly recognize any internal winding short-circuit condition, similar to the fast detection requirement for the other four ac machine types. The requirement for this fault detection feature resulted in the "intermediate" ranking for the switched reluctance machine shown in the table.

2.2.5 Concluding remarks

This section has shown that modular motor drives offer significant advantages for some important classes of motor drive applications. It is likely that these modular drive architectures will take on greater significance in the coming years as applications such as electrified aircraft propulsion and accessories (e.g., pumps, actuators, etc.) move into production. Designing the modular motor drive as an integrated assembly incorporating both the machine and the power electronics is likely to be mandatory because the negative impact of having large numbers of machine windings excited by modular inverters in separate enclosures is nearly eliminated by the integration.

One of the major outcomes of this discussion is that machine types that are compatible with concentrated windings offer some valuable advantages for achieving the manufacturability and fault tolerance advantages that are associated with modular motor drives. More specifically, PM synchronous machines with fractional-slot concentrated windings and switched reluctance machines both stand out as appealing candidates for highly modularizing the motor to achieve either or both of these objectives.

This discussion of fault-tolerant modular motor drives has highlighted both the opportunities and challenges for achieving high levels of fault tolerance that are posed by each of the five machine types that were investigated. Both FSCW PM synchronous motors and switched reluctance motors draw special attention because of their high ratings for achieving the four desired isolation characteristics. However, the special advantages offered by switched reluctance machines for offering the desired isolation characteristics as well as suppressing the serious risks posed by short-circuit faults are unique among the five machine types. In contrast, the higher vulnerability of the PM synchronous machines to short-circuit faults highlights the serious technical challenges that must be overcome by researchers in industry and academia who are seeking to harness the superior power density of PM synchronous machines in demanding applications such as aerospace motor drives that require superior fault tolerance as well.

2.3 Winding materials and wire types

Electrical machine windings are normally manufactured using a conducting wire (or magnet wire) with a very thin insulating coating. The conducting materials are generally unalloyed metals, predominantly oxygen-free high purity copper. Copper is readily available, has a high electrical and thermal conductivity and has good mechanical properties (annealed) for ease of winding. The conductivity of electrical conductors is normally defined as a percentage of the International Annealed Copper Standard (IACS) which is 1.724 micro-ohm-cm at 20°C.

The magnet wire can be purchased in a range of standard wire diameters defined by International Standards – for Europe and Asia this is the IEC standards (IEC 60317). Copper magnet wire can be obtained over a wide range of diameters from several mm down to 5–10 microns for high-frequency applications.

Other conductor materials are also available, such as aluminium, brass, silver, gold, nickel, or copper alloys with particular properties, for example minimum bend radius. Aluminium is the most common alternative to copper and is found in applications that require low cost and low mass. Aluminium also has particularly interesting properties at very low temperatures (cryogenic temperatures) and this has become of interest with the introduction of hydrogen fuels stored in cryogenic liquid form – this is discussed in more detail later in this section. The conductivity of pure aluminium is approximately 61% IACS which means you need more aluminium compared to copper for a given per unit resistance. However, the density of aluminium near room temperature is 2,710 kg/m^3 compared to 8,960 kg/m^3 for copper. In relation to the mass, this means that aluminium is almost twice as good as copper and this is one of the key reasons that aluminium is still preferred to copper in certain applications.

The other key benefits of using aluminium as a wire material are:

- it is a cheaper material
- it is more flexible than copper
- when exposed to air, it forms a thin oxide layer that protects the material from air and water, reducing the risk of corrosion
- the higher resistivity can also reduce the effect of eddy current in higher frequency applications.

The key benefits of copper are:

- its higher conductivity which reduces the losses and increases efficiencies for the same volume
- it has a lower thermal expansion and because thermal cycling is a known failure factor of the insulation, this would suggest that copper wire would have a longer insulation lifetime.

The key disadvantage of aluminium is related to the benefit of creating an oxide coating when exposed to air: this also makes aluminium magnet wire difficult to solder connections to and usually requires special welding equipment or processes that can remove the oxide layer from the material. An interesting variant to overcome this problem is copper clad aluminium, which has a thin layer of copper attached to the surface of the aluminium. This makes solder connections straightforward, even though the bulk of the conductor is aluminium.

In terms of the conductor material selection for integrated drives: winding or coil connections are one key factor and variable frequency operation with Pulse-Width Modulation (PWM) waveforms is another. Copper conductors are certainly easier to ensure a low-resistance connection to the coils or windings and any connection must be low-resistance to avoid a hot-spot at the joint, leading to thermal runaway and failure of the connection. Connections are normally mechanical (e.g. crimp) or solder jointed. For copper wire, this is usually reasonably straightforward: remove the insulation coating (chemical, mechanical, or heating) and then solder (dip, flow, hand, etc.). It is also possible to solder without removing the insulation first, but this requires more specialist equipment (fusion/spot or water welding).

Aluminium connections are more difficult. One particular issue worthy of specific mention here is the situation of joining an aluminium wire to a copper connection. One has to be a little careful in this situation because copper and aluminium can form an electrochemical reaction in the presence of an electrolyte, for example condensation, with the copper leading to galvanic corrosion of the aluminium. This can lead to an increased contact resistance and joint overheating, leading to failure. Encapsulation of the joint to eliminate water/condensation ingress is one possible solution.

The conventional process for connecting aluminium to aluminium is to remove the insulation using a chemical, mechanical, or heating process, then using a flux and solder appropriate for aluminium. Readers should remember that bare aluminium creates an oxide layer on the surface very quickly in air, which can cause problems for the joint. Aluminium joints can also be created using specialist welding equipment to ensure a good electrical joint, particularly if stranded or Litz type coils are being used. Stir-friction techniques are also being introduced. For integrated motor drives, this is likely to be fairly common, particularly with higher speed, higher frequency drives. In high-frequency motor drives (e.g. >300 Hz), Litz conductors commonly use a large number of insulated small diameter strand wires and it is important that the connection joint ensures a good electrical connection to all the strands, otherwise the supply current would transfer into the remaining 'connected' strands and this can lead to overheating in the conductor bundle. In this case, it is possible to get strand insulation that allows soldering using a solder dip/flow that dissolves the insulation to facilitate a good joint.

Figures 2.11 and 2.12 show a typical high-frequency copper Litz conductor and a bundled multi-stage Litz [12]. Litz wire was traditionally developed for early radio applications with frequencies in the range 10 kHz–5 MHz. The use of Litz wire is becoming more common for integrated drives which are operating at fundamental frequencies in the range 500 Hz–2,000 Hz. Litz wire is an alternative to multi-stranded conductors and offers the advantage of reducing the ac losses in comparison to multi-stranded conductors but with the disadvantages of higher cost, lower copper fill factor and a higher DC resistance. Litz wire is formed by bunching a number of thin insulated strands into a bundle and twisting these along the length of the wire. A number of these bundles are then grouped together and then twisted together along the length to form the final wire. The twisting of the strands reduces the AC losses but also increases the DC resistance. The AC losses are caused by eddy current and proximity losses in the wire.

Eddy currents can be controlled by choosing the strand size to suit the frequency and a rule of thumb is to choose a strand-size one-third of the skin depth at

Figure 2.11 Basic profiled copper Litz wire. Courtesy of Elektrisola Dr. Gerd Schildback GmbH & Co., Germany.

Figure 2.12 Multi-staged Litz wire with 5 bundles of 26 strands. Courtesy of Elektrisola Dr. Gerd Schildback GmbH & Co., Germany.

the operating frequency. Proximity losses are normally the main component of AC loss as a result. Internal proximity loss is caused by eddy currents in one strand induced by the current in a strand next to it, i.e. in close proximity to it. Other proximity loss elements to be aware of are the external proximity losses caused by external fields passing through the wire, such as those from rotor magnets. These proximity effects are effectively a redistribution of current within any strand but the total current in each strand is unchanged. The last source of AC loss is sometimes known as a coupling loss: it is caused by magnetic flux passing between strands, resulting in an induced emf which drives an eddy current that circulates through the connection joint at each end of the wire around each pair of strands that have a net flux linking them. Litz wire largely eliminates this by using the twist pitch to reduce the net flux linking any pair of strands. The perfect Litz wire would ensure that each strand in the wire occupies the same average position in the wire along its length. This ensures that there are no circulating eddy currents and each strand carries the same current. Litz wires come in different types: there is basic Litz as described above; compressed and profiled wire; strain-relief (central core) wire and even self-bonding Litz.

For aluminium Litz wires, the requirement that all strands need to be connected poses additional issues. Specialist welding techniques are usually needed, such as thermo-compression which heats the connection and compresses it to form a weld or ultrasonic welding which uses ultrasound energy to vibrate the strands causing friction, heat, and ultimately a welded joint. Ultrasonic welding is usually more appropriate for connections with a smaller number of larger strands. Thermocompression bonding makes an excellent joint for Litz wires with high strand numbers and is a cost-effective solution. This method uses localized heating to remove the insulation on the strands (for certain insulation grades) and mechanical compression to ensure a very high percentage of the strands are electrically connected.

The thermal behaviour of conventional conductors, such as copper or aluminium, is relatively straightforward if the operating temperature remains within the normal range of approximately –50°C to 200°C. The variation of the electrical resistivity in this range is closely linear and modelled using the standard expression:

$$\rho(T) = \rho(T_{ref})(1 + \alpha(T - T_{ref}))$$

where T is the operating temperature, T_{ref} is the reference temperature (normally 20°C or 0°C), and α is the temperature coefficient for resistivity. Table 2.4 shows the temperature coefficient at 20°C for several common metals.

If operation is expected over a wide temperature range, it is sensible to check the material data and use a polynomial function if necessary. National laboratory websites such as www.nist.org are a good resource for this and can also provide higher order polynomial expressions for curve-fitting temperature data. Low-temperature conductivities (cryogenic temperatures) are particularly non-linear. Figure 2.13 shows the behaviour of copper and aluminium down to 1K.

Table 2.4 Temperature coefficient of resistivity for some common materials (20 °C)

Metal	**Temperature coefficient of Resistivity**
Copper	0.00393
Aluminium	0.004308
Iron	0.005671
Gold	0.003715
Silver	0.003819
Nickel	0.005866
Tungsten	0.004403

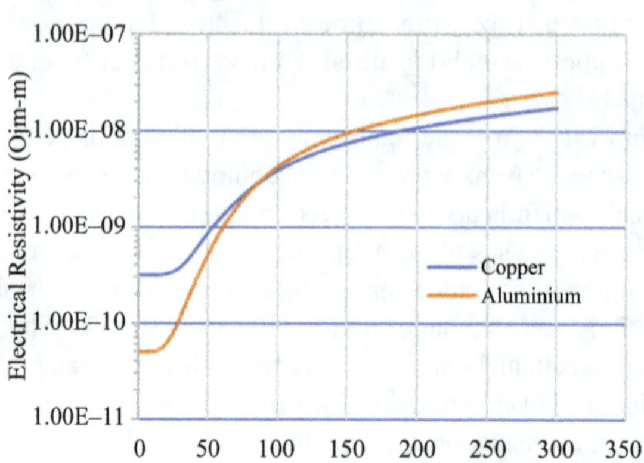

Figure 2.13 Electrical resistivity of copper and aluminium as a function of temperature in degrees kelvin, down to 1K

For high-temperature applications (e.g. >250°C), aluminium conductors have limited suitability because of the low melting point (660°C). The softening temperature is obviously lower, so this would reduce the operational temperature range further. Copper has a melting point of 1,080°C, so it is the most common conductor choice for high-temperature applications. The other key issues at high temperature are oxidation and corrosion. Increasing the temperature increases oxidation and the surface of copper wire will oxidise. This tends to be an unstable surface layer, with cracking and flaking occurring. Copper wire is commonly protected by coating the copper wire with nickel. The melting point of nickel (1,455°C) is much higher than that of copper, so nickel-coated copper wire is the most common high-temperature conductor choice. Oxidation occurs at a much lower rate (approx. 20 times) and the nickel coating also improves wear and corrosion resistance. The thickness of the nickel coating depends on the operating temperature. There are also some other interesting features of nickel-coated copper wire:

- Nickel is a mild ferromagnetic material so this can influence the field distribution in motor coils.
- Diffusion between the nickel and the copper occurs: nickel into the copper and vice versa. The rate of diffusion increases with temperature. The impact is to increase the conductor resistivity.
- Skin effect in AC conductors tends to push the current out to the surface of the conductor. The presence of the higher resistivity nickel on the surface might suggest that this is beneficial; however, the current distribution is quite complex in a two-layer material with current able to move across the boundary. Nickel has a high magnetic permeability, which reduces skin depth and therefore increases ac loss. Published data tends to confirm that the nickel coating does significantly increase the ac resistance of nickel-coated wire at high frequencies.

For very high-temperature applications (>750°C), pure nickel wire is the common choice. However, it is worth noting that nickel has a significantly higher resistivity than copper and also has a higher temperature coefficient of resistivity.

Cryogenic low-temperature applications are also of interest. This has been mainly focused on superconducting materials for high-field magnets (e.g. MRI scanners). However, conducting materials such as copper and silver are often used in conjunction with superconducting material to provide thermal stabilisation.

Hydrogen, stored in liquefied form at 20.3K, has been proposed as a fuel for future aircraft. This has created interest in aircraft electrical propulsion drives, operating with cryogenic cooling. This can utilize the very low resistivity of some materials at low temperature, without resorting to superconducting systems.

Figure 2.13 shows how the resistivity of aluminium reduces faster than that of aluminium as the temperature drops and below approx. 75K, aluminium becomes a better electrical conductor than copper. At very low temperatures, the purity of the material has a significant effect on the electrical resistivity. One common method of defining material purity is to use the residual resistivity ratio (RRR) value. This is ideally defined as the ratio of the electrical resistivity at 0K to that at room temperature. However, it is impossible to measure the resistivity at 0K so the value

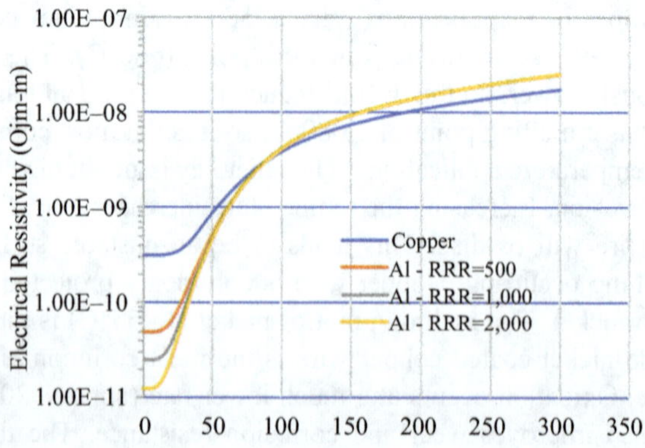

Figure 2.14 Electrical resistivity of copper and aluminium of varying purity v temperature in degrees Kelvin

of liquid helium – 4.2K – is used instead. Figure 2.14 shows the impact of increasing the purity of aluminium on the electrical resistivity below 100K.

Figure 2.14 shows that increasing the purity (increasing the RRR number) of aluminium can significantly improve the resistivity at low cryogenic temperatures. For applications using liquid hydrogen, for example, the liquid temperature is 20K so operation at these temperatures is certainly achievable. The impact for machines is a reduction in the electrical resistivity of the conductors of potentially 4 orders of magnitude and significant gains in operating efficiency.

2.4 Insulation

Electrical machine winding insulation is a critical issue for integrated drives because the majority of motor failures are insulation failures. Insulation failure normally occurs either through over-voltage electrical insulation breakdown or through thermal overload and/or cycling, leading to an insulation breakdown.

In general, machine insulation systems can be divided into two categories: high-voltage machines (typically >2–3 kV) and low-voltage machines (<2–3 kV). Most modern integrated drives are low-voltage systems so this section will focus on low-voltage motor drives. Low-voltage machines generally have multi-turn/strand random wound coils although there are number of hairpin winding coil arrangements appearing now particularly in the automotive sector. The insulation management of random wound machines is generally split into three main areas: coil–ground; coil–coil (in a slot or in the end-winding) and turn–turn insulation between the turns in a coil.

2.4.1 Resonant voltage overshoot at the motor terminals

One of the key aspects in insulation design for integrated drives is the PWM switching waveforms used in the converter. These are known to induce more winding failures than traditional sinusoidal line-fed machines. There has been a large number of papers published in recent years, devoted to an understanding of winding failures in converter-fed motors. One of the main factors is the 'transmission line' effect leading to resonant over-voltages caused by the PWM waveforms at the motor terminals.

A summary of the basic issues will be provided here for completeness because it has an impact on the trend towards faster device switching in converter technology. PWM waveforms are characterized by a string of pulses with sharp edges (leading and trailing edges) having a short rise time t_r or high dv/dt. Rise times for conventional Si-based converters are typically 100 ns, but with wide bandgap devices, such as SiC or GaN, the rise time can be two orders of magnitude faster. The short rise time leads to the cable connecting the motor to the converter behaving as a transmission line and this can lead to a significant voltage resonant overshoot at the motor terminal on the leading and trailing edge of each pulse. Figure 2.15 shows an example of a resonant overvoltage response at the motor terminal to a step voltage from the converter [13].

The cable between the motor and the converter can be modelled as a transmission line, with a cable represented as a two-port distributed impedance network. R and L represent the series resistance, R, and inductance, L, of the cable per metre length, and G and C represent the parallel conductance, G, and capacitance, C, per meter length. If a step pulse with rise time t_r is applied to the input terminals of the cable, the resulting wave equations can be solved in the time domain in terms of forward and backward travelling waves along the cable. It is very common to ignore the resistance of the cable R and G and consider the case of a lossless cable. In this situation, the equations can be greatly simplified to provide a good understanding of the behaviour of the motor and cable interactions. This shows that the forward and

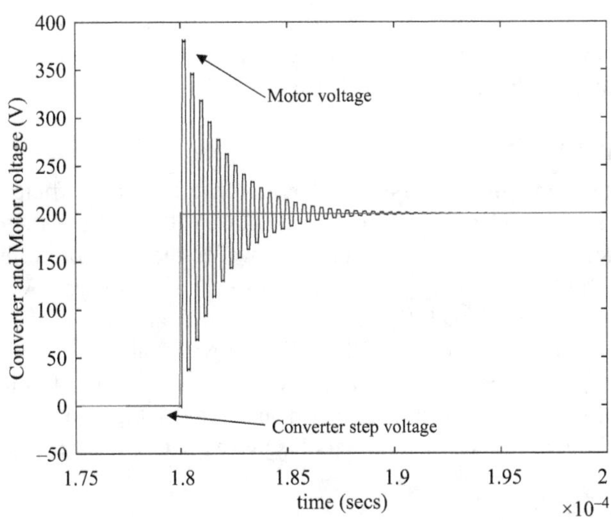

Figure 2.15 Resonant overvoltage waveform at motor terminals, 5 m cable length

backwards waves propagate at a velocity v_p and the cable has a characteristic impedance Z_c given by the simple equations:

$$v_P = \frac{1}{\sqrt{LC}} \text{ and } Z_C = \sqrt{\frac{L}{C}}$$

The characteristic impedance or surge impedance defines the terminating load impedance that would produce no reflection at the load. Any other load impedance results in a reflected wave along the transmission line. The magnitude of the reflected wave depends on the extent of the mismatch between the load and the cable surge impedance and is best modelled by a reflection coefficient that can be defined at the load (motor) and the supply (converter):

$$r_{mot} = \frac{Z_{mot} - Z_c}{Z_{mot} + Z_c} \text{ and } r_{con} = \frac{Z_{con} - Z_c}{Z_{con} + Z_c}$$

In general terms, the motor impedance tends to be significantly higher than the cable characteristic impedance so if we consider Z_{mot} tending to be very large, then r_{mot} tends to +1, which would represent a full positive polarity reflected wave. The DC link capacitor generally means the converter impedance appears as a short-circuit $Z_{con} = 0$ to a reflected wave, so r_{con} is -1, which represents a negative polarity fully reflected wave from the converter terminals.

If a voltage pulse appears at the input to the transmission line, a forward travelling pulse will travel along the cable of length, l, at a velocity, v_p taking a time $t_p = (l/v_p)$. Assuming the motor impedance is not equal to the cable characteristic impedance ($r_{mot} = 0$), a reflected backward travelling wave will be created. This returns to the supply where it is reflected back again as a forward travelling wave ($r_{con} = 0$) .... and so forth.

The typical resonant over-voltage appearing at the motor terminals is shown in Figure 2.15 and the magnitude U_1 of the initial overshoot for a pulse voltage steady-state magnitude of U is:

$$U_1 = (1 + r_{mot})U$$

If we assume that in the worst case, the motor reflection coefficient tends to 1, then we can see the 'worst' case peak initial overvoltage at the motor terminals would $2U$, i.e. double the pulse voltage. However there is some mitigation for pulses with a finite rise time, t_r. There will be a reduced voltage overshoot if the reflected wave can travel back to the converter and then return to the motor terminals in less than the rise time, t_r. If the reflected wave takes longer to return than the rise time, t_r, then the initial voltage overshoot at the motor terminals will be double the pulse voltage. This defines a critical cable length, l_c, to reduce a resonant voltage overshoot at the motor terminals for pulses with a rise time t_r:

$$l_C = \frac{t_r}{2\sqrt{LC}}$$

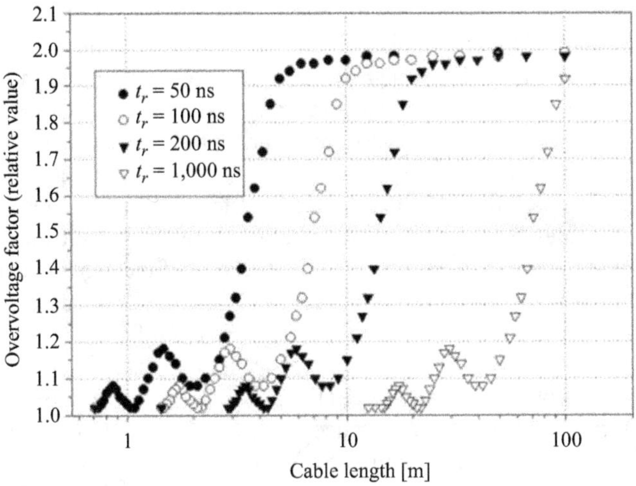

Figure 2.16 Voltage overshoot factor for varying pulse rise time and cable length. © 2023. Reproduced, with permission from, IEEE [14].

For integrated drives this is an important criteria. Figure 2.16 shows the voltage overshoot factor for varying pulse rise time and cable lengths [14]. The factor varies between 1 – no overshoot – and the worst case factor of 2 – double voltage. This is one of the advantages of a closely coupled integrated drive – the cable length between the motor and the converter can be kept very short and potentially reduce the voltage overshoot at the motor terminals. The trend for converter technology on the other hand is towards faster switching devices such as SiC and GaN and this leads to much lower rise times compared to Si devices and lower critical cable lengths.

If we return to the voltage overshoot as shown in Figure 2.15, the voltage overshoot creates a resonant voltage that decays with time. The decay is related to the motor reflection factor, the second, third, etc. peak have magnitudes:

$$U_2 = \left(1 - r_{mot}^2\right)U$$

$$U_3 = \left(1 + r_{mot}^3\right)U$$

The period of the resonant voltage is $4t_\mathrm{p}$ so the resonant frequency is:

$$f_r = \frac{1}{4l\sqrt{LC}}$$

2.4.2 Voltage distribution in the motor windings

The resonant voltage overshoot at the motor terminals described in the previous section relates to the voltage that develops at the motor windings terminals caused by high dv/dt pulses and a certain length of cable between the converter and the motor acting as a transmission line.

The voltage distribution through the motor coils also behaves like a transmission line but with a modified transmission line model representing the motor coils. Conventional motor insulation design assumes the voltage is uniformly distributed through the windings and these voltage levels are used to design the motor insulation: coil to ground (core); coil to coil (phase separators in the slots and end-windings) and turn to turn insulation. However, the motor winding can also be regarded as a transmission line and a modified line model used to determine how the resonant voltage overshoot propagates into the coils from the terminals. It is well known that any resonant voltage overshoot propagates into the first coil of the winding coil chain and decays over the first few turns of the first coil and this has the greatest impact on the turn–turn insulation around these turns. Winding insulation failures in this area of the winding have been observed in converter supplies motor drive systems.

The transmission line model used in the previous section has to be modified to include the series turn to turn capacitance per metre (K) found in motor coils systems. This is inserted in parallel with the turn inductance and resistance. The main assumption here is that the turn to turn capacitance is constant throughout the coils – however with random wound coils where the turns are randomly distributed, this is clearly not a reliable approximation. If we adopt the same procedure as the cable and assume a lossless model to gain an insight into the voltage propagation through the motor coils, the propagation constant becomes:

$$\gamma = \frac{j\omega\sqrt{LC}}{\sqrt{1-\omega^2 LK}} = \alpha + j\beta$$

This has an interesting feature in that the denominator can be real or imaginary depending on the frequency, ω. This allows us to define a critical frequency, ω_c:

$$\omega_c = \sqrt{\frac{1}{LK}}$$

The behaviour of the voltage propagation through the motor winding depends on the frequency in relation to the critical frequency:

$$\text{(a) } \omega < \omega_c \ \alpha = 0; \beta = \frac{\omega\sqrt{LC}}{\sqrt{(1-\omega^2 LK)}}$$

$$\text{(b) } \omega > \omega_c \ \alpha = \frac{\omega\sqrt{LC}}{\sqrt{(\omega^2 LK - 1)}}; \beta = 0$$

$$\text{(c) } \omega \gg \omega_c \ \alpha = \sqrt{\frac{C}{K}}; \beta = 0$$

If we look at case (a) – the voltage propagation represents a non-attenuated (lossless) travelling wave through the coils propagation at velocity, $v_p = \omega/\beta$. If the frequency is greater than the critical frequency – case (b), the voltage distribution becomes an attenuated standing wave. If the frequency increases further – case (c) – the voltage distribution remains an attenuated standing wave but the attenuation constant is simplified and is determined purely by the coil to ground and turn to turn capacitances. The attenuation constant in cases (b) and (c) is exponential $e^{-\alpha z}$, where z represents the distance measured from the motor terminal along the winding.

These equations can be used to determine the non-linear voltage distribution into the motor winding and determine the insulation stress around the first winding turns of the coils attached to the motor terminals. PWM waveforms with high values of dv/dt can produce high electric field values around the initial turns in the winding leading to the potential for partial discharge which over a prolonged period can lead to an insulation failure.

2.4.3 Mitigation methods for waveforms with high dv/dt values

Following the discussion in the previous sections, the obvious technique to avoid resonant voltage overshoots at the motor terminals is to keep the motor cable to the converter as short as possible. Fully integrated drives clearly have an advantage in this area.

Other options include pulse-shaping converter waveforms to reduce high dv/dt values. An alternative would be to modify the winding insulation to accommodate the non-linear voltage distributions in the motor coils.

The remaining options involve the use of filtering at the output of the converter: dv/dt and sine-wave filters for both differential and common mode voltages. However, these do tend to be large, heavy, and expensive.

2.4.4 Motor coil insulation systems

Conventional winding wire or magnet wire is drawn down to its bare conductor diameter and then coated with an insulating enamel (liquid resin) and then cured to form the wire insulation. The insulation coating provides a significant number of important functions besides the electrical insulation: mechanical stresses during winding (bending, etc.) and thermal expansion/contraction; chemical and environment protection; oxidation of the conductor; corrosion and many more. The technical performance of the many commonly used insulation grades is defined by standards such as IEC 60317.

The enamel coatings are normally polymer film coatings and the specific type of polymer defines its temperature rating. The temperature rating is defined in the standards as the maximum operating temperature to provide an average lifetime of 20,000 h. This assumes a degradation of the insulation caused by the breakdown of the polymer chains accelerated by increasing temperature based on the Arrhenius chemical reaction rate theory. Most wire manufacturers provide a graph of the expected lifetime versus temperature using this relationship. As a rule-of-thumb, the Arrhenius model tells us that the expected lifetime is halved for a 10 °C rise in temperature.

The relationship of lifetime to the temperature rating of insulation is based on constant or slowly changing temperature. There is growing evidence that thermal cycling can significantly reduce these lifetimes. The Arrhenius model is based on the influence of temperature on breaking the polymer chains in the insulation –

however, thermal cycling produces significant mechanical stresses on the insulation through cyclable thermal expansion and contraction leading to a failure caused by cracking or delamination of the insulation.

Over the years, the development of the polymer coatings has increased the maximum operating temperature up to a maximum of around 240°C – these are based on the polyamide-imide and polyimide insulation coatings. Nickel-plated copper wire with a polyimide coating can take the maximum temperature up to 280°C. However, to go beyond these temperatures, the organic polymer coatings have to be replaced by inorganic coatings (commonly ceramic based) and there are a few examples of high-temperature wire in the market – for example Ceramawire HT (500°C). These generally use a nickel-coated copper conductor and a ceramic based coating for insulation. Specific problems associated with these high temperature wires are related to flexibility, bendability and porosity to moisture.

The typical electric strength of the polymer-based coatings is approximately 200 V/μm so the wire coatings are very thin (10–30 μm). The manufacturing process of the coating can lead to tiny pin-holes in the coating and these are a potential issue for partial discharge and failure. The standards (IEC 60317) give guidelines on recommended maximum number of pin-holes per unit length. Insulation is also provided in three Grades (1–3) and this corresponds to the number of enamel coatings on the wire (one to three). This increases the insulation depth that provides extra voltage withstand and also reduces the likelihood of any pin holes through the full insulation.

There are also an increasing number of manufacturers offering winding wire specifically designed for inverter-fed motors. These are wires that are modified to accommodate the high d*v*/d*t* pulse waveforms discussed in the previous sections. Voids in the insulation materials or delamination can be subjected to locally high electric fields leading to internal partial discharges (PD) within the voids and not generally visible inside the insulation. External PD can also occur around the external surfaces of the wire insulation if there exist high electric field concentrations that can ionize the air near the insulation surface and these are often visible as sparking or the release of UV light (purple glow). White or black powder can also be deposited on the insulation surfaces. This can occur if overvoltages exist between wires near one another. Repeated PD can lead to erosion of the insulation material and ultimately a complete electrical breakdown between conductors around the void area. PD is characterised by a partial discharge inception voltage (PDIV) which corresponds to the increasing voltage level needed to start the electrical discharge and a partial discharge extinction voltage (PDEV) which is the reducing voltage needed to extinguish the discharge. The PDIV reduces with temperature but the change is relatively slow – for example a PDIV of 700 V with a temperature reduction of approx. 4 V/°C.

One of the other interesting environmental factors that influence PDIV/PDEV is pressure. This is very relevant to electrical insulation in aerospace applications – the pressure at 30,000 ft is 0.3 bar. Reducing air pressure reduces the breakdown strength of air as described by Paschen's curve for air shown in Figure 2.17.

Paschen's law describes the breakdown of a gap between two electrodes for different gases [15]. The curve predicts the breakdown voltage as a function of the

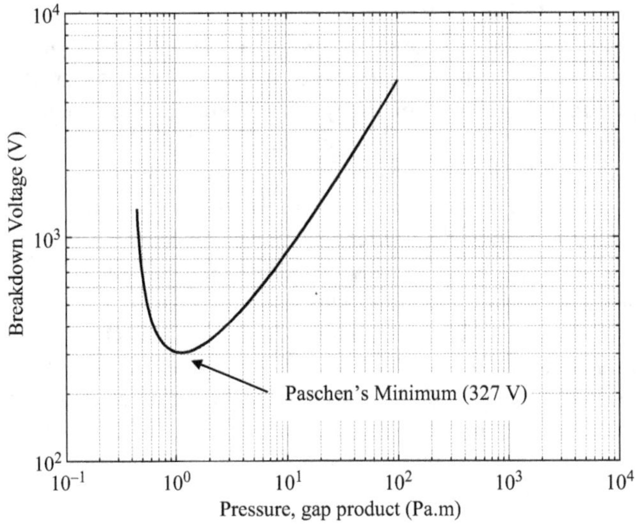

Figure 2.17 Paschen's curve of breakdown voltage versus pressure x distance

pressure times the gap distance. It is clear from Figure 2.17 that the breakdown voltage for air reduces as the pressure reduces to a minimum value (Paschen's minimum) of 327 V. This is very important for electrical systems in aerospace and explains why most of today's aircraft electrical systems are designed to operate below 300 V. However, as the electrification of aerospace gathers pace, power levels are increasing and it is increasingly difficult to operate at such low voltages, which is an important aspect in the design of high-power, aerospace electrical systems.

The inverter rated wire tends to address the problem of PD in two different approaches. Modifying the insulation material by adding nano-particles to prevent erosion of the material by PD. This can be facilitated by using the multiple coatings of different insulation materials. This does not stop the PD but it does prevent or slow down erosion of the insulation to ultimate breakdown. The other approach is to increase the PDIV of the insulation material and this aims at preventing PD completely (PD-free wire).

2.5 Magnetic cores

2.5.1 Soft magnetic materials

The function of the soft magnetic core in an electrical machine is to provide a very low reluctance path for magnetic flux to maximise the magnetic flux capability of the machine. This needs to be achieved whilst introducing the minimal loss. The magnetic core typically takes up over 60% of the active mass of the machine and so its design is critical in maximizing power density.

2.5.1.1 Laminations

An ideal core has a high magnetic permeability, a high saturation flux density and a low iron loss. In almost all cases, the core is made up of a set of laminations to reduce eddy current loss. There are many texts discussing the general properties of electrical steels [16,17] and so only a general summary will be given here, along with the current trends in materials.

A common feature of integrated drives is the need to minimize mass and volume, often accompanied by a high rotational speed. The soft magnetic core generally comprises the teeth and the core back and it is usually the core back which is the largest single component. This can be reduced in size by increasing the pole number of the machine, thereby reducing the flux per pole. However, a high pole number and a high rotational speed require a high electrical frequency. Whilst a motor connected directly to the grid will usually have an electrical frequency 50 or 60 Hz, it is now common for high performance drives to operate at over ten times this. For example, a 12,000 revs/min, 10 pole automotive drive will require an electrical frequency of 1 kHz.

The two major components of iron loss in the soft magnetic core are hysteresis loss and eddy current loss. The former is directly proportional to frequency, whilst the latter is a function of frequency squared. Compared to a 50 Hz main connected machine, a machine operating with a frequency of 1 kHz will therefore have an eddy current loss in the core of $(1{,}000/50)^2 = 400$ times greater.

To minimise eddy current loss, there are two measures usually taken:

- Silicon is usually added. This increases the resistivity of the material, though with a small reduction in saturation flux density. Electrical steels now commonly have up to 3.5% silicon added. Higher silicon content is difficult to achieve though, notably, one manufacturer, JFE Steel, is producing silicon content of 6.5% through a chemical deposition effect, producing a class leading loss density at high frequency. However, this is at the expense of reduced saturation flux density and poorer mechanical properties: the material becomes harder and much more brittle, making it difficult to punch. Other alloys are also possible, such as that including aluminium, but are much less common.
- Reducing the thickness of the laminations. Eddy current losses are proportional to the square of the thickness of the laminations: consequently there is a trend to use thinner laminations in high speed drives. 0.35 mm is the standard, but 0.2 mm is now also commonly used, with 0.1 mm also available.

As the need for increased efficiency and higher electrical frequencies continues, there continue to be investigations into the use of more exotic materials. Amorphous iron falls into this class [18,19]: it is an iron alloy, containing both boron and silicon. The material is very thin: typically only 0.025 mm thick, giving it exceptionally low eddy current loss. It is produced by a process of very fast cooling of the molten material, which does not give time for the usual crystalline structure to form. This results in an exceptionally low coercive force and consequently very low hysteresis loss. At first sight, it seems to be the ideal material, but in practice, there are a number of issues. It has a low saturation flux density and is exceptionally brittle so that it generally cannot be punched in the manner of production of normal laminations. It also suffers strongly from magnetostriction – an effect where the loss in the material increases rapidly when the material is under compressive stress. Most electrical machine cores are shrunk into a frame and consequently it is very difficult to avoid compressive stress – new methods of construction are therefore called for. Finally, because there are of the order of ten times more laminations required, there are ten times more gaps between laminations. This inevitably decreases the core 'stacking factor' – the ratio of lamination to overall axial space taken up by the core – to typically 80%, which in turn further reduces the peak flux which can be carried in the core.

A final material option is that of cobalt–iron [20], which is an alloy of iron comprising up to 50% cobalt, along with small amounts of other metals, such as vanadium. It has up to 30% higher saturation flux density than silicon iron, but this comes at a price – literally! The presence of the rare earth metal cobalt makes it an order of magnitude more expensive than silicon steels and, to produce optimum properties it has to be annealed for an extended period at up to 900 degrees. The cost has made it prohibitive for most applications: however, it has found a niche in the aerospace market.

Modular machines and segmentation were discussed earlier in the chapter. Another, related topic arises from the recent growth of cutting technologies to rival conventional stamping of laminations. Stamping is an excellent process for the production of large volumes of identical laminations, but it has a very high initial cost in the production of the punch and die, and it does not have any flexibility of shape. The shearing process of the stamping introduces stresses into the material in close proximity to the cut edge, which can significantly increase the iron loss [21]. The magnitude of this effect will be magnified in small components because a higher proportion of the lamination is close to an edge: it is also a function of the sharpness of the punch and die: an old, worn die will tend to produce greater additional loss than a new one. Manufacturers commonly introduce an empirical multiplication factor of up to 2.0 to account for this effect upon iron loss. Other cutting processes available include wire erosion, water cutting and laser cutting, all of which avoid any high initial cost of a punch and die, but all have been too slow for mass production and so have been used for prototyping or very low batch numbers.

In recent years, there has been the introduction of rapid laser cutting, which has created the possibility of mass production of lamination stacks in a flexible manner, without the need for any initial cost. Furthermore, the laminations in a stack no longer need to be the same. For example, tooth tips can be gradually changed along the length of the stack to produce skew; the air-gap length at the axial ends of the machine can be increased to control fringing flux at the ends of the cores. The opportunities which arise have yet to be fully investigated: interlocking features can be introduced into segmented designs to facilitate the jointing between segments; slits can be introduced into the lamination to reduce end-region loss and into rotors to introduce saliency. Features such as this can generally be accomplished with much smaller dimensions than is possible with stamping. There is still more to be understood concerning the effect of rapid laser cutting upon material properties, but it is certainly opening up a new world of flexible, customised design.

2.5.1.2 Soft magnetic composites

Soft magnetic composites form a competing technology to laminations. Instead of sheets of laminations, fine particles of iron, typically about 0.1 mm in diameter, are coated with an insulation layer and then isostatically pressed in a punch and die at up to 800 MPa to produce a net shape component, as shown in Figure 2.18. This component is then annealed to produce a net shape. The process is similar to that of sintering, but the annealing is performed at relatively low temperature to preserve the integrity of the insulation between particles. The result is an apparently solid component produced with almost no scrap. The composite material has very low eddy current loss because the grains are so small but can have significant hysteresis loss due to internal stresses within the grains. Saturation flux density is typically a little lower than that of laminations because it is hard to produce full density of iron and the unsaturated relative permeability is typically under 1,000 because the flux must travel across the boundaries between grains.

Unlike laminations, soft magnetic composites have the advantage of allowing three-dimensional paths for the magnetic flux. This allows more efficient use of space and innovative use of the region around the end-windings to make maximum use of the enclosed volume [22]. This is illustrated well in Figure 2.19, which shows a stator made from six compressed components.

Although there is insulation between grains, the finished component displays a low, but significant electrical conductivity. In small components this has negligible effect, but in larger components at high frequency, there is a bulk eddy current effect which must be taken into account.

The dimensional accuracy of components in the plane normal to the main direction of pressing is very good, with consistency down to a few microns,

Figure 2.18 An idealized set of pressed iron powder particles, alongside photographs of actual particles. (Courtesy Hoganas AB, Sweden.)

Figure 2.19 A soft magnetic composite stator, constructed from six components. The core back is able to extend axially beyond the teeth. (Courtesy Hoganas AB, Sweden.)

however, there are larger tolerances in the direction of pressing. However, this must be set against laminations, which will have ± 1 lamination variation in stack length. The aspect ratio of components must also be taken into account at the design stage. Aspect ratios of 10:1 or more, in which the longest dimension is in the direction of pressing, are not generally possible, as they result in reduced compaction rates, leading to poor mechanical strength and poor magnetic performance. This can also prevent the production of fine parts, such as narrow tooth tips.

Overall, soft magnetic composites tend to have higher iron loss than silicon iron laminations at low frequencies due to higher hysteresis loss. The lower magnetic permeability which occurs in the unsaturated region results in an increase in reluctance. This has negligible impact in machines with large air-gaps or in permanent magnet machines, where the reluctance to armature flux is already large, but it does significantly increase the armature inductance in machines with small air-gaps, such as induction and reluctance machines. For this reason, induction and synchronous reluctance machines made with soft magnetic composites tend to have lower magnetizing inductance, resulting in higher magnetizing current and a lower power-factor. Similarly, switched reluctance machines have lower unsaturated aligned flux-linkage, resulting in lower torque. At high frequencies, soft magnetic composites start to gain over laminations because of their low eddy current loss [23]. Figure 2.20 shows how above 4–600 Hz, some soft magnetic composites have a lower loss density than 0.35 mm laminations, but cannot quite match that of 0.2 mm laminations.

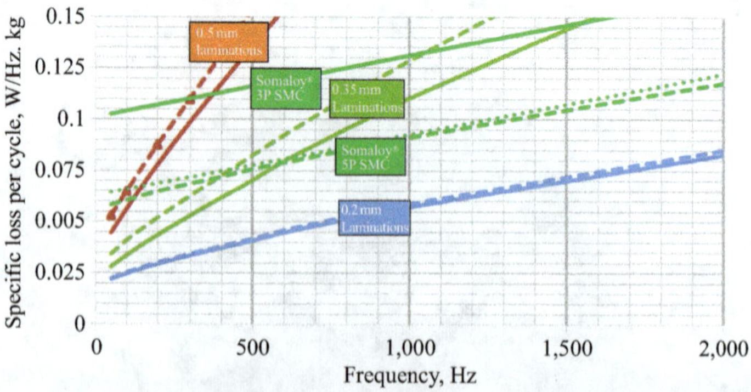

Figure 2.20 Curves showing the variation of loss density with frequency for various lamination steels and SMC. All curves are t a flux density of 1.0 T. Solid lines are manufacturers' data and dashed lines are experimental test. (Courtesy Hoganas AB, Sweden.)

Figure 2.21 An illustration of how, with an SMC core, crushing the machine at the end of life makes material separation much easier. (Courtesy Hoganas AB, Sweden.)

Recycling of electrical motors at their end of life can be particularly problematic because of the difficulty of separating the laminations from the copper windings. Soft magnetic composites are claimed to offer an advantage: unlike laminations, when they are crushed they go back into powder form. Figure 2.21 illustrates this, showing both a laminated motor and a soft magnetic composite motor after crushing.

The above discussion leads to the conclusion that soft magnetic composites are suitable for permanent magnet machines and for machines which have a significant three-dimensional flow of magnetic flux, such as transverse flux machines. The design must be undertaken in the knowledge of what is possible from a pressing perspective to ensure a good quality, manufacturable result.

2.5.2 *Hard magnetic materials – permanent magnets*

Over the last few decades, permanent magnet motors have become dominant in high performance drives where mass, volume and efficiency are paramount. Sintered rare earth magnets provide the highest energy density and are chosen: neodymium–iron–boron is used in most traction applications and its application is now extending up to large direct drive wind turbines, each of which uses several tons of magnet. In aerospace applications, samarium cobalt is generally employed because of its higher temperature capability and resistance to demagnetization. There are many texts focusing on permanent magnets and their characteristics, so it is not proposed to repeat this here. Instead, we will briefly discuss some of the challenges encountered in their use within integrated drives.

There is a strong centripetal force on the magnets, which becomes particularly large as the tip velocity of the rotor rises. In high-speed machines some form of mechanical containment is necessary: in surface mounted magnet rotors, this is most commonly achieved using a carbon fibre sleeve, whilst with inset magnet configurations a thin, saturated performs the retaining function.

2.5.2.1 Magnet loss

Sintered magnets, placed upon the rotor, rotate synchronously with the main field and so in an ideal machine they have no voltage induced in them and no eddy current heating. However, in a real machine drive, there are three sources of time varying fields seen by the rotor: stator slotting, asynchronous space harmonics produced by the winding and time harmonics of current produced by the switching action of the power converter. All three can be important, with the winding space harmonics becoming particularly large in machines with single tooth windings, as frequently used in modular integrated drives. Sintered neodymium–iron–boron has a typical conductivity of 0.7×10^6 Sm^{-1}, which is about one-third that of lamination steel. Consequently, the time varying magnet fields can result in magnet eddy currents and significant loss, particularly in high speed, high pole number machines.

The commonest method of magnet loss reduction is to laminate the magnets, generally in the axial direction. For high-speed machines, the lamination may have to be as thin as 2 mm slices and can prove to be very expensive.

The losses due to very high-frequency winding current harmonics tend to be greater in high-speed drives because the winding inductance is low. These losses can be reduced with a highly conducting screen placed round the rotor but, when there are also low-frequency winding space harmonics present, the sleeve is not generally effective and can sometimes exacerbate the problem.

2.5.2.2 Demand, supply and price volatility

The demand for neodymium–iron–boron magnets is rising exponentially as more and more permanent magnets are used in traction and generator applications. According to Statistica, the demand has risen from 6,700 tonnes in 2018 to a predicted 16,100 tonnes in 2022.

Two of the largest customers are electric vehicles and offshore wind, where direct drive wind turbine generators are increasingly being developed using permanent magnet generators. It is interesting to see how much magnet material could potentially be used by each of these applications. There are over 1 billion cars worldwide and an electric car has typically 2 kg of magnet within its traction motor. If all the world's cars were electric, then this would correspond to about 2 million tons of magnet. Direct drive wind turbine generators require of the order of 0.8 tonnes of magnet per MW of output power. One large offshore windfarm, such as that in Dogger Bank off the east coast of England, could generate 3.6 GW of power, which would require of the order of 29,000 tonnes of magnet if all the generators were direct drive permanent magnet machines. It is clear that if we are to continue to use rare earth permanent magnet machines, the supply of neodymium will have to rise by more than a factor ten.

The above should be seen in the context that the worldwide reserves of neodymium are of the order of 8 million tons, so there is enough to supply demand, if it can be mined and supplied at an acceptable financial and environmental cost. Over the last five years the cost has been very volatile. Figure 2.22 illustrates the raw material cost of neodymium over the last four years. Within this time, the cost has varied by a factor of four or more.

A further complication arises because of where most of the neodymium is to be found. China currently supplies 70% of the world's demand. This, in tandem with environmental concerns, has led to a desire for many western countries, which have no or small natural reserves, to seek alternative technologies which reduce or circumvent its use [24,25]. Examples of this include Renault, who are using wound

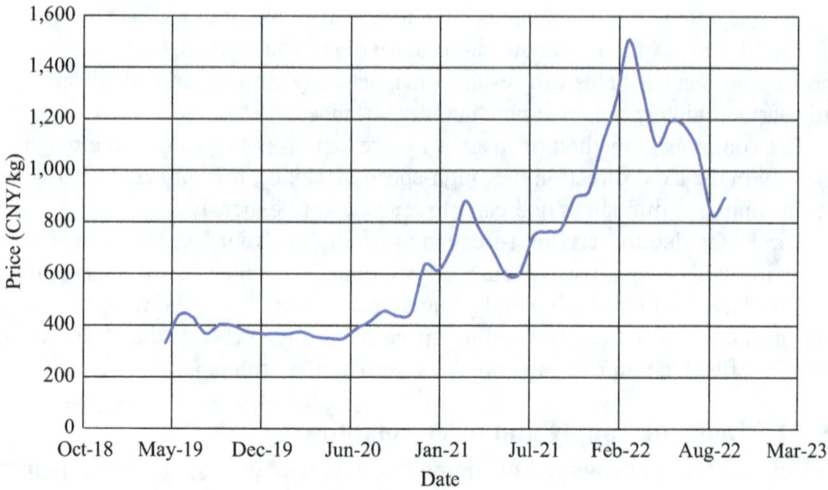

Figure 2.22 The raw metals price variation of neodymium in Chinese Yuan per kilogram (data taken from Trading Economics)

field machines in their electric vehicles, and BMW, who are considering this technology for future vehicle drives. Recycling of rare earth magnets has also become important, with some major initiatives in this field [26,27].

If neodymium iron boron magnets are to be used at elevated temperature (typically greater than 80°C), then another rare earth metal, dysprosium, is added to prevent demagnetization [28]. Most high-performance motors operate at temperatures well in excess of this and so their magnets contain a small percentage of dysprosium. As the percentage of dysprosium rises, then the residual magnetism falls. Hence, magnets capable of high-temperature operation typically have a lower residual magnetism. Dysprosium is central to the cost of finished magnets because it is approximately five times more expensive than neodymium. Its price has been exceptionally volatile: at the time of writing in 2022 it stands at $600 per kilogram, but it has been as high as $3,000 per kilogram: the world production is about 100 tonnes per annum. It is has much smaller world reserves than neodymium and may prove critical in future supply chains.

Research and development of magnets is intense, with emphasis on removal of critical materials. This is beyond this text, but readers are pointed to three papers discussing the prospects of success [29–31].

References

[1] A. G. Sarigiannidis, M. E. Beniakar, P. E. Kakosimos, A. G. Kladas, L. Papini, C. Gerada, "Fault tolerant design of fractional slot winding permanent magnet aerospace actuator," *IEEE Transactions on Transportation Electrification*, vol. 2, no. 3, pp. 380–390, 2016, doi:10.1109/TTE.2016.2574947.

[2] N. R. Brown, T. M. Jahns, R. D. Lorenz, "Power converter design for an integrated modular motor drive," *2007 IEEE Industry Applications Annual Meeting*, New Orleans, LA, USA, 2007, pp. 1322–1328, doi: 10.1109/07IAS.2007.205.

[3] System Design and Analysis, U.S. Dept. of Transportation, Federal Aviation Administration, FAA Advisory Circular AC No. 25.1309-1A, June 21, 1988.

[4] J. K. Tangudu, T. M. Jahns, "Comparison of interior PM machines with concentrated and distributed stator windings for traction applications," in *2011 IEEE Vehicle Power and Propulsion Conference*, Chicago, IL, USA, 2011, pp. 1–8, doi: 10.1109/VPPC.2011.6043171.

[5] B. C. Mecrow, "New winding configurations for doubly salient reluctance machines," *IEEE Transactions on Industry Applications*, vol. 32, no. 6, pp. 1348–1356, 1996, doi:10.1109/28.556638.

[6] N. Bianchi, S. Bolognani, M. D. Pre, and G. Grezzani, "Design considerations for fractional-slot winding configurations of synchronous machines," *IEEE Transactions on Industry Applications*, vol. 42, no. 4, pp. 997–1006, 2006, doi:10.1109/TIA.2006.876070.

[7] C. M. Spargo, B. C. Mecrow, and J. D. Widmer, "Application of fractional slot concentrated windings to synchronous reluctance machines," in *2013*

International Electric Machines & Drives Conference, 2013, pp. 618–625, doi: 10.1109/IEMDC.2013.6556159.

[8] N. D. Veeramaneni and A. Saxena, "Synchronous reluctance motor design with multi-layer fractional slot concentrated winding," in *2021 National Power Electronics Conference (NPEC)*, 2021, pp. 01–05, doi:10.1109/NPEC52100.2021.9672541.

[9] T. M. Jahns, "Improved reliability in solid-state AC drives by means of multiple independent phase drive units," *IEEE Transactions on Industry Applications*, vol. IA-16, no. 3, pp. 321–331, 1980, doi:10.1109/TIA.1980.4503793.

[10] B. C. Mecrow, A. G. Jack, and J. A. Haylock, "Fault tolerant permanent magnet machine drives," in *IEE Proceedings*, Port B, November 1996.

[11] T. M. Jahns, V. Caliskan, "Uncontrolled generator operation of interior PM synchronous machines following high-speed inverter shutdown,", *IEEE Transactions on Industry Applications*, vol. 35, no. 6, pp. 1347–1357, 1999, doi:10.1109/28.806049.

[12] ELEKTRISOLA Dr. Gerd Schildbach GmbH & Co KG: www.elektrisola.com

[13] D. D'Amato, J. Loncarski, and V. G. Monopoli, "Impact of PWM voltage waveforms in high-speed drives: a survey on high-frequency motor models and partial discharge phenomenon," *Energies*, vol. 15, 1406 pp., doi: 10.3390/en15041406.

[14] J. C. G. Wheeler, "Effects of converter pulses on the electrical insulation in low and medium voltage motors," *IEEE Electrical Insulation Magazine*, vol. 21, no. 2, pp. 22–29, 2005.

[15] I. Christou, "Optimisation of high voltage electrical systems for aerospace applications," PhD thesis, University of Manchester, 2011.

[16] A. Krings, M. Cossale, A. Tenconi, J. Soulard, A. Cavagnino, and A. Boglietti, "Magnetic materials used in electrical machines," *IEEE Industry Applications Magazine*, vol. 23, pp. 21–28, 2017.

[17] A. T. Sankara Subramanian, P. Meenalochini, S. Suba Bala Sathiya, and G. Ram Prakash, "A review on selection of soft magnetic materials for industrial drives," in *Materials Today*, New York, NY: Elsevier, Proceedings 45, 2021, pp. 1591–1696.

[18] A. Hirzel, "The basics of amorphous-iron motors", *Motors/Drives Magazine*, April 2014, pp. 38–42.

[19] Z. Wang, Y. Enomoto, M. Ito, *et al.*, "Development of a permanent magnet motor utilizing amorphous wound cores," in *IEEE Transactions on Magnetics*, vol. 46, no. 2, pp. 570–573, 2010, doi:10.1109/TMAG.2009.2033350.

[20] N. Fernando, G. Vakil, P. Arumugam, E. Amankwah, C. Gerada, and S. Bozhko, "Impact of soft magnetic material on design of high-speed permanent-magnet machines," *IEEE Transactions on Industrial Electronics*, vol. 64, no. 3, pp. 2415–2423, 2017, doi:10.1109/TIE.2016.2587815.

[21] M. Bali and A. Meutze, "Influences of CO_2 laser, FKL laser and mechanical cutting on the magnetic properties of electrical steel sheets," *IEEE Transactions on Industry Applications*, vol. 51, no. 6, pp. 4446–4454, 2015.

[22] A. G. Jack, B. C. Mecrow, P. G. Dickinson, *et al.*, "Permanent magnet machines with powdered iron cores and prepressed windings," *IEEE Transactions on Industry Applications*, vol. 36, no. 4, pp. 1077–1084, 2000.

[23] D. Gumbleton-Wood, G. J. Atkinson and L. Sjöberg, "Electromagnetic properties of soft magnetic composites and electrical steels at high frequencies considering material manufacturing techniques," in *2019 IEEE International Electric Machines & Drives Conference (IEMDC)*, 2019, pp. 2027–2034, doi: 10.1109/IEMDC.2019.8785071.

[24] T. Jahns, "Getting rare earth magnets out of EV traction machines," *IEEE Electrification Magazine*, vol. 5, no. 1, pp. 6–18, 2017.

[25] B. Poudel, E. Amiri, P. Rastgoufard, and B. Mirafzal, "Toward less rare-earth permanent magnet in electrical machines: a review," *IEEE Transactions on Magnetics*, vol. 57, no. 9, 2021.

[26] M. C. Bonfante, J. P. Raspini, I. B. Fernandes, S. Fernandes, L. M. S. Campos, and O. E. Alarcon, "Achieving sustainable development goals in rare earth magnets production: a review on state of the art and SWOT analysis," in *Renewable and Sustainable Energy Reviews*, New York, NY: Elsevier, 2020.

[27] K. Binnemans, P. McGuiness, and P. T. Jones, "Rare-earth recycling needs market intervention," *Nature Reviews, Materials*, vol. 6, pp. 459–461, 2021.

[28] S. Constantinides, "The important role of dysprosium in modern permanent magnets", in Arnold Magnetics Website, Accessed October 2022.

[29] K. Hono and Sepehri-Amin, "Prospect for HRE-free high coercivity Nd-Fe-B permanent magnets," in *Scripta Materialia*, New York, NY: Elsevier, 2018, pp. 6–13.

[30] F. Mazeleyrat, "Getting rid of critical raw materials in hard magnets: is it feasible," *IEEE Transactions on Magnetics*, vol. 58, no. 2, 2022.

[31] J. M. D. Coey, "Perspective and prospects for rare earth permanent magnets," in *Rare Earth Permanent Magnets Review*, New York, NY: Elsevier, 2019.

Chapter 3
Power converter and control techniques

Xu Deng[1] and Judith Apsley[2]

In Chapter 2, the authors have discussed the fundamentals of electrical machines and their application in integrated motor drives (IMDs). In order to serve electrical machines and help them achieve their designed performance, associated power electronics, sensors, controllers, and the advanced control techniques will be introduced in this chapter.

3.1 Overview of electric drives

The general arrangement of an electric drive system is shown in the block diagram in Figure 3.1. In addition to the electrical machine, a standard drive system consists of power converters, sensors for current/voltage/position measurement purposes, and a control unit where microprocessors are commonly employed. The red arrows represent the power flow in an electric drive system. The purple arrows show that the control signals flow from sensors to the control unit, producing switching signals to control the power converter.

The purpose of the power converter is to convert the electrical energy from the form of the provided power source into the desired form for the electrical machine. For example, a brushless direct current (BLDC) motor drive supplied by a battery inverts the dc input voltage into ac output with a fundamental frequency which is the same as the fundamental frequency of the motor. Sensors collect the rotor position from the shaft and current/voltage information from power converter, while the control unit then utilizes this information to generate the optimal control signals for power converters.

In this chapter, we will first review the most widely used semiconductor devices in power converters including suitable applications according to their VA rating and switching frequency, followed by some converter topologies that are widely used in IMDs. Control methods for BLDC machines will be discussed, including the conventional and optimized torque and speed control strategies.

[1]School of Engineering, Newcastle University, UK
[2]Department of Electrical and Electronics Engineering, University of Manchester, UK

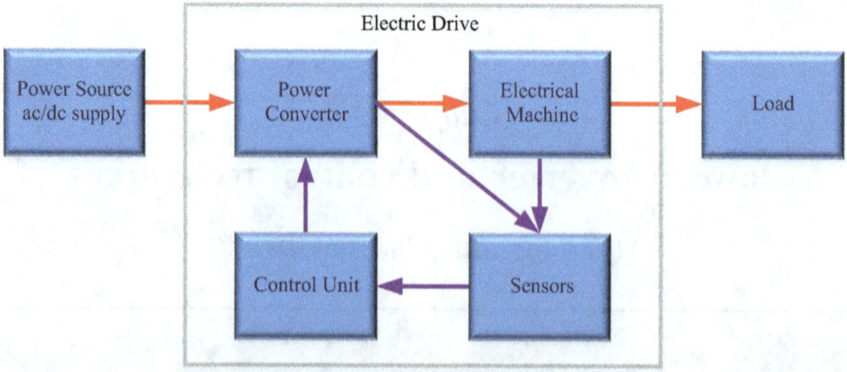

Figure 3.1 General arrangement of an electric drive system

Associated microcontrollers will be discussed at the end of this chapter, giving guidance on each controller family's strength and suitable application.

3.2 Power converters

3.2.1 Power semiconductor switching devices

Constituting the heart of modern electric drives, the evaluation of power semiconductor switching devices is leading the pace of the development of power converters in IMDs. Insulated gate bipolar transistors (IGBT) and metal oxide semiconductor field effect transistor (MOSFET) are two of the most widely used semiconductor switching devices in industry. Due to their superior switching characteristics, reduced device power losses and simplified gate control ability, IGBTs and MOSFETs are no doubt the most competitive candidates for different power range IMD applications.

The device choice varies from application to application, and a wide range of factors, such as motor fundamental frequency, power rating, volume, and cost, all play a role in determining the right choice [1]. Obviously, a classification based on power rating and switching frequency helps to narrow the range of which switching device is the better choice in a specific application. The power rating and switching frequency of common semiconductor switching devices in IMDs are shown in Figure 3.2. Generally, MOSFETs are more widely employed in low-voltage, low-current and high switching frequency applications, whilst IGBTs are more suitable for high-voltage, high-current, and low switching frequency applications.

3.2.2 Converter topologies

Figure 3.3 shows a typical power circuit of an electric drive powered by a three-phase voltage source. Depending on the power rating and its application, the topology shown in Figure 3.3 varies. The inverter can also be powered by single-phase voltage or in many cases, by a battery.

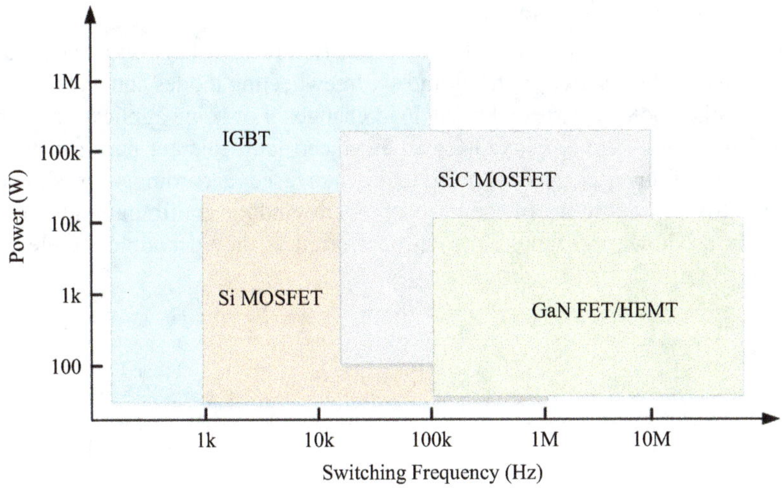

Figure 3.2 Switching frequency and power rating of common semiconductor switching devices in IMDs

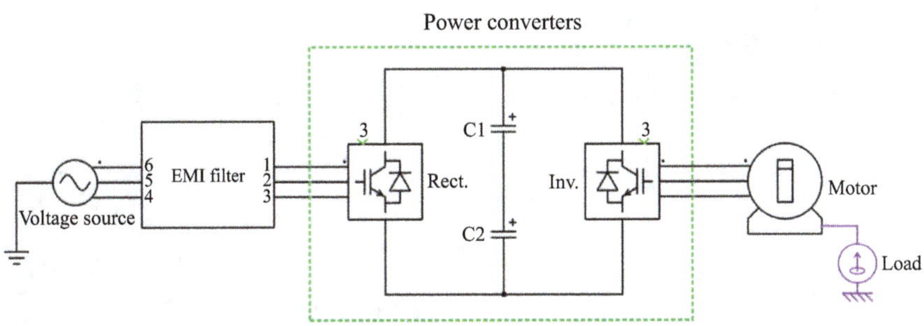

Figure 3.3 Three-phase ac–dc–ac converter system for electric motor drives

The majority of the IMDs are ac drives due to the higher power density of PMSMs. According to the applications, there are many different types of power converters used in the ac drives. In this section, the voltage source inverters are discussed, since they are the most commonly used converter topology in the IMD systems. The two-level three-phase inverters are first discussed in this section, which are widely used in commercial servo drives and domestic products, followed by discussion on multi-level inverters for low-voltage drive applications. These inverters are a source of variable voltage variable frequency ac supply when the output of the inverters is pulse width modulated. In addition, this chapter will also show how the power electronics have been integrated with the electric machines.

3.2.2.1 Two-level VSI in IMDs

Figure 3.4(a) shows a conventional three-phase two-level voltage source converter, which consists of six active switches and six freewheeling diodes connected in anti-parallel. This topology varies when its load changes. For some applications, where multi-phase motors are employed for an enhanced fault tolerant performance [2], the number of inverter phases need to be increased accordingly as shown in Figure 3.4(b). In addition, in the case of open-winding configuration, H-bridge converters as shown in Figure 3.4(c) are preferred as they decouple the electrical

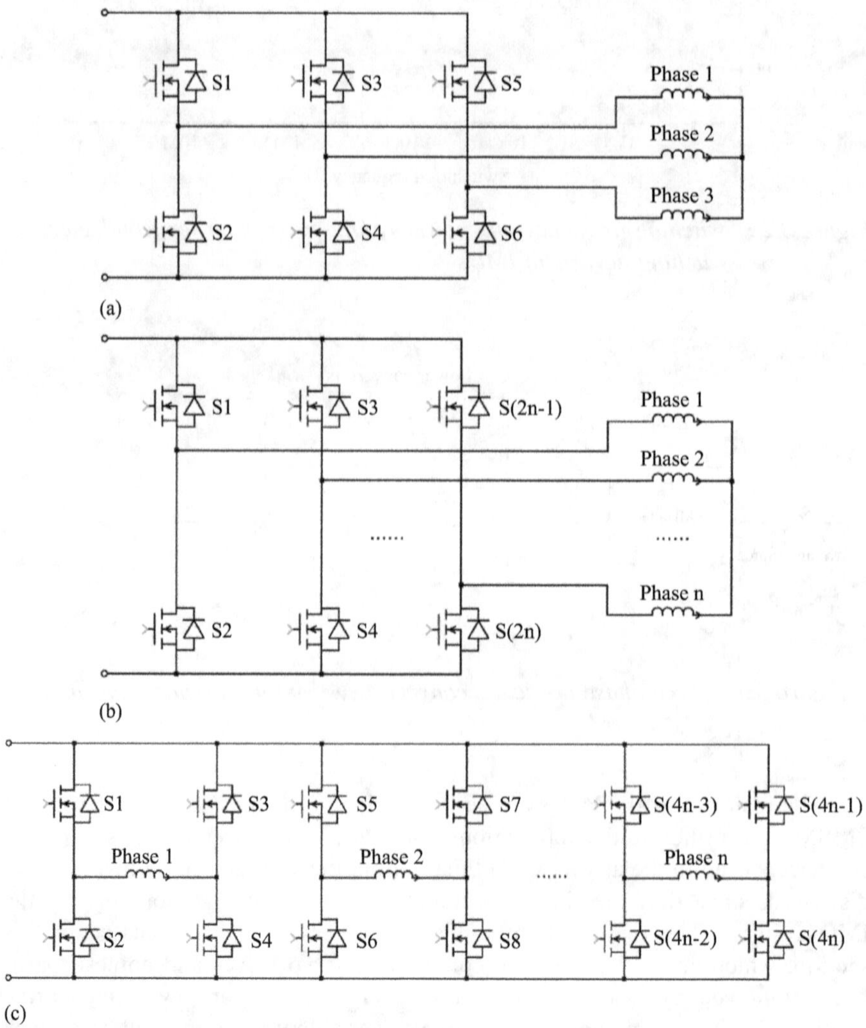

Figure 3.4 Two-level voltage source converters. (a) three-phase two-level inverter, (b) n-phase full bridge inverter, and (c) n-phase H-bridge inverter

connection among output ends [3]. In this case, phase windings and the inverter are decoupled with each other physically, which enables fault tolerant control, and simplifies the fabricating procedures.

3.2.2.2 Multi-level VSI

In high-power propulsion systems, the power requirements cannot be efficiently fulfilled by using a typical two level VSI; the multi-level converters represent a better solution for this application. The most commonly used three-level converters are T-type converter and neutral point clamped (NPC) converter as shown in Figures 3.5 and 3.6. Comparing with two-level VSIs, they both provide an

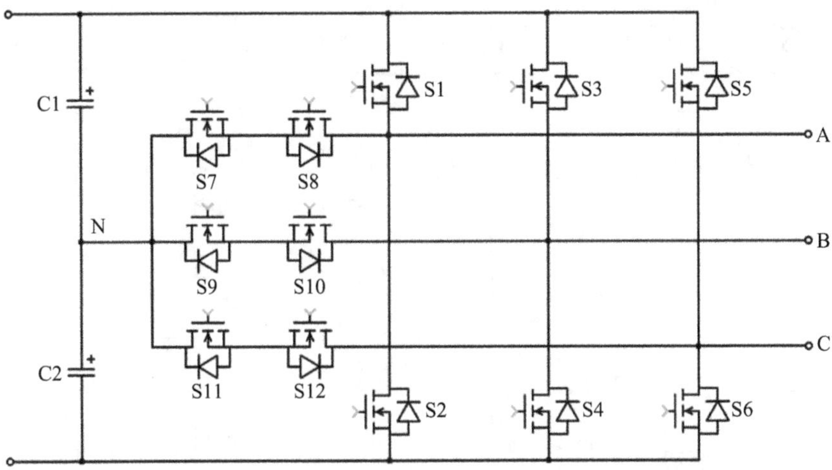

Figure 3.5 Three-phase three-level T-type converter

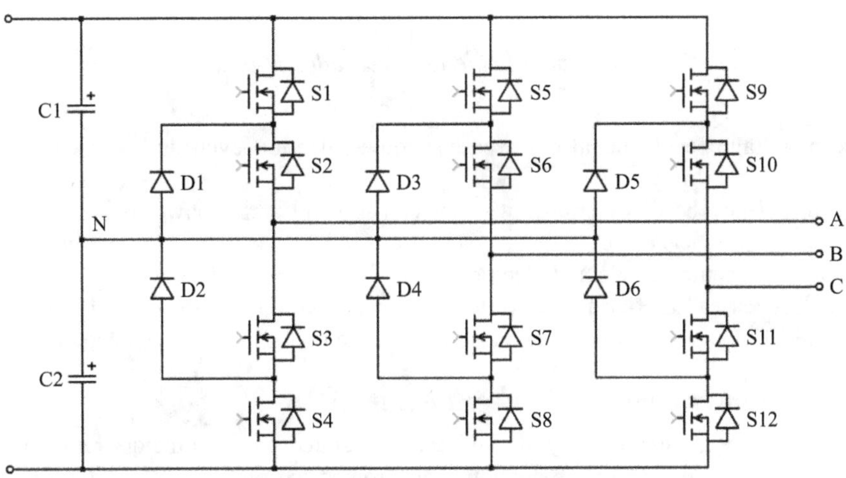

Figure 3.6 Three-phase three-level neutral point clamped converter

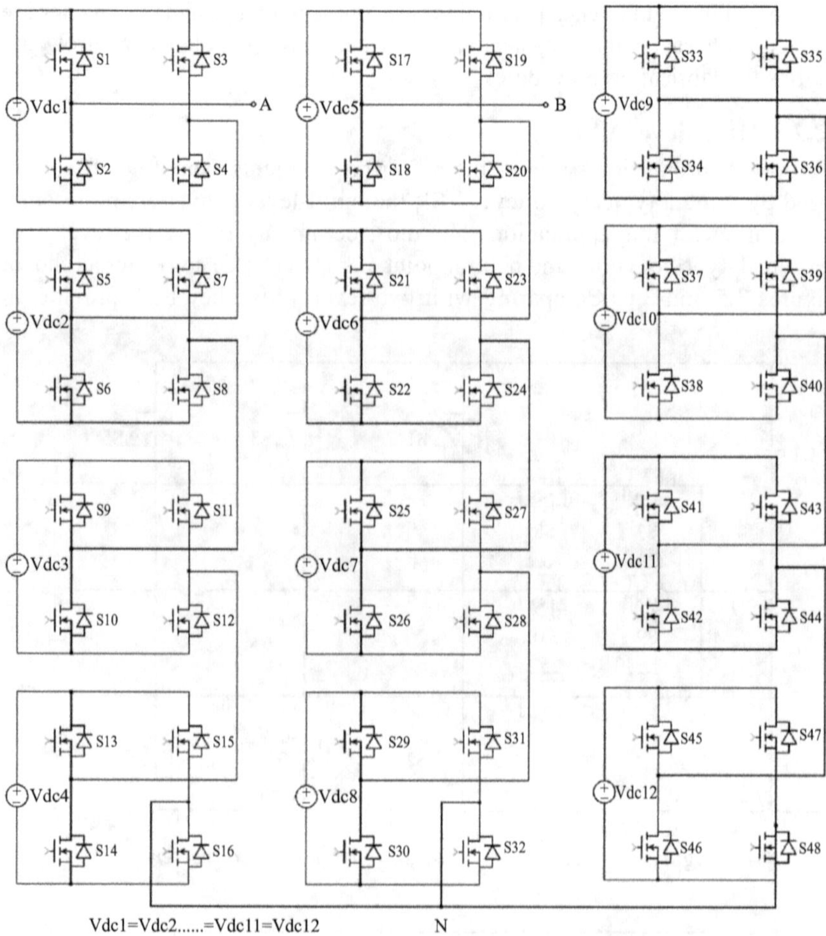

Figure 3.7 Three-phase nine-level cascaded H-bridge converter

extra voltage level and also lower harmonic injection even at low switching frequency.

Efficiency becomes critical when it comes to a battery-powered system. To achieve high efficiency, cascaded topologies are widely adopted in high power rating drives for electric vehicle and electric aircraft [4,5]. Figure 3.7 shows a three-phase nine-level cascaded H-bridge converter. Comparing with the two-level VSI, it offers reduced THD and lower dv/dt, and also enables the use of low voltage rated devices.

3.2.3 Power converters integration in IMDs

The most straightforward way of making an integrated drive from a discrete motor drive system is to put all the main power electronics on or into the motor. It reduces the system size, eliminates the connection cables, and eliminates the need for separate housings. Figure 3.8(a) shows a prize-winning 34.0 kW high-speed IMD prototype [6]. All the main power electronics are closely assembled on the motor housing, and part of the passive components are integrated into the machine housing. The system diagram is shown in Figure 3.8(b).

This research prototype employed a three-phase LCL filter which is connected between the power supply and the active rectifier. The inductors of the LCL filter are integrated within the machine housing through careful electromagnetic design.

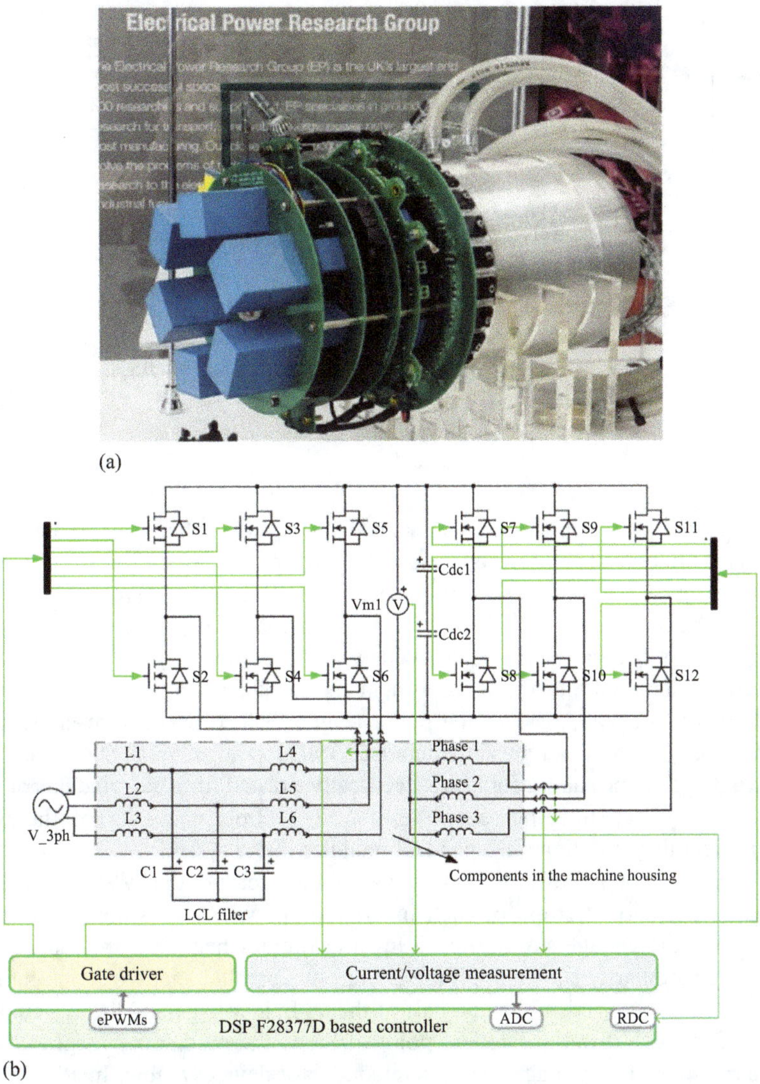

Figure 3.8 A three-phase high-speed IMD research prototype. (a) Photo of the research prototype, (b) system diagram of the prototype

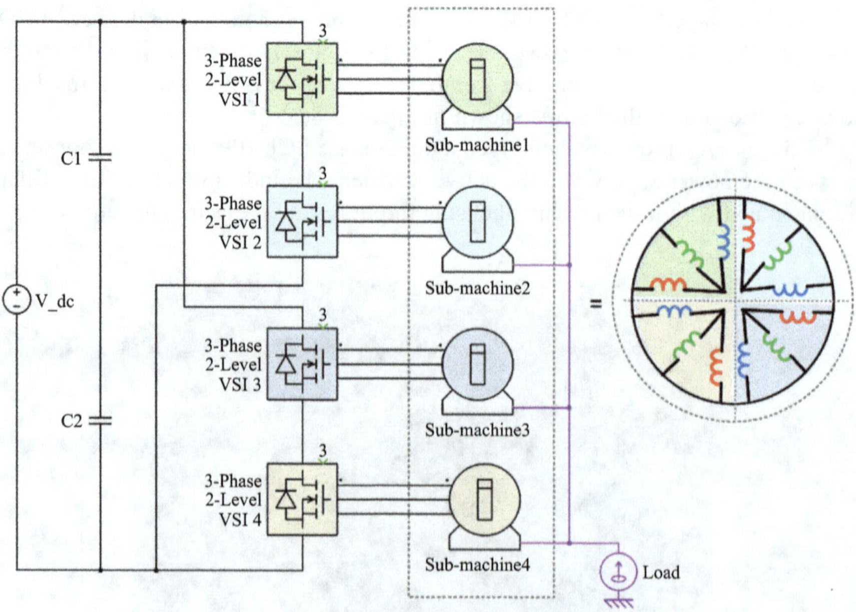

Figure 3.9 Modularity of power converters in IMDs

The inductors L1–L6 share part of the magnetic paths and cooling path with the motor stator. This prototype is a very good example, showing the level of power electronics integration that a grid connected commercial servo drive can achieve nowadays.

To further reduce the overall size of the system, modularity of power converters is another concept that has been adopted widely [7–9]. A more compact system can be achieved through splitting the drive system into modules. Each power electronics module serves a single phase of the motor (Figure 3.4(c)) or a sub-machine of the motor. Figure 3.9 shows a typical modularized IMD topology. Since the windings are split, windings of one sub-machine are electrically isolated from the windings of other sub-machines. Each three-phase two-level VSI controls a quarter of the whole machine, and they constitute independent modules in the system.

There are couple of different ways to connect all the VSIs to dc link. Generally, the current stress of each inverter drops when VSIs are connected in parallel, and the voltage stress of each inverter drops when VSIs are connected in series. The modularity shown in Figure 3.9 has two parallel and two series paths, so that each VSI only needs to handle half of the dc link voltage and the system input current. Obviously from a practical point of view, the modularity of power electronics can make low-voltage power semiconductor devices viable. In addition, the modularity of power converters enables interleaving control to be implemented, which can further reduce the requirement for dc link capacitors [10].

3.3 Sensors

In order to conduct the desired control algorithm, most of the electric drives require accurate current sensing and position sensing technology. When it comes to choosing sensors for IMDs, the fundamental sensor requirements on magnitude, accuracy, resolution, and bandwidth are not different from the sensor requirements of discrete electric drives. Instead, the size and robustness of the sensors are the key features, needing careful consideration. As IMDs are aiming to get significant size reduction from fitting all the drive components in a small volume, apart from the largest components in most power converters, the passive components and the heat-sinks, the size reduction of sensors in IMDs is always worth careful consideration.

3.3.1 Current sensing technology

A current sensor is a device that detects electric current in a conductor and generates an analog voltage/current signal or a digital signal that is proportional to the measured current. The output signal is then used for the purpose of control. Table 3.1 shows the most commonly used current sensing technology. When it comes to IMDs, generally speaking, current transducers are more widely employed

Table 3.1 Current sensing technology

Fundamental principles	**Isolation**	**Sensing technology**	**Measuring signal**	**Advantage/*disadvantage***
Ohm's Law	Non-isolated	Shunt resistor sensing	ac and dc	Low cost High reliability *Power loss* *Require electrical isolation*
		Trace resistance sensing	ac and dc	Low cost Space saving No additional power loss *Large thermal drift* *Require electrical isolation*
Faraday's Law of induction	Isolated	Current transformers (CTs) [11]	ac	No extra driving circuitry Low power consumption while measuring high current *Core saturation by dc current* *Demagnetization needed once the core's magnetized*
		Rogowski coils	ac	Coreless *Low sensitivity*
Magnetic fields sensors	Isolated	Open-loop Hall-effect sensors [12]	ac and dc	Low power consumption Wide sensing range Simple circuit and low cost High energy efficiency *Large thermal drift* *Poor accuracy and linearity* *Slow response speed*
		Close-loop Hall-effect sensors	ac and dc	High precision Fast response speed Low-temperature drift

(Continues)

Table 3.1 (*Continued*)

Fundamental principles	Isolation	Sensing technology	Measuring signal	Advantage/*disadvantage*
				Good linearity Strong anti-interference ability *Narrow sensing range* *High cost* *High energy consumption*
		Fluxgate sensors (also known as zero flux sensors)	ac and dc	Extremely high resolution and accuracy Low offset and gain drift Large bandwidth *Expensive* *High secondary power consumption* *Limited bandwidth for simpler design*
		Magneto-resistive (MR) current sensors	ac and dc	Coreless High sensitivity *Nonlinear behavior* *Distinct thermal drift* *High cost*

in high power rating IMDs, while shunt and trace resistance sensing technology fit better with low power rating IMDs when taking power loss and thermal drift effect into consideration.

In some cases where the current rating is not excessively high and surface mounted current sensors must be used to obtain the minimum vertical size, current sensing ICs are good choices. However, the Hall-effect current sensing ICs and other magnetic field sensing-based current sensors are very sensitive to the external magnetic field. The external magnetic field must be lower than the maximum permissible value, when these sensors are placed between the machine end plate and the end windings. Otherwise, the normal operation of power electronics will be affected, thus the replacement solution or magnetic shielding must be implemented.

3.3.2 Rotational position and speed sensing technology

Rotational position sensors in electric drives are devices used to measure rotary displacement of the rotor. They work by transforming mechanical rotary movement into electrical signals. Three types of common position sensors that have been widely used in electric drives as listed in Table 3.2.

Resolvers have a longer history and are still used in military and aerospace applications. They are also adopted for industrial use due to their robust structure. Optical encoders (absolute and incremental) have long been favored by motor control system engineers for their superior accuracy, high resolution, and especially their standard digital outputs that can be easily handled by microcontrollers. Incremental encoders are used in applications where velocity and direction are required. However, incremental encoders do not provide absolute position of the rotor, instead they produce pulse signals, measuring the change of the position. Therefore, in some of the

Table 3.2 Rotational position and speed sensing technology

	Type	Output	Advantage/*disadvantage*
Encoder	Contact Optic	Digital	Ease of use in control systems High accuracy and resolution Do not require extra complex electronics or circuitry Lower mass and inertia *Not suitable for use in harsh environment* *Direct light source interference High cost*
Resolver	Contact Inductive	Analog	Suitable for extremely harsh applications Withstand higher temperature, vibrations and shock loads Resistant to electrical disturbances *Require Resolver to Digital Converter (RDC)*
Hall-effect position sensor	Contactless Magnetic	Analog	Function in harsh environment Compact and almost unlimited lifetime Available at low cost *Require separate magnet* *Vulnerable to external magnetic fields* *Dynamic angle error at high rotation speed*

applications where the absolute position is needed when the system is switched on, the absolute encoders are preferred. Hall-effect position sensors are becoming popular in the application of electric drives due to their low cost, compact size and viability in harsh environments. However, they only offer high accuracy when the rotor shaft is stationary or is rotating at relatively low speed.

3.4 Control methods

The control of an integrated drive is very much the same as that of a conventional machine of similar dynamic performance. Steady-state efficiency may be slightly improved in integrated drives through the use of more than three phases, which also offers improved fault tolerance [13,14], but performance is dominated by the thermal design, as covered in Chapter 6. This section uses a brief overview of conventional control schemes as a basis for identifying the specific features arising from the integration of the power electronics with the machine, which can enhance performance, or need additional design considerations.

3.4.1 Torque and speed control

The main control variable in an electrical drive is either angular speed or torque. In the latter case, cascaded control can be used. The faster inner current loop controls torque and a slower outer loop controls speed. Further slower outer loops may be added for control of additional variables such as flow rate, for a pump or fan, temperature, for a heating or ventilation system, or position for an actuator. This nested loop structure is shown in Figure 3.10. Details on configuring and tuning the conventional Proportional, Integral, Differential (PID) control can be found in textbooks such as [15].

Integrated low-power actuation motors are widely available for hobby radio controlled (RC) vehicles, valve actuation and robotics, packaging the control, power electronics, machine and multistage, high-ratio gear train in a single unit, powered by a low voltage supply, with a third wire for the control signals, as shown in Figure 3.11. Units may be configured for speed or have a built-in rotary potentiometer for closed-loop position feedback. For hobby RC applications, these units do not allow adjustment of PID gains and can be unreliable due to the dc machine brushes and plastic gears. The response rate is constrained by the use of pulse width modulation (PWM) with a variable duty cycle reference signal on a 50 Hz carrier, limiting the reference update rate to 20 ms.

For valve actuators, units may be higher power, with epicyclic gears, and BLDC rather than brushed dc machines, incorporating Hall sensors for commutation. A serial communications interface allows the user to set PID gains and current

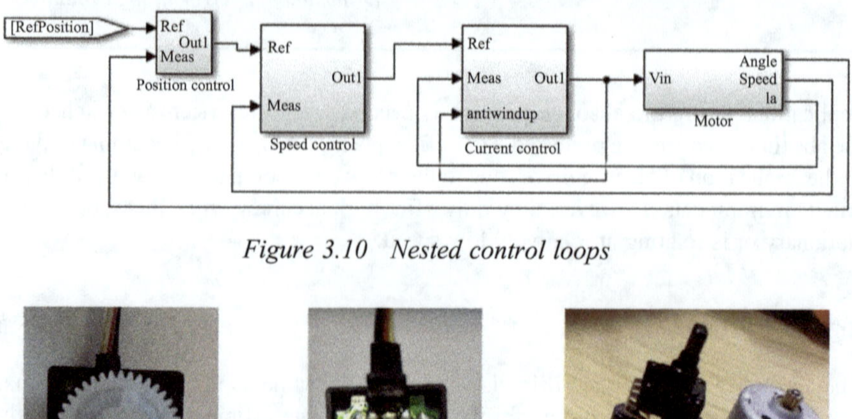

Figure 3.10 Nested control loops

Figure 3.11 Integrated RC drive: (a) gearbox, (b) power electronics, and (c) motor with covers removed

limits. Measurement and reference-following capabilities are possible, within the limits of the serial data transfer rate, typically slower than 1 ms per message. The hardware will include commutation, overcurrent limiting and thermal protection.

High performance, integrated, robotic drives may use trapezoidal BLDC or sinusoidal BLAC machines, with the two-level inverter circuit in Figure 3.4(a), built-in resolver or encoder position feedback, and a high bandwidth data link, such as EtherCAT. Such units allow reference following and sensor feedback, and may have sophisticated control algorithms, for example for self-tuning, parameter tracking and protection. However, the algorithms are generally proprietary, with no access to voltage or current control signals. This restricts the use of modern multi-axis control methods such as computed torque control [16].

3.4.2 Open loop voltage/frequency control

For an AC machine, terminal voltage can be used can be used to approximately set motor speed, provided that the voltage/frequency (*v/f*) ratio is held constant. More advanced schemes use a current-dependent voltage boost at low speeds to increase starting torque and may reduce the *v/f* ratio at light loads to improve efficiency. *V/f* control is mainly used for induction machines with low dynamic requirements, for example pumps and fans, since the control is based on steady-state theory. Further details can be found in textbooks such as [17].

Early research on integrated drives proposed integrated *v/f* controlled IM drives, with a single-phase mains input, for low power domestic equipment [18]. Alternative drives with a direct, single-phase mains connection include: universal machines, single-phase, capacitor-start IMs, and shaded pole machines, all of which have low efficiency and high starting currents. In the domestic market, power converters were seen as expensive and unreliable. However, the EU Ecodirective requirement for new products to display energy ratings [19] has led to a much wider adoption of converter-controlled variable speed drives in domestic items such as washing machines and vacuum cleaners, for example the Dyson drive in Chapter 9. The desire for dual use battery/mains operation has done the same for hand tools and kitchen equipment. In this market, the "plug and play" aspect of integrated drives is attractive, but the challenge is to keep costs down. The drive would normally be part of a larger system, so environmental protection is less important than space envelope.

3.4.3 Dynamic torque control

There are two main types of algorithms for fast dynamic control of torque (the inner loop in Figure 3.10) – field-oriented control (FOC) and direct torque control (DTC), introduced below.

3.4.3.1 Field-oriented control (FOC)

In FOC, the torque is controlled via stator currents, by imposing field orientation and transforming motor variables to a synchronously rotating reference frame. The theory assumes sinusoidal windings, so is only applicable to IM, SM, and VR

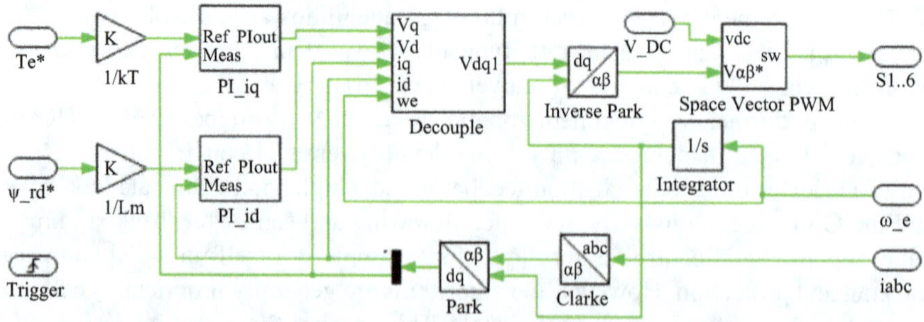

Figure 3.12 FOC

machines. The Clarke transform is used to map from 3-phase (*abc*) to an equivalent 2-phase ($\alpha\beta$) system in the stationary reference frame. The Park transform is used to map from the $\alpha\beta$ system to a second 2-phase (*dq*) system in a synchronously rotating reference frame. Orientation with the rotor flux linkage (IM) or magnet flux linkage (PM) simplifies the machine equations such that torque is proportional to torque current i_q. Flux current i_d is proportional to rotor flux linkage in an IM and can be used to offset the magnet flux linkage in a PMSM. Further details can be found in standard textbooks, for example [20].

The resulting generic controller is shown in Figure 3.12 for an IM. Reference torque and flux currents are related to torque and flux through the control laws. Classical PI control is implemented in the synchronous reference frame. Estimated back EMF and decoupling terms are added to cancel speed-dependent cross-coupling terms. The resulting reference voltages are transformed to $\alpha\beta$ and space vector modulation (SVM) is used to generate inverter switch commands for control of terminal voltages. Measured currents are mapped to *dq* providing feedback for the current controllers. The orientation angle θ_e is either derived from a flux observer (direct method), or taken as the integral of the electrical frequency ω_e, which is the sum of the electrical angular speed of the rotor and the slip (indirect method), where the slip is estimated from reference currents. PM control is similar, except that *d*-axis current is usually set to zero, since the field is provided by the magnet. In the PM machine, the orientation angle can be taken from the measured angular position of the rotor.

FOC algorithms need speed measurement for back EMF cancellation and decoupling and angle measurement for orientation. Decoupling terms, torque and flux gains and observers for sensorless control are sensitive to machine parameter variations. The advantage of integration is that the control can be tailored to a specific machine and converter, with detailed parameter mapping to improve accuracy. When combined with on-line parameter tracking, the information can also be used to detect degradation or unusual load conditions.

3.4.3.2 Direct torque control (DTC)

DTC has two control objectives: to choose a switch combination that keeps the flux magnitude constant, so the locus of the flux is circular, and to advance the rate of

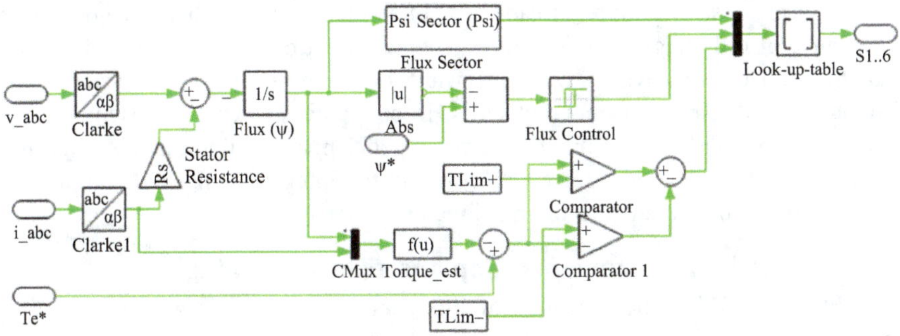

Figure 3.13 DTC

rotation of the flux to increase torque or decrease it to reduce torque [21]. As shown in Figure 3.13, stator flux is estimated from measured voltages and currents, using the Clarke transform to map to 2-phase $\alpha\beta$ values. In classical DTC, hysteresis control is used, and switch vectors are only changed when the flux exceeds the boundaries of the desired flux envelope. A set of switching laws can be derived according to the sector number of the flux vector. In a conventional 3-phase machine, and six-switch inverter, there are eight viable combination of inverter switches, referred to as switch states (assuming one switch per leg is closed and the other open). DTC has been applied to IM, PM, and SR machines.

DTC is less parameter-dependent than FOC and does not need speed or position measurements. However, flux estimation from voltage and current is challenging at low speeds, so DTC is generally used for speed control rather than position holding at zero speed. Modifications to reduce switching ripple, improve low-speed performance, implement with a fixed switching frequency or maximize torque per amp all tend to reintroduce parameter dependency [22]. Again, the advantage of integration is that the control can be tailored to the specific machine and converter.

3.4.3.3 Control of flux-switched machines

In flux-switched machines, at low speed, current is injected into the winding over a commutation interval determined by integrated position sensors. The magnitude of the current pulse is adjusted to control the torque. According to the machine design, one or more phases may conduct at any time. Commutation angles are fixed, and the current control over the on period is similar to that of a dc machine [23].

As the machine speed is increased and the current commutation time becomes a significant proportion of the on-time, torque control is achieved by adjusting of the turn-on and turn-off angles of the machine. Timings are based on commutation maps as a function of machine speed and load, derived from experimental measurement or finite element analysis. For SR machines operating into saturation, gain scheduling, with fuzzy logic control, may be used to manage the non-linear torque–current relationship and the transition low and high-speed operating modes [24].

Because these machines already use integrated sensors, there is good user acceptance of full integration. SR drives have unidirectional currents, which means that they can be used with a reduced switch count in an integrated system with discrete, bare-die, power semiconductors. In conventional separate machines and converters, the market is dominated by packaged 3-phase bridge modules, and the reduced switch count does not reduce costs. In the integrated machine, commutation timing can be matched to the machine characteristics.

3.4.3.4 Maximum torque per amp and field weakening

Both FOC and DTC allow control of flux magnitude, so flux weakening (FW) control can be used to improve efficiency at light loads, exploit saliency torque (in SMPM) and extend the speed range. In IM, *d*-axis stator current determines the rotor flux linkage. In SMPM, negative *d*-axis current is injected to partially cancel the magnet flux linkage and the speed range extension depends on magnet flux linkage, stator inductance and rated current [20,25]. In a PM machine, maximum torque per amp control (MTPA) optimizes the *d*-axis current to minimize copper losses. In an IM, this formal optimization is less common, but is more likely to include other loss components. While this control is no different to a conventional motor-drive system, integration again allows more detailed characterization of losses and saturation.

3.4.4 Multiphase and modular machines

The use of machines with more than three phases offers moderate efficiency gains, through better use of the magnetic circuit, as well as enhanced fault tolerance through redundancy [13,14]. Multiphase is attractive for integrated machines because the connections are internal to the machine so do not add to the complexity of the external wiring. In modular machines, the phases are divided into multiple 3-phase sets, as shown in Figure 3.9.

The main advantage of integration is the ability to share power through more inverter legs, reducing the power rating per leg, in applications where parallel combinations of devices and windings would be necessary to meet the voltage and current rating in a three-phase configuration. Multiple phases offer better current sharing (through explicit control of each current path) and reduced switching current ripple (through the use of interleaving) than the equivalent three-phase topology.

Multiphase machines are attractive for electric vehicle (EV) propulsion applications, due to the relatively low operating voltage (600–800 V), high current, and high efficiency requirements of road vehicles. This has been due to safety concerns about high dc voltages, and to the relatively low-voltage ratings of GaN devices. Added to that, the unit cell voltage of lithium batteries is 2–4 V [26], so large numbers of cells in a series/parallel combination are needed for direct connection to an EV dc bus, with associated challenges for current and voltage sharing. High current and low-voltage converters typically use multiple parallel connections, for example the 2015 Tesla S85PD Series S inverter used 2 × 14-parallel IGBTs per leg.

3.4.4.1 Multiphase control

In a sinusoidal ac machines (PM or IM), the most common method of control is to use an extended Clarke's transform. Further details can be found in [13,14] according to the layout of the phase windings, together with other forms of control. The extended Clarke transform maps the multiphase machine to a 2-phase $\alpha\beta$ set, which produces torque and a multiple higher order xy sets which are usually controlled to give zero current. The higher order windings offer redundancy, and in a fault condition, control of currents in the higher order winding sets enables continued operation with de-rated torque [27].

Figure 3.14 shows a typical multiphase machine with FOC. The torque control in the d–q sets is identical to the three-phase machine in Figure 3.12. Additional control is required to force the current to zero in the additional xy sets. Because these higher order sets are often linked by different higher-order harmonics, control of the xy sets is usually implemented in the stationary reference frame using proportional resonant (PR controllers) rather than conventional PID control, as shown in Figure 3.14.

In SR machines, where mutual inductance can be neglected, the control is much easier. In this case, both current-shaping and commutation timing controls can be performed per phase, so with more phases, the algorithm is repeated for more inverter phase legs with a $2\pi/m$ phase shift for m symmetrically arranged phases per phase. The same is true of weakly coupled modular winding machines, such as fractional slot PM machines, where the self-inductance is dominated by the leakage term. In this case, conventional 3-phase control can be implemented per submachine. If the coupling is significant, then MMF terms can be passed between submachines as implemented by [28].

3.4.4.2 Voltage synthesis and interleaving

Space vector PWM is complex in multiphase machines, because for a 2-level converter, the number of switch combinations is 2^m switch for an m-phase machine. So although SVM has been implemented, most researchers prefer sine-triangle PWM. One reason for this is that the introduction of a phase shift between triangle waveforms increases the effective switching frequency, reducing the DC link filter requirements, with extensive analysis by the authors of [29]. The effect of multiple phases and interleaving is illustrated in Figure 3.15, which shows the dc side current for the two-level converter for one output cycle of sine-triangle PWM, with a resistor–inductor load of 80 kW, a power factor of 0.85, 650 V ideal dc voltage source, 600 Hz output frequency, and 25 kHz carrier frequency. The average dc current is 123.3 A in each case. The 3-phase case has higher rms and peak-peak current ripple than either the 9-phase system or the set of 3 interleaved 3-phase systems, with more significant power and switching frequency components.

There are some drawbacks to interleaving. If the triangle reference is phase shifted, peak-and-trough current sampling will not reject switching disturbances in other phase legs. Oversampling followed by digital filtering is recommended. Interleaving can also introduce a phase delay between when outputs are changed in the different phase legs, which needs to be allowed for in the controller design.

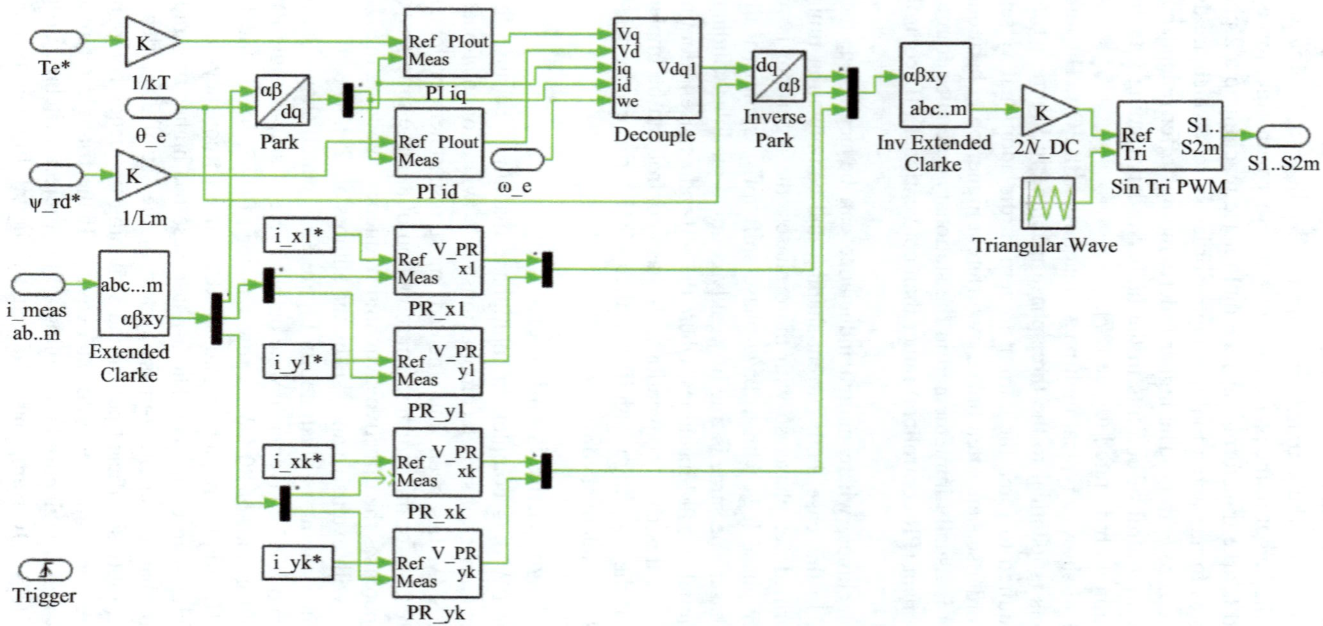

Figure 3.14 FOC extension to multiphase

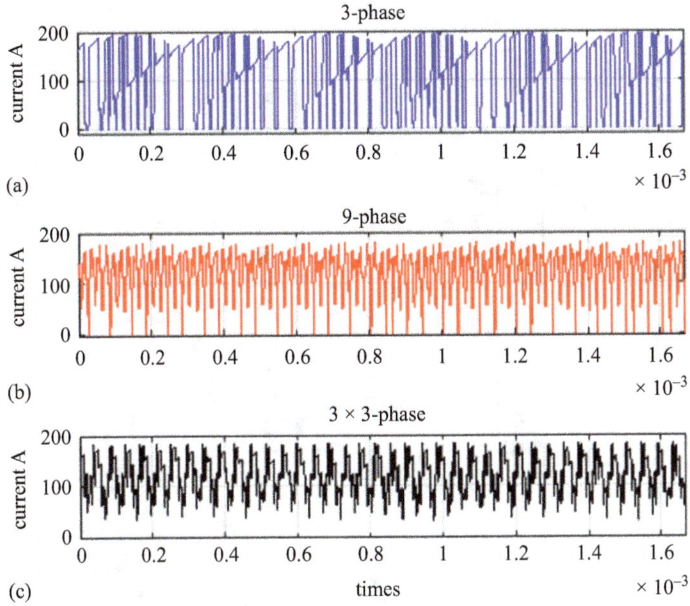

Figure 3.15 Effect of interleaving on dc current ripple (a) 3-phase, (b) 9-phase, (c) 3×3-phase with interleaving.

Interleaving can also be used to reduce motor phase current ripple and control circulating harmonic currents [30].

3.5 Control hardware

The control platform, peripheral requirements, and prototyping systems, outlined in this section, are fundamentally the same as for a conventional drive. The additional consideration for integration is the interface to external systems. Multiphase designs, which lend themselves to a distributed modular control structure, also need internal interfaces between subsystems. In both cases, this implies some form of data network.

3.5.1 Control platforms

The four main categories of control platform currently in use are summarized in Table 3.3. Microcontrollers have been in use for decades for industrial IM, PM, and BLDC drives and offer a good range of peripherals, support libraries, and development boards. The same is true of digital signal processors (DSP), but these are more expensive, and generally offer fewer peripherals. With fast, optimized floating point mathematics, they are used in higher performance products, with more demanding numerical algorithms (e.g., Kalman filters, Fourier transforms, model predictive control, sensor-less speed estimation) and for prototyping systems. Field

Table 3.3 Summary of digital control units available

	Description	Peripherals	Communications	Libraries	Numerics	Code
Micro-controller	Micro-processor with dedicated hardware peripherals	QEI, PWM, ADC, DIO, Timers, Interrupts	SPI, I^2C, CAN, USB, USART	Park, Clarke, Modulation, Control	Fixed point, Floating point in software so slow	C, C++, Mbed, Arduino
DSP	Micro-processor optimized for fast maths	Maths coprocessor. Other peripherals as above, but fewer channels	USART, USB, SPI, others, but fewer channels	Park, Clarke, Modulation, Control, Optimization, Filters	Fast hardware floating point	C, C++
FPGA	Logic array	DIO. Specialist peripherals can be configured using libraries.	Can be configured using libraries	Low level, for peripherals, telecoms, filters	Fixed point	VHDL
System on a chip (SoC)	Logic array with integrated micro-processor	Full range, but quantity limited by number of pins	Full range, but quantity limited by number of pins	Park, Clarke, Modulation, Control	Fixed point, Floating point in software so slow	VHDL with C++, Mbed, Arduino

ADC, analog to digital conversion; CAN, controller area network; DIO, digital input/output; I2C, inter-integrated circuit; PWM, pulse width modulation; QEI, quadrature encoder interface; SPI, serial peripheral interface; USART, universal asynchronous receive transmit; USB, universal serial bus.

programmable gate arrays (FPGA) provide fast parallel channels, but have limited numerics, so may be used in combination with a microcontroller, either as separate devices, or by configuring some of the gate array as a microcontroller as a system-on-chip (SoC). Gate arrays are used on their own for DTC and finite control set predictive current control, and as client units for signal conditioning and PWM generation and are configured using VHDL.

3.5.2 Peripherals

Most microcontrollers include the peripherals required for control of a dc or 3-phase ac machine as illustrated in Figure 3.16. The exception is the resolver interface, which may require an external integrated circuit, usually available with a local bus interface such as I^2C or SPI. Other standard peripherals include a hardware watchdog and high priority interrupts for protection, but these are the same as for a conventional drive. For machine with more than three phases, options include:

- Separate integrated circuits on a bus interface, for example current sensing, sample and hold (S/H), DAC, ADC. This allows synchronization of data acquisition and more choice in quality measures such as accuracy, linearity, resolution, and repeatability.
- An add-on gate array. These are good for PWM generation, digital filters and start-up, shutdown and fault protection logic sequencing.
- Additional low-level microcontrollers, in a server–client configuration.

3.5.3 Prototyping systems

Control algorithms will usually be developed in simulation. The prototyping system assists with transfer to the target platform and commissioning of the power converter. This is more important for an integrated drive, because once the drive has been assembled, there is no opportunity to put a probe on internal signals for

Control Hardware	**Duty cycle** ↓ Dead time, Min pulse ↓	**Current** ↑ Digital filter ↑	**Speed** ↑ Phase locked loop ↑	 ↑	**Position** ↑ ↑ ↑ ↑
Peripheral	PWM counter/timer	S/H DAC	QEI	SPI	Digital Input
Circuit	↓ Isolation ↓ **Gate drive**	↑ Filter ↑ **Sensor**	↑ ↑ ↑ **Encoder**	↑ IC ↑ **Resolver**	↑ ↑ **Hall sensor**

Figure 3.16 Information flow between the circuit and the control hardware. ADC, analog to digital conversion; PWM, pulse width modulation; QEI, quadrature encoder interface, S/H: sample and hold, SPI: serial peripheral interface.

development or fault diagnosis. Monitoring needs to be designed in from the start, with appropriate scaling and bandwidth.

- *Automatic code generation*: Third-party libraries are available for many microcontrollers and DSPs, providing automatic code generation from simulation packages such as MATLAB© and PLECS®. The quality of code can be excellent, but should be used with care especially between a floating point simulation platform and fixed point target. Generally, the target code can be modified by hand, but this is easier in platforms that retain the variable naming used in the simulation.
- *Control hardware-in-the-loop platforms*: Physical control boards can be connected to representative high-speed dynamic models of the power devices and circuit, to allow for development and debugging of the finalized controller code and testing of simulated faults in the power hardware. This is a useful feature for an integrated drive.
- *Power hardware-in-the-loop platforms*: Power hardware can be interfaced to a development platform from physical sensors and switch commands to allow the control algorithm to be tested on the physical power hardware. Interfacing to power hardware is difficult in an integrated drive. Activities such as sensor calibration would have to be done before integration.

3.5.4 Distributed architectures and data networking

Data networks may need to be considered at 2-levels. A system interface may be needed for

- device programming, unless this is pre-programmed before assembly,
- reference following,
- parameter setting and fault diagnostics,
- on-line monitoring.

Domestic and hobby applications are plug-and-play, with a 3-wire connection for the reference, power, ground, where the reference may be speed or position, as a PWM signal or one-wire bus. CAN is the standard for vehicle systems, and there are a multitude of proprietary Fieldbus protocols with MTConnect for industrial systems. The drive is usually represented as in the same way as other peripherals, as an addressable device with a set of addressable registers. Future applications will need to consider FlexRay (for vehicles) and EtherCAT (for robotics) and Industry 4.0 standards including 5G, (for manufacturing and building services) for IoT connectivity.

Within the drive, especially for multiphase designs, a local bus such as I^2C or SPI can link bus-based data acquisition and distributed control. The choice depends on the number of connected devices, the physical distance and timing, and synchronization requirements.

Timing and synchronization are illustrated in Figure 3.17, which shows the timing constraints to compute a new voltage reference value within one cycle of the PWM period. Sampling is usually synchronized with the output of a new PWM

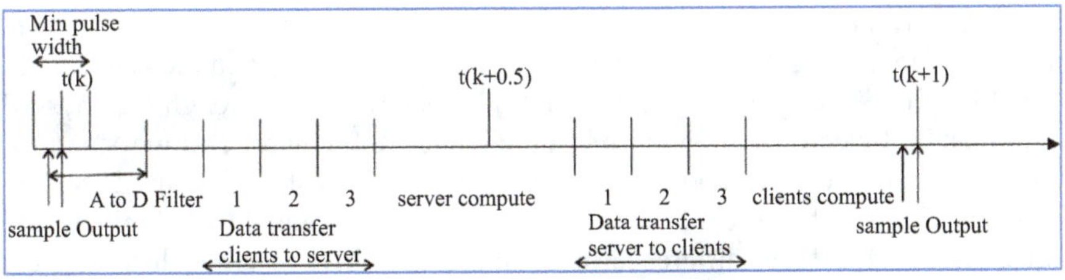

Figure 3.17 Typical timing requirements for single cycle reference computation

reference. Sometime called "peak or trough sampling," synchronized sampling has two advantages; it captures the average value of the current over the period and provided it can be completed within the minimum pulse width constraint of the power switch it avoids switching transitions, eliminating switching noise. Sampling should be synchronized across phases to prevent sample delay phase lag. The full sequence is listed below:

1. sample (with output of the previous reference),
2. analog to digital conversion and local processing,
3. data transfer from the clients to the server,
4. server computation,
5. data transfer of reference values from the server to the clients,
6. local computation at the clients,
7. output the new reference.

3.6 Thermal, environmental, and EMI constraints

In most integrated drives, the controller is located under the casing of the machine, for example in the terminal box, around the outside of the stator, in the end cap, or beyond the endcap [12,31]. The control is either isolated from the converter and machine thermal paths, or cooled first. Nevertheless, the controller may experience a wider temperature range than in a conventional converter. Controllers for automotive or aerospace (e.g. AEC-Q100 [32]) are certified for specified temperature ranges and reliability measures. However, some components can be hard to source, for example wide temperature range optical devices. Conformal coatings or epoxy resin potting can be used to improve thermal conductivity and mechanical integrity of the circuit.

Conformal coatings or epoxy resin potting are also important for withstanding the high vibration environment of an electrical machine. Again, automotive (AEC-Q100 [32]) and aerospace standards address component and circuit construction requirements. There is increasing concern about audible noise and vibration (ANV) for electric vehicle drives, identifying that without proper design, mounting locations for the controller such as within terminal boxes and cowling can be particularly prone to vibration.

In a conventional converter, the control board is usually mounted as close to the power modules as possible to minimize parasitic inductance in the gate circuit. The effect of device switching on the controller is minimized through the physical separation and electrical isolation of gate circuits from control and measurement circuits and the use of ground planes. Anti-aliasing filters should be used on electrical signals, except where the device response is naturally bandwidth limited (e.g., Hall sensors). The main difference in the integrated drive is the inclusion of the electrical machine. Most integrated drives locate the power electronics between the controller board and the machine. In this arrangement, the physical separation distance may be sufficient to minimize effects of stray flux on the control [12]. Iron keepers at the ends of the machine to contain the flux and shield the control board may increase motor eddy currents loss.

3.7 Summary

The choice of converter topology, control hardware, and associated control technique for an integrated drive is not unique. The choice is often a result from holistic consideration on motor, drive, and other parts. The main benefits of converter integration are modularity, short connections, and simplification of external wiring. The benefits of the integrated controller are the close coupling of the machine, converter and drive, which allows accurate parameter mapping. However, prototyping becomes more complicated, with no direct access to internal measurements. Additional consideration must be given to environmental protection and EMI.

References

[1] B. J. Baliga, *Fundamentals of Power Semiconductor Devices*, New York, NY: Springer Science & Business Media, 2010.

[2] S. Wu, C. Tian, W. Zhao, J. Zhou, and X. Zhang, "Design and analysis of an integrated modular motor drive for more electric aircraft," *IEEE Transactions on Transportation Electrification*, vol. 6, pp. 1412–1420, 2020.

[3] Z. Gao, D. Jiang, W. Kong, *et al.*, "A GaN-based integrated modular motor drive for open-winding permanent magnet synchronous motor application," in *2018 1st Workshop on Wide Bandgap Power Devices and Applications in Asia (WiPDA Asia)*, 2018, pp. 73–79.

[4] M. Quraan, P. Tricoli, S. D'Arco, and L. Piegari, "Efficiency assessment of modular multilevel converters for battery electric vehicles," *IEEE Transactions on Power Electronics*, vol. 32, pp. 2041–2051, 2017.

[5] V. Patel, C. Buccella, and C. Cecati, "Analysis and implementation of multilevel inverter for full electric aircraft drives," *Energies*, vol. 13, p. 6126, 2020.

[6] X. Deng, M. A. S. Mohamed, S. Lambert, and B. Mecrow, "Development of a high-speed, permanent magnet, SiC-based drive with integrated input filters," *IEEE Transactions on Energy Conversion*, vol. 35, pp. 863–874, 2020.

[7] M. Ugur and O. Keysan, "Design of a GaN based integrated modular motor drive," in *2018 XIII International Conference on Electrical Machines (ICEM)*, 2018, pp. 1471–1477.

[8] J. Wang, Y. Li, and Y. Han, "Evaluation and design for an integrated modular motor drive (IMMD) with GaN devices," in *2013 IEEE Energy Conversion Congress and Exposition*, 2013, pp. 4318–4325.

[9] P. Han, G. Heins, Y. Zhang, and D.M. Ionel, "Integrated modular motor drives based on multiphase axial-flux PM machines with fractional-slot concentrated windings," in *2021 IEEE International Electric Machines & Drives Conference (IEMDC)*, 2021, pp. 1–6.

[10] J. Wang, Y. Li, and Y. Han, "Integrated modular motor drive design with GaN power FETs," *IEEE Transactions on Industry Applications*, vol. 51, pp. 3198–3207, 2015.

[11] S. Haghbin, K. Khan, S. Zhao, M. Alakula, S. Lundmark, and O. Carlson, "An integrated 20-kW motor drive and isolated battery charger for plug-in vehicles," *IEEE Transactions on Power Electronics*, vol. 28, pp. 4013–4029, 2013.

[12] T. Chen, C. Caicedo-Narvaez, H. Wang, M. Moallem, B. Fahimi, and M. Kiani, "Electromagnetic compatibility analysis of an induction motor drive with integrated power converter," *IEEE Transactions on Magnetics*, vol. 56, pp. 1–4, 2020.

[13] E. Levi, R. Bojoi, F. Profumo, H.A. Toliyat, and S. Williamson, "Multiphase induction motor drives – a technology status review," *IET Electric Power Applications*, 1(4), pp. 489–516, 2007.

[14] E. Levi, F. Barrero, and M.J. Duran, "Multiphase machines and drives – revisited," *IEEE Transactions on Industrial Electronics*, 63(1), pp. 429–432, 2015.

[15] C.D. Richard and R.H. Bishop, *Modern Control Systems,* Upper Saddle River, NJ: Pearson Prentice Hall, 2008.

[16] F.L. Lewis, D.M. Dawson, and C.T. Abdallah, *Robot Manipulator Control: Theory and Practice*, London: CRC Press, 2003.

[17] N. Mohan, T.M. Undeland, and W.P. Robbins. *Power Electronics: Converters, Applications, and Design*, New York, NY: John Wiley & Sons, 2003.

[18] I.A. Viorel, L. Szabo, and R.C. Ciorba, "Integrated motor and control unit for industrial variable speed drive," http://citeseerx.ist.psu.edu/viewdoc/download?doi=10.1.1.301.2182&rep=rep1&type=pdf, Accessed November 2021.

[19] Directive 2009/125/EC of the European Parliament and of the Council of 21 October 2009, Establishing a framework for the setting of ecodesign requirements for energy-related products (recast) EUR-Lex – 02009L0125-20121204 – EN – EUR-Lex (europa.eu).

[20] S.K. Sul, *Control of Electric Machine Drive Systems*, New York, NY: John Wiley & Sons, 2011.

[21] M. Ahmad, *High Performance AC Drives: Modelling Analysis and Control*, New York, NY: Springer Science & Business Media, 2010.

[22] M. Preindl and S. Bolognani, "Model predictive direct torque control with finite control set for PMSM drive systems, Part 2: field weakening operation," *IEEE Transactions on Industrial Informatics*, vol. 9, no. 2, pp. 648–657, 2013, doi:10.1109/TII.2012.2220353.

[23] T.J.E. Miller (ed.), *Electronic Control of Switched Reluctance Machines,* New York, NY: Elsevier, 2001.

[24] C. Mademlis and I. Kioskeridis, "Gain-scheduling regulator for high-performance position control of switched reluctance motor drives," *IEEE Transactions on Industrial Electronics*, 57(9), pp. 2922–2931, 2009.

[25] Y. Inoue, S. Morimoto, and M. Sanada, "Comparative study of PMSM drive systems based on current control and direct torque control in flux-weakening control region," *IEEE Transactions on Industry Applications*, vol. 48, no. 6, pp. 2382–2389, 2012, doi:10.1109/TIA.2012.2227134.

[26] Y. Miao, P. Hynan, A. Von Jouanne, and A. Yokochi, "Current Li-ion battery technologies in electric vehicles and opportunities for advancements," *Energies*, 12(6), p. 1074, 2019.

[27] J.M. Apsley, "Open-circuit fault mitigation for multiphase induction motors with a unified control structure," in *5th IET International Conference on Power Electronics, Machines and Drives* (*PEMD 2010*), 2010 Apr 19, pp. 1–6, IET.

[28] B. Wang, J. Wang, A. Griffo, and B. Sen, "A general modeling technique for a triple redundant 3 × 3-phase PMA SynRM," *IEEE Transactions on Industrial Electronics*, vol. 65, no. 11, pp. 9068–9078, 2018, doi:10.1109/TIE.2018.2793229.

[29] M. Diana, R. Ruffo, and P. Guglielmi, "Pwm carrier displacement in multi-n-phase drives: an additional degree of freedom to reduce the dc-link stress," *Energies*, 11(2), p. 443, 2018.

[30] N. Bodo, E. Levi, and M. Jones, "Investigation of carrier-based PWM techniques for a five-phase open-end winding drive topology," *IEEE Transactions on Industrial Electronics*, 60(5), pp. 2054–2065, 2012.

[31] A. Shea and T.M. Jahns, "Control challenges and mitigation techniques for sensor errors in modular motor drives with weakly-coupled distributed control architectures," in *Proc. IEEE Int. Electr. Mach. Drives Conf. (IEMDC)*, May 2015, pp. 890–896.

[32] http://www.aecouncil.com/AECDocuments.html

Chapter 4
Wide bandgap semiconductor switching devices in integrated motor drives

Thomas M. Jahns[1] *and Bulent Sarlioglu*[1]

4.1 Wide bandgap power semiconductor devices

4.1.1 WBG (SiC and GaN) material properties and benefits for integrated motor drive applications

4.1.1.1 Introduction

Since the introduction of silicon-based thyristors in 1957, researchers in the field of power semiconductor devices have achieved remarkable progress. In the late 1960s, the gate turn-off thyristor (GTO) made it possible to turn off the latching thyristor devices using their gate terminals, enabling switching frequencies exceeding 1 kHz. In the mid-1970s, new silicon bipolar junction transistors and metal-oxide-semiconductor field-effect transistors (MOSFET) were introduced, pushing switching frequencies upward above 10 kHz. The insulated-gate bipolar transistor (IGBT) was developed in the 1980s with combined features of the MOSFET and bipolar transistor that made it an appealing choice for many power electronics applications, including motor drives. Today, IGBTs are the dominant power switching devices for power electronic converters operating at voltages greater than 600 Vdc.

Silicon-based power devices (e.g., MOSFETs, IGBTs) have dominated power electronics applications ever since thyristors were commercialized in the 1950s, and their performance has approached or reached the theoretical limits imposed by silicon material properties in many cases [1]. Alternative semiconductor materials with larger valence bandgap voltages than silicon offer significant potential performance benefits that have been recognized for many decades, with increasingly serious research initiatives beginning in the 1970s. Among the many candidate wide bandgap (WBG) materials, silicon carbide (SiC) and gallium nitride (GaN) eventually emerged as the most appealing choices for commercial development. However, major technical obstacles had to be overcome to prepare SiC and GaN for use in commercial power semiconductor devices. The first commercial SiC wafers became available in the 1990s, with the first commercial SiC and GaN

[1]Wisconsin Electric Machines and Power Electronics Consortium (WEMPEC), Department of Electrical and Computer Engineering, University of Wisconsin – Madison, USA

power switches appearing in 2008 and 2010, respectively [2]. Tremendous progress has been made since that time to expand the voltage and current ratings of commercially available WBG switches. Many power electronics professionals are convinced that the rapid expansion and maturation of WBG power switch offerings is heralding a disruptive revolutionary advance in power electronics technology because of the major performance advantages offered by WBG power devices in a wide range of power electronics applications. Integrated motor drives is one of the significant emerging applications that stand to benefit greatly from the availability of these new devices.

The purpose of this chapter is to explore the potential impact of WBG power switches on the future development of IMD technology. This examination will begin with a concise discussion of the advantages offered by the key material properties of WBG semiconductor materials compared to silicon for IMD applications. This section will be followed by a brief discussion of the major types of WBG power semiconductors that are currently available as candidates for use in IMDs.

The next section of this chapter focuses on a balanced discussion of the advantages and special challenges associated with the introduction of WBG power switches into IMD systems currently using voltage–source inverter (VSI) technology. VSI technology has dominated adjustable-speed motor drive applications since the 1980s when IGBTs and MOSFETs became the leading choices for new motor drive applications. Candidate techniques for addressing the identified challenges are discussed.

The final section of the chapter is devoted to exploring the opportunities opened by WBG switch technology for their application in IMD systems using current–source inverter (CSI) technology in the place of conventional VSIs. Opportunities are discussed for taking advantage of the CSI approach to overcome the significant challenges imposed by inserting WBG power switches into conventional VSI topologies as drop-in replacements for IGBTs and MOSFETs.

4.1.1.2 WBG material properties and benefits for power semiconductor devices

Bandgap refers to the energy difference between the conduction and valence bands of the material as illustrated in Figure 4.1. Wide-bandgap (WBG) semiconductors

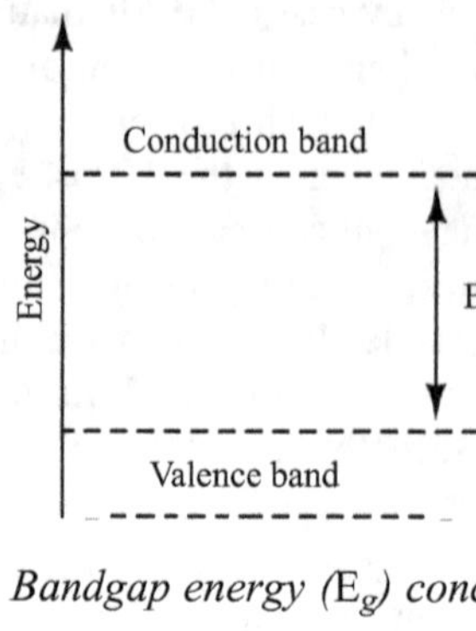

Figure 4.1 Bandgap energy (E_g) concept diagram

are semiconductor materials that have a larger bandgap than conventional semiconductors. Conventional semiconductors such as silicon have a bandgap energy in the range of 1–1.5 electron volts (eV), whereas wide-bandgap materials have bandgaps in the range of 2–4 eV [3]. For example, silicon carbide (SiC) has a 3.26 eV bandgap, and the corresponding bandgap for gallium nitride (GaN) is 3.43 eV. Generally, wide-bandgap semiconductors have electronic properties that fall in between those of conventional semiconductors and insulators.

Wide-bandgap semiconductors have material properties that allow them to operate at much higher voltages, frequencies, and temperatures than conventional semiconductor materials including silicon and germanium. Their intrinsic qualities make them suitable for a wide range of applications, and they are leading contenders for next-generation power semiconductor devices for use in future power electronics equipment.

Table 4.1 lists several key material properties of Si, SiC, and GaN that are most relevant to power semiconductor applications [4]. Five of these key material properties are plotted as a radar chart in Figure 4.2 for these three semiconductor materials to provide a more convenient visual comparison [5]. An important insight to draw from the table and figure is that it is the higher bandgap energy in

Table 4.1 Key electronic properties of Si, 4H-SiC, and GaN [4]

Property	Si	4H-SiC	GaN
Band gap, E_g (eV)	1.12	3.26	3.43
Critical breakdown electric field, E_{crit} (MV/cm)	0.25	2.2	3.5
Electron mobility, m_n (cm^2/V s)	1,400	800	2,000
Electron saturated velocity, v_{nsat} ($\times 10^7$ cm/s)	1.0	2.0	2.5
Thermal conductivity, Θ (W/cm K)	1.5	3.0–3.8	2.3
Intrinsic carrier concentration, n_i (cm^{-3}) at 300 K	1.5×10^{10}	5×10^{-9}	1.9×10^{-10}

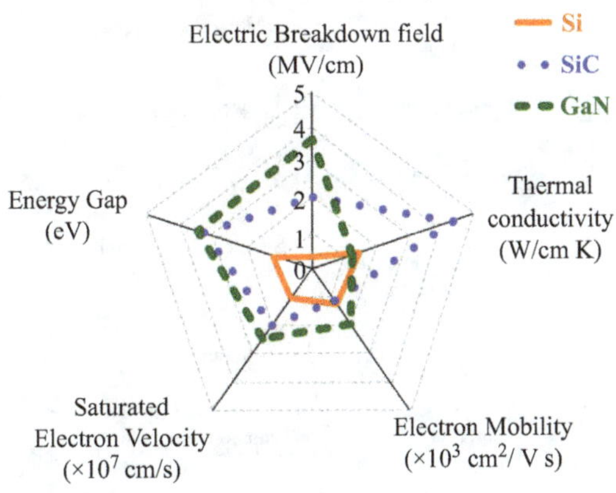

Figure 4.2 Summary of Si, SiC, and GaN material properties. © 2023 Reproduced, with permission, from IEEE [5].

combination with several other desirable material properties that make SiC and GaN appealing for power semiconductor device applications. Some of these other properties are directly linked to the material's bandgap energy and others are not.

The significance of each of the key properties to the resulting power device properties and performance characteristics, and, ultimately, to the associated IMD benefits will be briefly discussed in the following paragraphs. Figure 4.3 provides a visual summary of these relationships in the form of a flow map to accompany these descriptions.

High electric breakdown field: This important material property, which is linked to the material's bandgap energy, is significantly higher for both SiC and GaN than for Si by factors that reach 14 for SiC and 8.8 for GaN. This property is one of the keys to achieving high device breakdown voltages in majority-carrier devices such as MOSFETs without being forced to a high thickness of the device's drift region (W_N in the vertical MOSFET cross-section view in Figure 4.4) or a low doping concentration. Keeping the drift region as thin as possible without significantly reducing the doping concentration is critical for minimizing the device's forward on-state voltage drop which, in turn, has a major beneficial impact on the IMD's inverter losses and efficiency. Both of these device parameters are also valuable for reducing the required device area, making it possible to reduce its parasitic capacitance, helping to suppress switching losses as well as to reduce the device's cost that is tightly linked to its area. The major breakdown advantage of SiC over Si and, to a lesser degree, over GaN, is one of the important factors that has made it possible to fabricate SiC MOSFETs with voltage ratings of 10 kW.

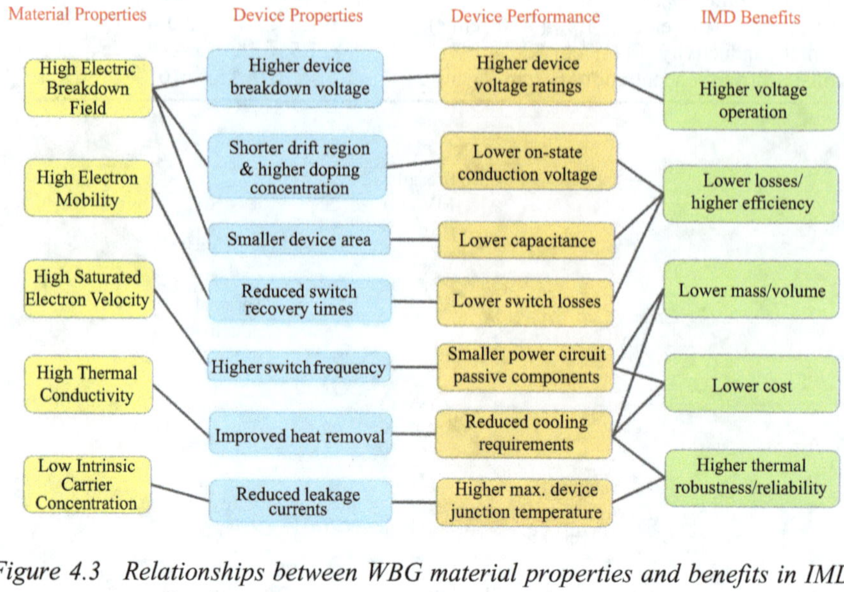

Figure 4.3 Relationships between WBG material properties and benefits in IMD applications

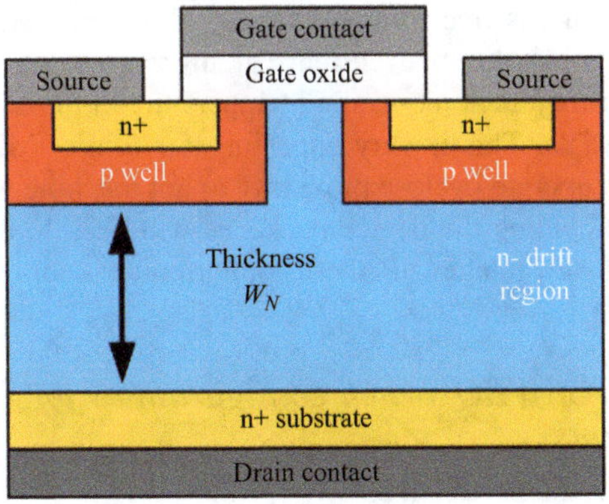

Figure 4.4 Cross-section view of a vertical MOSFET structure

High electron mobility: The higher electron mobility of GaN compared to both Si and SiC is advantageous for achieving high switching frequencies with lower switching losses. This is the case because high electron mobility helps to accelerate a device's recovery time following a switching event, which has the effect of minimizing the dissipated switch energy associated with the event. This feature contributes to the known advantages of GaN power switches over both Si and SiC devices for implementing very high switching frequencies that can extend into the MHz range.

High saturated electron velocity: Both SiC and GaN have advantages in their saturated electron velocity values over Si by at least 2:1. This is significant because high saturated electron velocity values contribute to the advantage in fast switching times that both of the WBG materials hold over silicon. The fast switching times reduce the energy loss per switching event, helping to enable higher switching frequencies with SiC and GaN that can be an order of magnitude larger than conventional Si-based devices in the same application. In addition to the efficiency advantages, the high switching frequencies make it possible to significantly shrink the sizes of passive components (i.e., capacitors, inductors) required in the inverters, yielding valuable mass and volume reductions in IMD applications.

High thermal conductivity: High thermal conductivity is highly desirable in IMDs to more easily remove the heat losses dissipated inside IMDs with less mass and energy needed for cooling systems including heatsinks, heat exchangers, and pumps/fans. The thermal conductivity of SiC is superior to that of both Si and GaN. In fact, the thermal conductivity of SiC is more than twice the value of GaN, identifying one of the reasons that SiC is known to be an appealing choice for power electronics applications (including IMDs) seeking to push power density limits.

Low intrinsic carrier concentration: The intrinsic carrier concentration values for SiC and GaN are both incredibly low compared to that of Si, a feature that can be linked to their higher bandgap energy values. The intrinsic carrier concentration measures the number of electrons that are free in the conduction band of intrinsic material that is free from any dopant. Low intrinsic carrier concentration values result in low leakage currents at high temperature, which helps to raise the maximum safe junction operating temperature of the devices that use these wide bandgap materials. This is very appealing for future IMDs that require the power electronics to operate in close proximity to hot machine windings. However, other components including the device packages will need to be thermally ruggedized to take full advantage of these opportunities for elevated operating temperatures.

4.1.2 Overview of WBG switch configurations for IMD applications

4.1.2.1 WBG device structures

The designs of wide-bandgap power semiconductor devices generally falls into one of two categories: vertical and lateral. Figure 4.5 shows simplified sketches of both of these device design classes that identify their key features, but it should be noted that the depicted layer width/thickness dimensions are significantly distorted in order to make it practical to identify all of the key features in each drawing. These drawings are intended only to show the key distinguishing features of each type since there are so many variants of both vertical and lateral device designs.

Nearly all SiC power devices that are manufactured today use the vertical design which is generally preferred because it requires significantly less chip area per device compared to the lateral design. This factor makes vertical devices more economical to manufacture since the cost of the device wafers is a major contributor to the final device cost. However, the vertical device design requires device wafers with the highest possible quality since even small defects in the wafer's crystalline lattice can ruin the device and degrade the number of good devices per wafer (i.e., the yield rate). This issue becomes increasingly important as the device's current rating increases, since larger device areas increase the probability that any given device on the wafer chip will include a destructive defect, thereby reducing the wafer's yield of good devices. The quality of SiC wafers has reached sufficiently high levels today – a major accomplishment – that it justifies adoption of the vertical device structure [7].

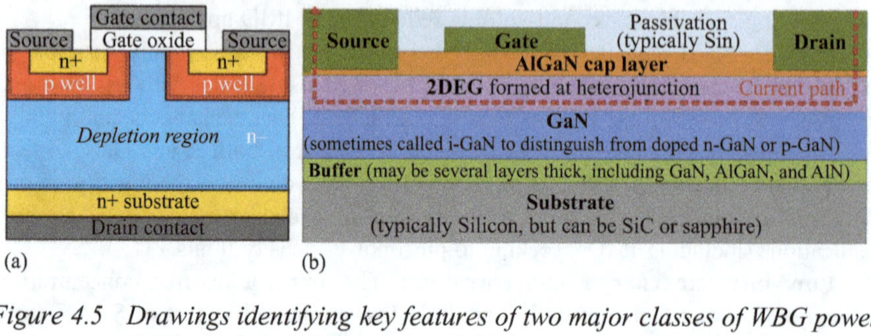

Figure 4.5 Drawings identifying key features of two major classes of WBG power device design: (a) vertical and (b) lateral. © 2023 Reproduced, with permission, from IEEE [6].

In contrast, the quality of GaN wafers at the time of this writing has not reached the same level as that of SiC. As a result, GaN material is typically grown as a thin epitaxial layer on a much thicker wafer substrate made of a different material which is typically either Si, SiC, or sapphire as indicated in Figure 4.5(b). Even though GaN material has a higher breakdown field value that SiC, the limitations imposed by the dielectric qualities of the epitaxial geometry typically limit the maximum breakdown voltage ratings of lateral GaN power devices to 600 and 900 V, in contrast to the higher maximum voltage ratings of commercially available vertical SiC devices which are 1,200 and 1,700 V. Vertical SiC devices with breakdown voltage values of 10 kV are in the early stages of reaching the marketplace, albeit at a much higher cost reflecting the early stage of their development.

Unlike many silicon power semiconductor devices, the designs of WBG power semiconductor switches are designed as majority-carrier devices with characteristics of field-effect transistors (FETs). In fact, the dominant class of SiC power devices that are commercially available today are vertical power MOSFETs that share many of the same basic features as their silicon counterparts. However, the highly desirable material properties of the WBG materials discussed earlier in this section give these SiC MOSFETS breakdown voltage ratings and on-state resistance values that are far superior to conventional Si-based MOSFETs. While GaN power devices are designed to share many of the same terminal behavior characteristics of enhancement-mode MOSFETs, they often use different semiconductor phenomena to achieve these characteristics. For example, many lateral GaN power devices are designed as high-electron-mobility transistors (HEMTs) that depend on the presence of a junction of two materials with different bandgap energies to form the conducting channel that can be controlled by the gate.

Despite the differences between their channel modulation techniques, both SiC MOSFETs and GaN HEMTs behave like very fast-switching FETs with resistive on-state voltage drops that increase linearly with the drain current amplitude. The temperature sensitivity of the on-state voltage for a given current is significantly lower for SiC MOSFETs than for GaN HEMTs, reflecting another advantage of the vertical channel in the SiC devices compared to the lateral channel in the GaN switches.

4.1.2.2 WBG device normal conduction state

WBG power switches can generally be categorized into two types based on their switching characteristics: depletion or enhancement mode. Depletion mode devices are normally on when the applied gate voltage is zero, while enhancement mode devices are normally off under the same zero gate voltage condition. The designs of the WBG devices determine whether they are candidates for depletion or enhancement mode operation. Enhancement mode device designs typically require the presence of a thin layer of electrically insulating material between the gate and the device body that is typically some type of oxide material (see Figure 4.5(a)).

Since the development of this gate insulating material for WBG power switches posed significant technical hurdles, the earliest generation of commercial WBG devices had depletion mode characteristics that did not require an insulating gate. However, the majority of power converter topologies used today have voltage–source characteristics that make enhancement mode characteristics much more desirable in order to prevent destructive short circuits from appearing in the converter circuit if the gate voltages are unexpectedly removed for even a short period. As result, the large majority of commercial WBG power switches (both SiC and GaN) that are available and sold today have enhancement mode characteristics so that they are normally off devices.

As a result of the differences between the material properties, device structure, and channel modulation techniques in SiC and GaN, the positive gate voltage for turning the enhancement mode devices fully-on is significantly higher in SiC devices (18–20 V) than in GaN devices (5–6 V).

4.1.2.3 WBG bidirectional switches

Although several points raised in the preceding sections highlight the disadvantages of lateral device configurations compared to vertical devices, there are also some features of the lateral devices that work to their advantage. One particularly notable example is the superior suitability of the lateral device configuration for developing monolithic bidirectional (BD) switch devices. Although various alternate names are associated with bidirectional switches (e.g., 4-quadrant switches), BD switches are distinguished by their ability to block high voltage in both polarities and achieve gate-controlled current flow in both polarities. A more detailed discussion of WBG-based BD switches in presented in Section 4.3.2 later in this chapter.

4.2 Application of WBG devices in integrated motor drives

4.2.1 Potential benefits of WBG devices for integrated motor drives

Both SiC and GaN devices are appealing candidates for use in future integrated motor drives. They are suitable for a wide variety of power converter topologies that can be used in IMD systems including both VSIs and CSIs. Some of the key advantages offered by WBG devices that are particularly attractive for IMD applications are discussed in the following paragraphs.

4.2.1.1 Higher switching frequencies yield smaller and lighter passive components

Inductor and capacitors are widely used in power electronics to form filters and to serve as passive buffers for power sources (e.g., dc-link capacitors for VSIs). Amplitudes of inductor ripple current and capacitor ripple voltage are inversely proportional to the switching frequency in nearly all cases. Silicon-based IGBTs in conventional hard-switched VSIs are typically limited to PWM switching frequencies of 20 kHz or less for motor drives with power ratings of 10 kW or higher. The PWM frequency typically drops into the range of 5–10 kHz when the inverter power rating climbs above 100 kW.

With the advent of WBG-based inverters, it is now possible to increase the PWM switching frequency in hard-switched VSIs to values of 50 kHz or higher for motor drive ratings up to 100 kW depending on the current and voltage requirements [8]. For lower power ratings of 10 kW or less, GaN power devices make it possible to reach switching frequencies above 100 kHz because of their ultra-fast switching capabilities [9].

By taking advantage of the exciting opportunities provided by WBG switches to raise PWM switching frequencies by factors of 5–10 times, the mass and volume of IMD inverter capacitors and inductors can be significantly reduced. However, it should be noted that designing high-frequency inductors and capacitors for high-power applications present technical challenges that are beginning to receive increased attention from power electronics engineers because of the increased demands being placed on them by fast-switching WBG switches [10].

4.2.1.2 Higher operating temperature yields reduced cooling requirements

A particular advantage of WBG devices that is extremely beneficial for the integrated motor drive is their ability to work at higher temperatures than Si devices. This takes on particular importance since high-performance electric machine typically operate with internal temperatures of 200°C or higher compared to maximum junction temperature of 150°C or, at most, 175°C that apply to Si-based power devices. Hence, WBG switch technology opens a pathway to mounting the drive power electronics in the same housing chamber as the motor in order to achieve higher power densities [11].

Mounting the power electronics close to the motor also opens opportunities to adopt integrated cooling topologies for the motor and power electronics. For example, in today's electric vehicles, engine oil is typically used to cool the hotter motor, and water-ethylene glycol (WEG) cools the traction inverter at lower temperatures. In the future, it will be possible to use a single integrated coolant loop that reduces its cost, mass, and volume, while improving its reliability.

There are significant challenges that remain to achieve elevated temperature operation of future motor drive power converters [12]. First, the maximum operating temperature of a power switch is often limited as much or more by its packaging as it is by the power device alone. While high-temperature packages for power devices are already available for special applications at premium prices, much work is required to make power switches widely available with maximum junction temperatures of 200°C or higher for high-volume, cost-sensitive applications.

Next, it important to note that all of the power converter components, and not just the power devices, must be compatible with higher-temperature operation to achieve the full potential of this vision. These parts can be classified into two categories. The first category is the high-power components. For VSIs, these passive components include the dc-link capacitor and output EMI filters, if they are used. This poses a serious challenge because both electrolytic capacitors and film capacitors that are typically adopted in commercial motor drive inverters have temperature ratings of 125°C or less.

The second category of components that must be thermally hardened consists of signal-level components including digital processors, op amps, sensors, instrumentation I/O components, and communications logic. Several of these components are already available for high-temperature operation at a significant premium cost for special applications such as down-hole drilling electronics. The challenges associated with making these components at much higher volumes and much lower cost are very real.

Finally, it is important to note that low-temperature operation is also highly important in some applications, and the ability to tolerate frequent thermal cycling between the temperature extremes in any application will be critical to achieving high reliability. The ultimate goal is to insure that *all* electronic components can tolerate the thermal environment associated with their particular IMD application in order to meet the demanding reliability requirements applicable to the complete integrated motor drive system.

4.2.1.3 Lower device loss yields high inverter efficiency and/or reduced cooling

The low switching losses and low on-state forward voltage drop of WBG switches make it possible to significantly reduce the total motor drive losses, raising its efficiency [13]. This opportunity is particularly important for electrified transportation applications that draw their power from energy sources with limited capacities so that the traction drive's efficiency has a significant impact on vehicle range. In addition, lower loss in the IMD's power electronics results in reduced heat that must be dissipated by the cooling system, resulting in a smaller and lighter cooling system that increases the total IMD power density and lowers cost. With some WBG-based traction drives reported to achieve efficiencies exceeding 99%, it is becoming increasingly plausible to discuss advanced cooling schemes for WBG-based drives that depend only on passive means such as heat pipes and convection to accomplish the required cooling without pumps or fans [14].

Engineering tradeoffs are inevitably present when deciding how to take best advantage of the benefits offered by WBG technology since some of these benefits compete with other. A good example is the desire to raise the switching frequency as high as possible to minimize the mass and volume of passive components (i.e., inductors, capacitors) while recognizing that the associated switching losses are increased, reducing the WBG efficiency enhancement benefits and increasing the mass and volume of the cooling system. Inevitably, these very real tradeoffs must be evaluated and optimized on a case-by-case basis.

4.2.2 Challenges caused by WBG devices in VSI-based IMDs

With the market arrival of new enhancement mode (normally off) WBG power switches offering appealing performance advantages, it is very natural and tempting to treat these switches as drop-in replacements for the Si-based IGBTs and power MOSFETs in VSIs that have dominated the power electronics landscape for the past 30+ years. Unfortunately, the same features that lie at the source of the benefits discussed in the preceding sub-section can give rise to serious complications when Si-based power devices in VSIs are simply removed and replaced with SiC and GaN counterpart switches. The nature of these complications and some available techniques that can be adopted to suppress them are discussed below.

4.2.2.1 Motor winding over-voltages caused by high excitation voltage *dv/dt*

The switching speed of WBG switches is so fast that the wire motor cables connecting the VSI to a motor load behaves like a transmission line that can interact with the motor phase impedances to cause serious transient over-voltages at the motor phase terminals. Impedance mismatches between the motor cables and the machine can result in voltage overshoot amplitudes that can double or even triple the VSI's dc link voltage [15]. Such high voltages create the risk of catastrophic damage to the motor cables attributable to insulation voltage breakdown and failure if the repetitive over-voltages are not suppressed.

This voltage overshoot problem was recognized and studied more than 30 years ago when then-new IGBTs were being adopted in hard-switched VSIs for motor drive applications. Generally, this phenomenon did not cause serious problems when using IGBTs unless the length of the motor cables exceeded 30 m. However, the approx. 10× switching speed increase provided by WBG switches over Si-based IGBTs results in motor terminal voltage overshoot problems with motor cables as short as 3 m in length [16]. Making matters worse, these overshoots are accompanied by large capacitive current pulses passing through the inverter switches that are caused by the parasitic common-mode capacitance to ground of the motor cables.

Figure 4.6 shows an example of the measured voltage overshoot at the terminals of a three-phase passive *L–R* load emulating a motor connected to a WBG-based hard-switched VSI with a 3-phase set of shielded motor cables that are 7.6 m long. Load terminal voltage overshoots that reach 2× the inverter dc link voltage occur at each switching event, and significant current pulses are absorbed by the inverter switches during each voltage pulse. The amplitude of the voltage *dv/dt* during each inverter switching event is approx. 14 V/ns, a value that is not uncommon with hard-switched VSIs unless special steps are taken to slow the device switching rate.

The adoption of an IMD architecture might seem at first like an effective approach for avoiding this problem by shrinking the motor cable length to very small values less than 0.2 m. However, closer analysis has shown that the voltage overshoot problem is not eliminated with IMD configurations, but instead transferred inside the machine's stator windings where individual winding turns experience turn-to-turn and turn-to-ground overshoot voltage values that are several times larger than their expected values without these undesired transmission line effects [17]. Figure 4.7 shows a cross-section view of the 7-turn stator tooth winding of a 1 MW PM machine and the calculated values of the turn-to-turn voltages developed in response to a step voltage applied to the winding's input terminals with a *dv/dt* value of 14 V/ns. The peak turn-to-turn voltages for a couple of the adjacent turns reach nearly 5 times their expected values without the transmission line effects caused by the ultra-fast switching events, demonstrating the seriousness of this voltage overshoot issue.

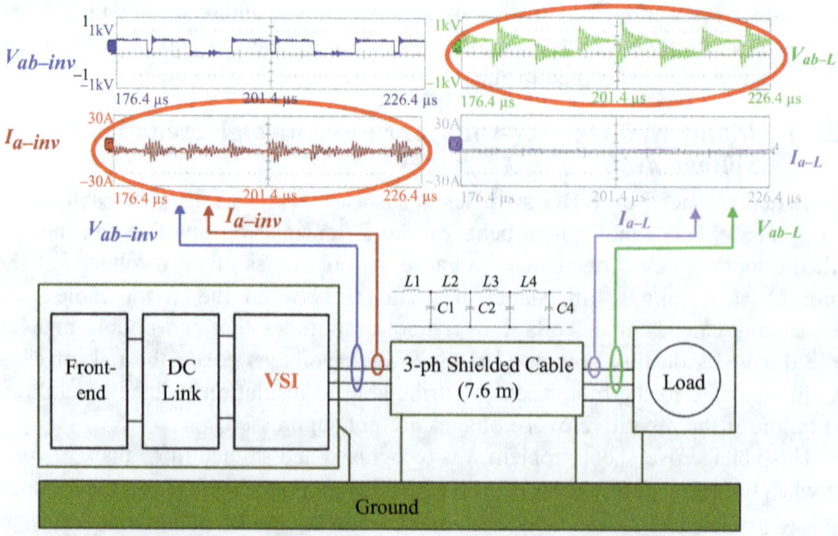

Figure 4.6 Examples of voltage and current overshoot waveforms for a hard-switched VSI using SiC power switches connected to a 3-phase L–R load using a medium-length (7.6 m) motor cable

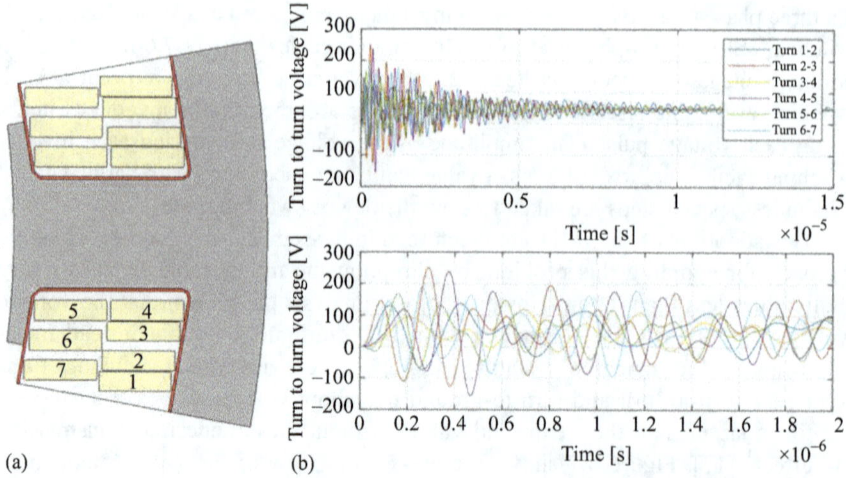

Figure 4.7 Calculated turn-to-turn voltages in response to a step phase voltage with dv/dt = 14 V/ns: (a) winding cross-section with turn number identification and (b) calculated turn-to-turn voltage waveforms with zoomed-in view of overshoot amplitude

4.2.2.2 High conducted and radiated EMI

The extremely high *dv/dt* switching rate of WBG devices (>10 V/ns) significantly aggravates the risk of elevated conducted and radiated electromagnetic interference (EMI) caused by the power electronics [18]. In particular, any parasitic capacitance between the earth ground and the power converter, motor cables, or motor components (stator and rotor) provide pathways for common-mode (CM) currents since the capacitor current is directly proportional to the values of both the capacitance *C* and *dv/dt*. The common-mode EMI emissions are also increased as the switching frequency is raised, creating a higher risk of exceeding the prevailing EMI maximum limits set by international standards for motor drives in different applications (e.g., industrial, commercial, residential, military, aerospace, etc.).

Figure 4.8 shows an equivalent circuit of a 3-phase WBG-based VSI (see Figure 4.6) that highlights the parasitic capacitances associated with the *RL* load as well as the *LC* transmission line model of the interconnecting cable [16]. Figure 4.9(a)

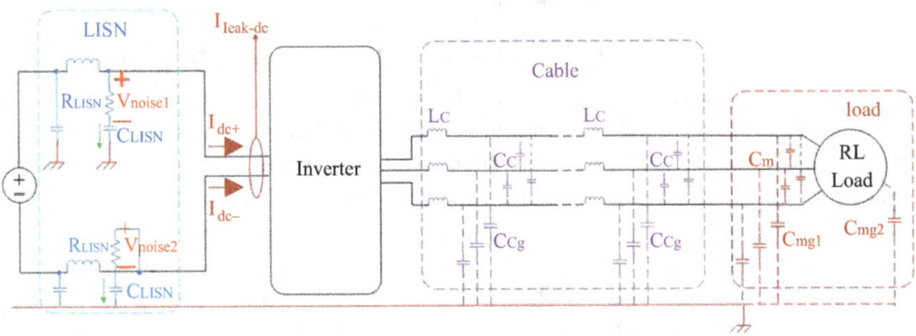

Figure 4.8 Common-mode EMI equivalent circuit for a 3-phase VSI showing parasitic capacitances-to-ground associated with the 3-phase RL load and cable. © 2023 Reproduced, with permission, from IEEE [16].

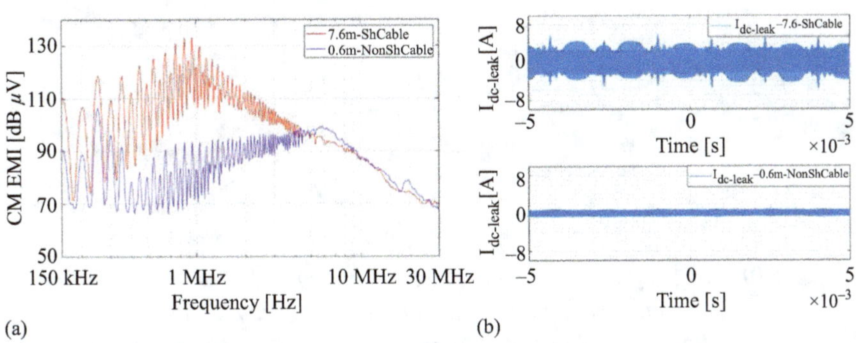

Figure 4.9 (a) Measured conducted CM EMI of VSI with two cable lengths and different shielding; and (b) measured CM leakage current for the two cable cases. © 2023 Reproduced, with permission, from Dr Hang Dai [19].

presents the measured common-mode EMI frequency spectra for this VSI with a long shielded cable (7.6 m) and a short non-shielded cable (0.6 m), demonstrating the significant degrading impact of the longer cable length on the CM EMI amplitude [19]. Both CM EMI spectra exceed the maximum amplitude limits set by IEC standard CISPR 22 for information technology equipment. Figure 4.9(b) compares the measured common-mode leakage current to ground for the two cable cases, confirming that the long shielded cable results in significantly higher CM leakage current to ground.

The ultra-fast switching of WBG switches in hard-switched VSIs also contributes to elevated levels of radiated EMI that can seriously interfere with other nearby electronics, causing serious malfunctions and failures [20]. Placing the power electronics inside tightly sealed metal container can help to suppress this radiated EMI, but power cables interconnecting the inverter to its load can behave like an antenna to broadcast the damaging EMI throughout buildings and vehicles. Adding shielding to the cables can help suppress this radiated EMI, but the additional mass and cost of the shielded cable is undesirable, and the conducted EMI can be aggravated by the shielding's presence.

4.2.2.3 Bearing damage attributable to rotor charge build-up

Another undesirable impact attributable to inverter common-mode voltage waveforms with ultra-fast-switching voltage pulses is the generation of dangerous current discharges inside the metal ball bearings of drive load motors. The arcing current discharges inside the bearings appear because of voltage build-up between the rotor shaft and grounded motor housing fed by the parasitic capacitance between motor windings and the rotor assembly (Figure 4.10(a)). Fast-rising common-mode voltage pulses with fast-rising edges cause the voltage to build up sufficiently on the rotor shaft to trigger current discharges inside the bearings (Figure 4.10(b)) that ultimately destroy them by damaging their internal races (Figure 4.11). The rate of bearing damage is aggravated by high *dv/dt* values

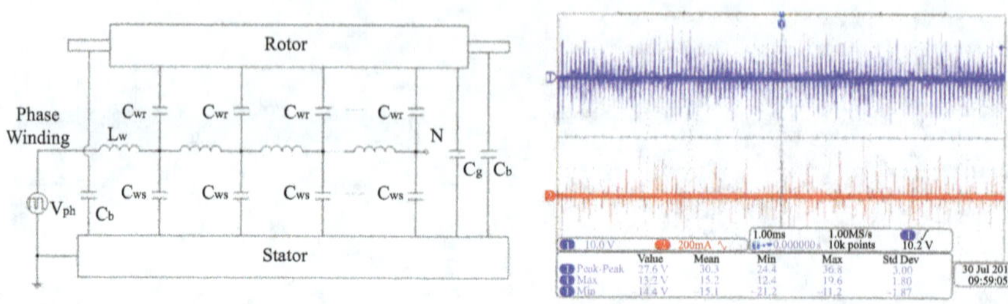

(a) (b)

Figure 4.10 (a) Simplified equivalent circuit depicting shaft voltage build-up mechanism resulting from motor parasitic capacitances and terminal common-mode voltage pulses (left); and (b) measured shaft voltage (upper) and bearing current (lower) generated by 250 kW, 460 Vrms SiC inverter operating at dv/dt > *14 V/ns switching rate. © 2023 Reproduced, with permission, from IEEE [21].*

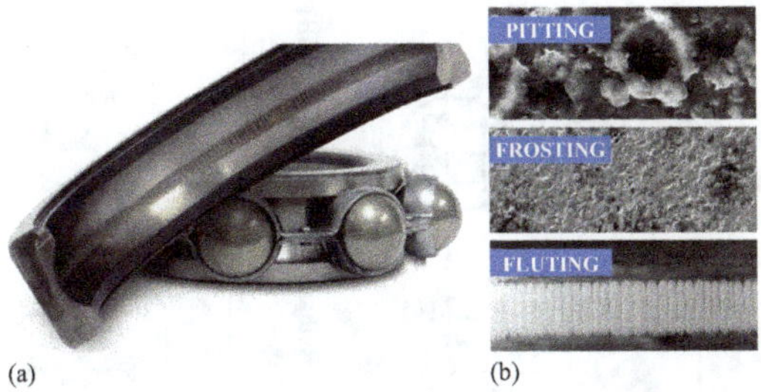

Figure 4.11 (a) Bearing race damage caused by internal discharge currents [22]; and (b) examples of bearing race pitting, frosting and fluting damage caused by bearing currents [23]. © 2023 Figures, courtesy of Electro Static Technology, an ITW Company.

combined with high switching frequencies, making inverters with WBG power switches more dangerous than their counterparts with slower Si devices that switch with lower *dv/dt* values at lower frequencies [21].

Depending on the details of the motor construction and voltage excitation waveforms, bearing destruction can occur after years, months, or, in some cases, only a few days of operation. A variety of factors associated with the motor drive installation influence the bearing current risk including the common-mode voltage amplitude, switching frequency, *dv/dt* value, temperature, motor construction, motor speed, etc. The adoption of ball bearings made with non-conducting ceramic balls or the addition of grounded brushes to remove the rotor charge before it can build up to dangerous levels are two of the ways that bearing currents can be eliminated, albeit with a negative impact on motor cost and maintenance.

4.2.3 Standard techniques for overcoming WBG switch challenges in VSI-based IMDs

4.2.3.1 Higher gate resistors and snubbers to reduce output voltage *dv/dt* and overshoot voltage amplitudes

Raising the gate resistor values is one of the most commonly applied techniques for reducing the *dv/dt* values during WBG switch turn-on and turn-off events in hard-switched VSIs. Elevated gate resistor values slow the charging/discharging of the WBG device gate capacitor, thereby slowing the switching transients at the devices's output terminals. Using two different resistor values at the gate drive circuit's output terminals – one for turn-on and the other for turn-off – makes it possible to independently adjust the turn-on and turn-off *dv/dt* values (Figure 4.12) [24]. Reducing the *dv/dt* values has a beneficial effect on all three of the challenges discussed in the preceding subsection: motor winding overvoltages, EMI, and bearing currents.

A second well-known technique for shaping the voltage transient waveforms during turn-on and turn-off events is the introduction of a snubber circuit across the output terminals (e.g., drain to source) of the WBG switches in the VSI phase legs (Figure 4.13) [25]. As shown in Figure 4.13(b), changing the values of the

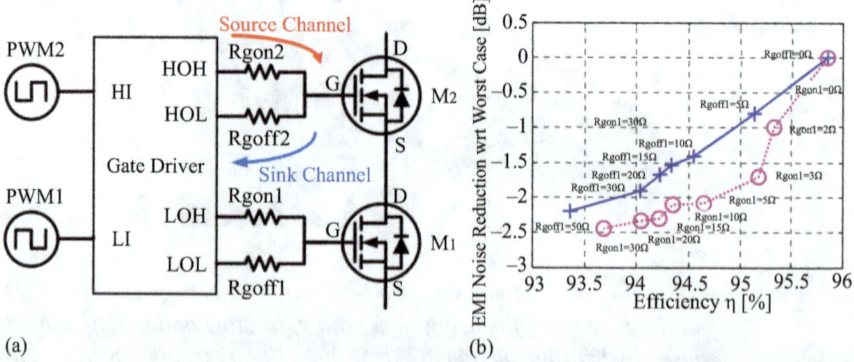

Figure 4.12 *(a) Simplified schematic of gate driver circuit with gate resistors; and (b) influence of gate resistances Rgon1 (magenta) and Rgoff1 (blue) on GaN converter efficiency and CM noise reduction with respect to the worst case (the GaN converter with zero gate resistance). © 2023 Reproduced, with permission, from IEEE [24].*

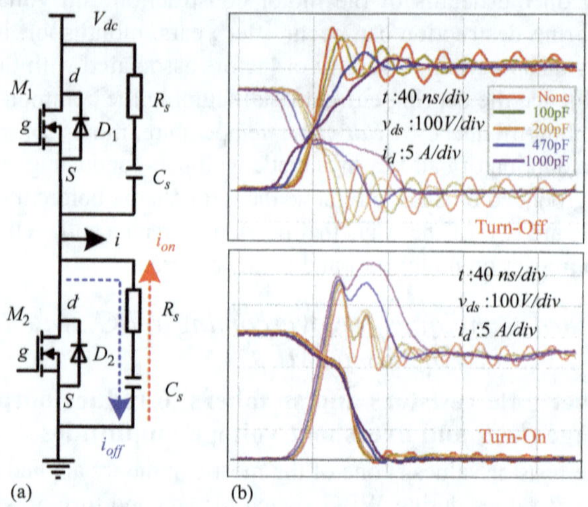

Figure 4.13 *(a) VSI phase leg that includes RC snubber circuits (R_s, C_s) across the drain–source terminals of both WBG switches; and (b) impact of changing capacitor value C_s in the two snubber circuits on drain–source voltages and currents of switch M_2 during both turn-off and turn-on events. © 2023 Reproduced, with permission, from IEEE [25]*

snubber component values provides a means of reducing the peak amplitude of the output voltage overshoot transient.

Unfortunately, there are serious negative consequences associated with both of these techniques. Increasing the gate resistance values extends the turn-on and turn-off times, increasing the inverter's switching losses. As a result, this technique negates the increased switching speed and lower switching losses offered by the WBG switches, neutralizing one of their most potent advantages over the conventional Si-based devices that they are replacing. Similarly, snubber circuits typically include resistors that accumulate additional undesired losses with every switching event, reducing the inverter's efficiency.

4.2.3.2 Filters to suppress high *dv/dt* and CM EMI in output voltage waveforms

One of the most well-established techniques to reduce the *dv/dt* amplitude in the output voltage waveforms of hard-switched VSIs is to introduce a three-phase low-pass filter. In addition to reducing the *dv/dt* amplitude, the filter also has the beneficial effect of attenuating the common-mode voltage delivered to the load.

Two versions of the inverter output filter are presented in Figure 4.14. The "sinusoidal filter" shown in Figure 4.14(a) places the corner ("break-point") frequency of the filter at a sufficiently low value that it significantly suppresses all high-frequency components including the PWM switching frequency component and all of its higher-frequency harmonics [26]. The major advantage of this filter design is that the output voltage waveforms are highly sinusoidal (i.e., low total harmonic distortion, THD) at the fundamental frequency (see Figure 4.15), and the common-mode voltage is substantially reduced. As a result, the motor terminal voltage overshoot, common-mode EMI, and bearing current problems highlighted in the preceding subsection are all significantly reduced or eliminated. However, the required inductor and capacitor components are typically quite large and heavy to accomplish this filtering, undesirably compromising the power density advantages associated with WBG switches that are so valuable in IMD applications. In addition, the losses in the filter components substantially compromise the efficiency advantages provided by the WBG switches.

The second type of output filter shown in Figure 4.14(b) is known as a *dv/dt* filter that shares the same basic second-order *LC* filter topology as the sinusoidal filter [27]. However, it is a "light" version of the sinusoidal filter that has a more modest objective of reducing both the leading- and trailing-edge *dv/dt* values of the PWM voltage pulses delivered to the load; the pulsed-nature of the output voltage waveforms delivered to the load is not suppressed. By moving the corner frequency

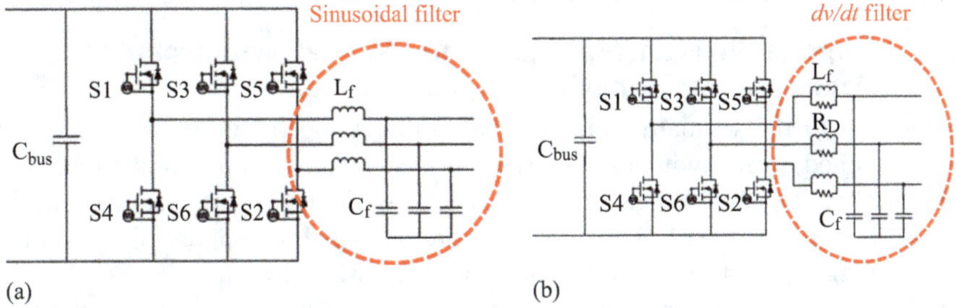

Figure 4.14 WBG-based VSI with three-phase output filter: (a) sinusoidal filter and (b) dv/dt filter

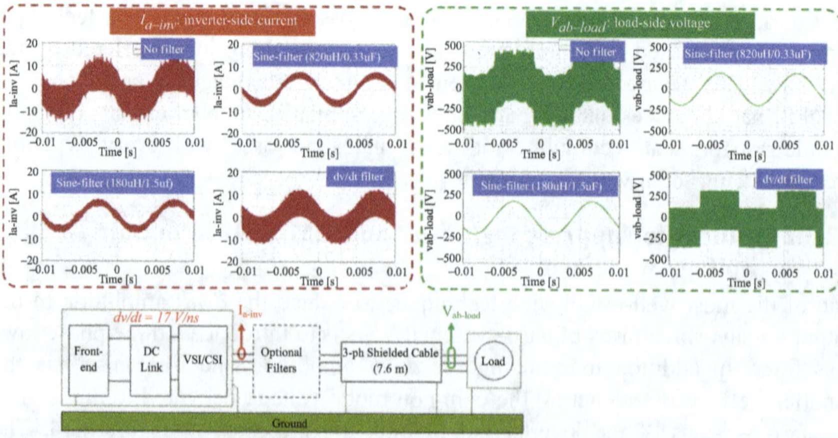

Figure 4.15 Current and voltage waveforms of VSI with different output filters. © 2023 Reproduced, with permission, from Dr Hang Dai [19].

of the *dv/dt* filter to a significantly high frequency value closer to the PWM switching frequency, the mass and volume penalties associated with the filter inductor and capacitor components are substantially reduced.

Unfortunately, the benefits provided by the *dv/dt* filter for reducing the load voltage overshoot, the common-mode EMI amplitudes, and the PWM switching frequency harmonic components are also much more modest than for the sinusoidal filter (see Figure 4.14(b)). As a result, engineering judgments must be made on a case-by-case basis to determine whether the sinusoidal or *dv/dt* output filters are better suited for a particular application.

If the desired objective is narrowed to reducing the amplitude of the conducted EMI, there are a wide variety of passive EMI filter topologies that are designed to specifically accomplish that result. Widely used examples include common mode choke inductors and Y-capacitors that are designed to block the transmission of CM EMI from an inverter to either the load or the inverter's dc power source. In addition to passive filter designs, there are many active EMI filters that use amplifiers together with feedback techniques to cancel the EMI current, thereby preventing it from reaching either the inverter's load or source, depending on their placement [28].

4.2.3.3 Special inverter topologies and PWM algorithms to cancel common-mode voltage generation

In addition to the standard techniques for suppressing the undesired effects of hard-switched WBG switches discussed in the preceding subsections, there are additional approaches that include special hard-switched inverter configurations as well as tailored PWM algorithms that are designed to suppress the common mode voltage content in inverter output waveforms.

An example of a hard-switched VSI that is configured to suppress the CM voltage via cancellation is shown in Figure 4.16(a) [29]. It uses a total of nine WBG switches to effectively form two overlapped six-switch VSIs sharing three

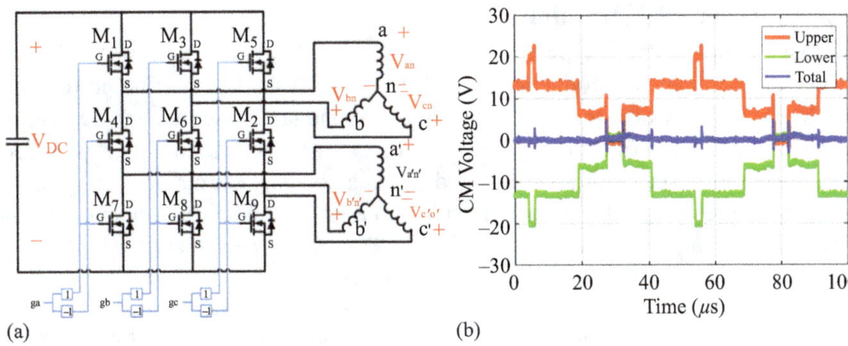

Figure 4.16 *(a) Nine-switch balanced inverter and its gating signals; and (b) measured common-mode voltage components of the output voltage waveforms for the upper and lower halves of one phase leg, and their summed total CM voltage. © 2023 Reproduced, with permission, from IEEE [29].*

switches. Each of the 3-switch phase-legs excites both halves of a split phase winding for a single motor. By controlling the two overlapped 3-phase inverters to apply identical PWM switching algorithms that are 180° out of phase, the phase windings receive their full fundamental voltage excitation, but the common mode voltage components in the two half-windings cancel each other (Figure 4.16(b)). The addition of the three switches to form this nine-switch balanced inverter eliminates the need for bulky and lossy output filters to achieve this common-mode voltage suppression. Care must be taken to carefully synchronize the switching of the upper and lower switches in each phase leg in order to achieve complete common mode voltage cancellation. Any differences between their switching times will result in progressive re-emergence of the undesired common-mode voltage component.

There have also been significant technical contributions to the development of PWM algorithms for conventional 6-switch full-bridge VSIs that significantly reduce but do not eliminate common voltage components in their output voltage waveforms. Although many candidate PWM algorithms with reduced common mode voltage characteristics have been proposed [30], many of them suffer from some form of performance degradation such as extra switching events each cycle that hurt efficiency, or loss of access to the full voltage vector space accessible with the conventional space vector PWM algorithm.

4.2.4 Current and future applications of WBG technology in VSI-based IMDs

Despite the challenges associated with applying WBG power switches in VSIs, some exciting applications of WBG-VSI technology in integrated motor drive configurations are already in commercial production or are being actively developed in industry and university research facilities. Two examples will be briefly introduced in this section. The first is an electric vehicle (EV) traction drive, and the second is a prototype IMD designed for electrified aircraft propulsion.

4.2.4.1 Tesla 3 EV traction drive inverter

The first example is the EV traction drive for the Tesla Model 3 passenger EV that has been in commercial production since 2018. The Tesla 3 is the first production battery EV in the world that has adopted wide-bandgap device technology in its traction drive. A view of the rear (main) Tesla 3 transaxle powertrain unit is provided in Figure 4.17(a). The inverter is mounted inside a compact aluminum casting highlighted with a red rectangle on the right side of the integrated powertrain unit that has the appearance of a machine endbell. The interior permanent magnet (IPM) synchronous motor is mounted on the left side of the powertrain, and the gearbox is sandwiched between the two. The fact that the inverter is separated from the motor housing by the gearbox raises questions about whether this powertrain should be classified as a true integrated motor drive, but it will be treated as an IMD for purposes of this discussion because of the many features it shares with a "pure" IMD.

A view of the Tesla 3 inverter power stage components is provided in Figure 4.17(b) after removal of its single printed circuit board containing all of the control electronics and gate drives. Inspection reveals that it uses a classic six-switch VSI topology, with four discrete SiC MOSFETs connected in parallel for each inverter switch. The film capacitors are potted in the black rectangular block component that is visible to the left of the inverter switches [31].

Table 4.2 provides a summary of some of the key electromechanical specifications of the inverter and its associated IPM motor that are relevant to this discussion. The inverter delivers >200 kW to the machine during peak power operation with a nominal dc link voltage of 350 Vdc. It is notable that the inverter phase current during peak output power operation approaches 1,000 Arms (i.e., ~238 Arms per MOSFET switch). The power density metrics for the inverter based on mass and volume are quite impressive compared to other commercial passenger EV traction drives at the time of this writing, exceeding 40 kW (pk)/kg and 40 kW (pk)/liter, respectively.

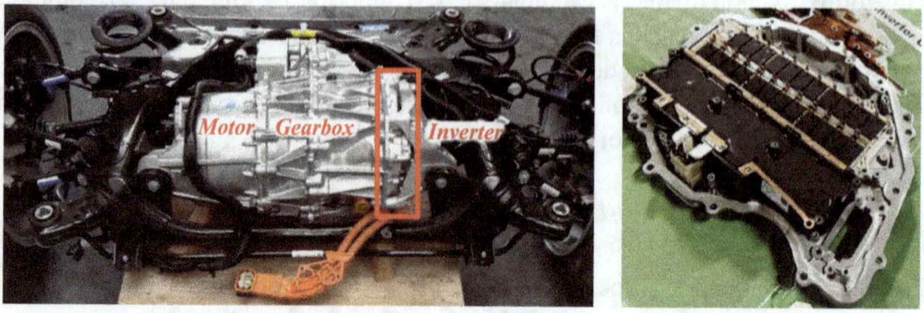

(a) (b)

Figure 4.17 (a) View of Tesla 3 transaxle powertrain identifying inverter location and (b) an internal view of SiC-based inverter power stage. © 2023 Figures, courtesy of Ingineerix (a), and MotorTrend Magazine (b).

Table 4.2 Tesla 3 traction drive inverter specifications and metrics [32–34]

Specifications and metrics	Value
Power rating (peak)	211 kW (peak)
Torque (peak)	450 Nm
Corner point speed	4,478 rpm
Nominal inverter DC link voltage	350 Vdc
Inverter peak current (RMS)	953 A rms
Inverter unit mass (approx.)	4.8 kg
Inverter unit volume (approx.)	5.0 liters
Inverter mass power density (approx.)	44 kW (pk)/kg
Inverter volumetric power density (approx.)	42 kW (pk)/liter

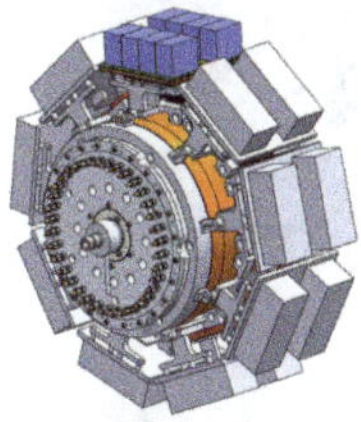

(a)

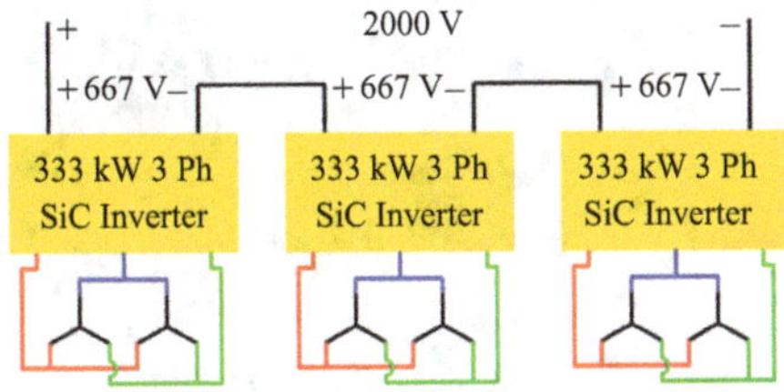

(b)

Figure 4.18 (a) Drawing of prototype 1 MW integrated motor drive for aircraft propulsion application; and (b) modularized inverter excitation schematic diagram for 2000 Vdc source voltage

4.2.4.2 1 MW prototype electrified aircraft propulsion motor drive

The second example is a prototype 1 MW IMD unit that is being developed for an electrified aircraft propulsion motor drive application. It is being developed by a group of four universities led by the Ohio State University under the sponsorship of the US National Aeronautics and Space Administration (NASA) as a part of their University Led Initiative (ULI) program. The objective of this R&D effort is to design and demonstrate a prototype propulsion drive that pushes the state-of-the-art in power density using an integrated motor drive configuration. The prototype motor drive hardware is in the process of being fabricated at the time of this writing, during the last year of a 5-year project.

A drawing of the prototype 1 MW IMD unit in its final form is provided in Figure 4.18a. As illustrated in this drawing, the inverter power electronics is distributed around the outer circumference of the surface permanent magnet (SPM) synchronous machine [35,36]. The machine has six 3-phase sets of windings, for a total of 18 machine phase windings to excite. The machine drive's power circuit schematic in Figure 4.18(b) highlights the modular nature of the motor drive architecture. The 2,000 Vdc that is supplied is divided into three 667 Vdc power buses that are connected in series. This configuration reduces the required voltage rating of the power electronics and machine windings which is appealing for high-altitude operation where the dielectric strength of the atmosphere is much lower than at sea level. Each of the six 3-phase windings has its own dedicated 167 kW inverter, and two of them are excited in parallel for each of the three 667 Vdc stages. Furthermore, each of the inverter phase-legs in the six inverters excites a single phase winding. The highly modular architecture chosen for this motor drive lends itself well to the future development of fault-tolerant operating modes.

Figure 4.19(a) shows a view of the prototype 1 MW IMD unit mounted on a dynamometer during its final assembly stage. Two inverter phase-legs are built into

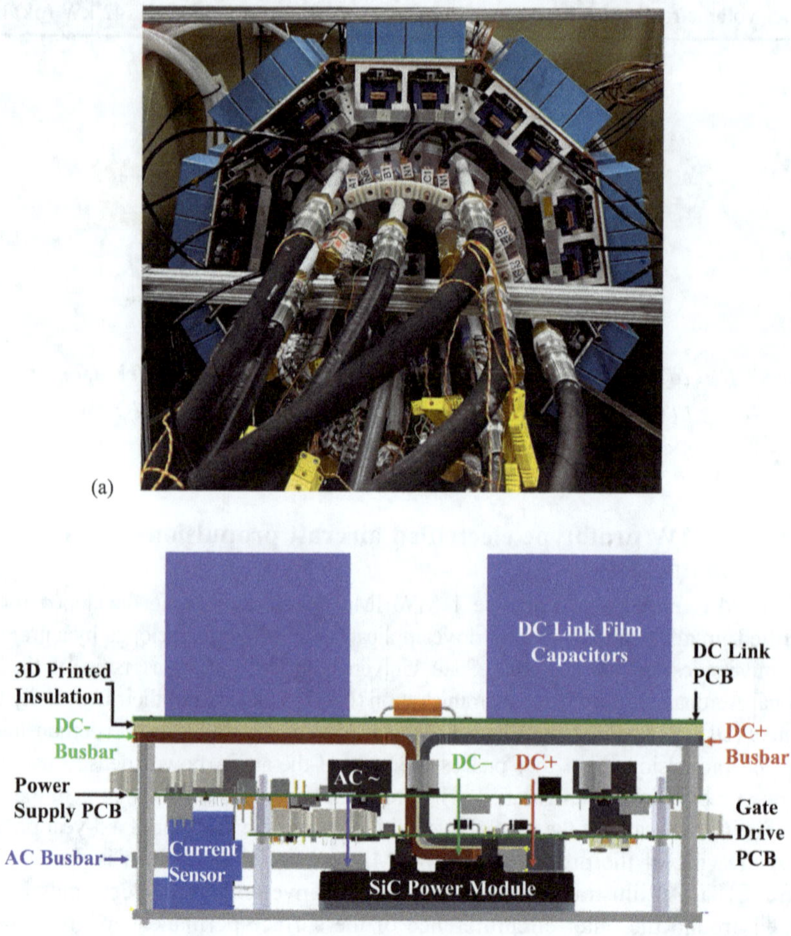

Figure 4.19 (a) View of assembled prototype 1 MW IMD unit mounted on dynamometer and (b) side view drawing of inverter module unit that excites 2 of 18 machine phase windings. © 2023 Figure, courtesy of Dr Jin Wang, The Ohio State University (b).

Table 4.3 Specifications and performance metrics of 1 MW prototype aircraft propulsion IMD

Specifications and metrics	Value
Total rated power	1 MW
Rated torque	477 Nm
Rated speed	20,000 rpm
Individual inverter power rating (1 of 6)	167 kW
Total DC supply voltage	2,000 Vdc
Individual inverter DC bus voltage	667 Vdc
Individual inverter current (RMS) @ rated power	624 A rms
Total combined inverter mass (approx.)	29.1 kg
Total combined inverter volume (approx.)	20.6 liters
Total inverter mass power density (approx.)	34.4 kW/kg
Total inverter volumetric power density (approx.)	48.6 kW/liter
Total IMD mass (approx.)	124 kg
Total IMD volume (approx.)	74 liters
Total IMD mass power density	8.1 kW/kg
Total IMD volumetric power density	13.5 kW/liter

each of the 9 inverter modules mounted around the outer circumference of the machine. Figure 4.19(b) provides a side view drawing of one of these inverter modules, highlighting the presence of two stacked printed circuit boards (PCBs) containing the gate drive electronics on top of each SiC power module and a third upper PCB for the dc-link bus capacitors. Table 4.3 provides a summary of key specifications and metrics for the IMD equipment, including power density values for the total inverter power electronics, as well as for the combined IMD including the motor and the inverter. The rated motor speed of 20,000 rpm plays a key role in reducing the mass of the machine. As shown in this table, the expected power density of the complete IMD is 8.1 kW/kg which is well on the way to reaching the power density values of modern turbofan jet engines which fall in the range of 11–19 kW/kg [37].

4.3 Unfolding opportunities for WBG devices in future IMDs

4.3.1 Introduction to CSIs as alternative to VSIs

As described in the first half of this chapter, WBG power switches offer some powerful advantages to new generations of integrated motor drives because of their impressively high switching frequencies and low on-state conduction voltages. While the potential improvements in IMD power density and efficiency are tantalizing, the very real technical challenges that WBG switch create when used in conventional VSIs can be daunting, to point of discouraging motor drive designers from adopting WBG switches in new designs. The objective of this section is to propose current source inverters (CSIs) as a promising alternative approach for applying WBG switches in future IMDs that overcomes several of the key obstacles that hinder their use in conventional VSIs. Furthermore, CSIs offer some intriguing application advantages made possible by the special features of the CSI topology.

Although CSIs were once the dominant inverter topology for motor drives during the 1960s and 1970s when thyristors were the major available power switches, CSIs have largely disappeared from view since the 1980s when MOS-gated silicon power switches (i.e., power MOSFETs and IGBTs) displaced thyristors and established VSIs as the dominant inverter topology. Three-phase two-level VSIs and CSIs (see Figure 4.20) are power converter topologies that are related to each other by the mathematical concept of duality for which the roles of voltage and current are reversed. Some of the major differences between them can be summarized as follows:

1. DC bus capacitors C_{bus} are the signature passive energy storage components in parallel with the dc input terminals of a VSI, while a series inductor L_{bus} plays a similar role in the CSI.
2. The six power switches in the VSI can block voltage only in the forward polarity while behaving as uncontrolled diodes in the reverse polarity. The corresponding six power switches in the CSI are duals of the VSI switches, blocking voltage in both polarities while conducting current under gate control only in the forward direction.
3. The CSI requires a three-phase capacitive filter C_f across its output terminals, serving as a buffer between the dc-bus inductor L_{bus} in the inverter's front-end and the inductive load impedance that is associated with all classes of electromagnetic electric machines.

The reason that CSIs lost their dominant position in adjustable-speed motor drives to VSIs in the 1980s and 1990s lies with the difference between the basic electrical terminal characteristics of thyristors and the MOS-gated switches that replaced them. Thyristors can be readily designed with symmetrical voltage characteristics so that they can block equal voltages in both forward and reverse polarities (Figure 4.21(a)). However, thyristors can only conduct current in the forward polarity which is controlled by triggering the gate terminal to initiate the device's conduction state. Importantly, a thyristor can only be returned to its off-state by having the external circuit apply reverse voltage to the thyristor's anode–cathode terminals. The

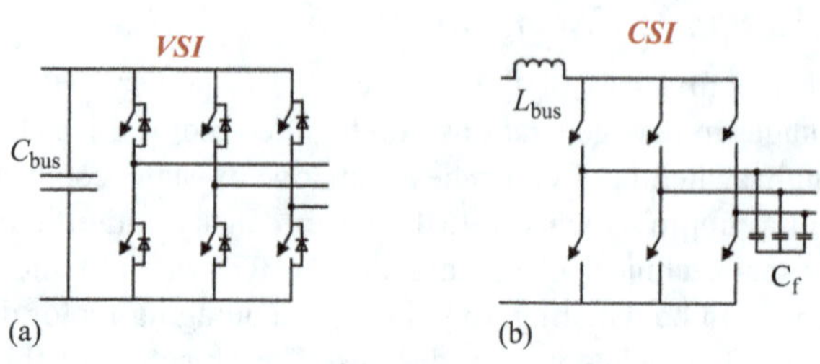

Figure 4.20 Comparison of two-level full-bridge topologies for VSI (left) and CSI (right)

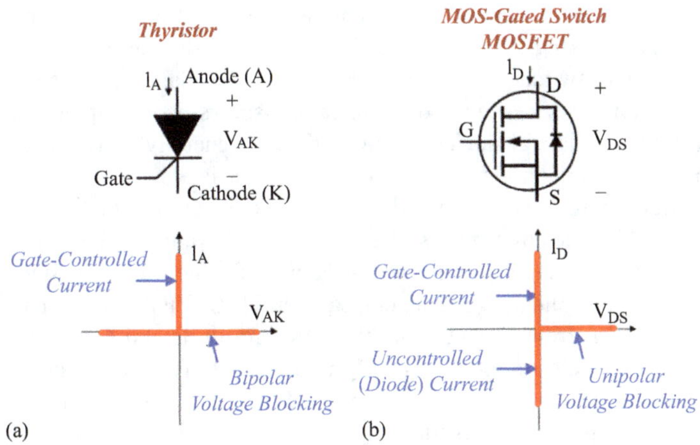

Figure 4.21 *(a) Comparison of idealized switch* I–V *locus lines for examples of reverse-voltage blocking switch; and (b) forward-only voltage blocking switch*

nature of the thyristor's volt-current characteristics with symmetrical voltage blocking and unidirectional current flow make it perfectly suited for CSIs.

In contrast, nearly all MOS-gated controlled switches can only block voltage in only a single (i.e., forward) polarity (see Figure 4.21(b)). The gate terminals of these devices fully control the forward polarity current to either flow (on-state) or stop flowing (off-state), offering a major advantage over thyristors. When the external circuit applies voltage to a MOS-gated controlled switch in the reverse polarity, it behaves as a diode corresponding to uncontrolled current flow in the negative (reverse) polarity. These volt–current characteristics with unidirectional voltage-blocking and bidirectional current flow make the MOS-gated controlled switches perfectly suited for VSIs.

Given all of the performance advantages held by MOS-gated switches compared to thyristors (i.e., full on-off gate control, much higher switching frequency, etc.), the MOS-gated switches progressively displaced thyristors, forcing nearly complete displacement of CSIs by VSIs which has persisted ever since the 1990s. While some researchers have explored the development of CSIs using MOS-gated switches by placing a diode in series with each MOSFET or IGBT to provide the required reverse-voltage-blocking (RVG) capability, the additional loss and cost of the diodes has significantly limited the appeal of this approach, preventing it from reaching commercial viability.

4.3.2 Impact of WBG switch devices on renewed CSI interest

The development of modern wide-bandgap power semiconductors has opened the door to reconsidering CSIs as an alternative to VSIs. There are at least two reasons why WBG switches are reviving interest in CSIs. The most important reason is that

WBG technology is generally more compatible with developing monolithic reverse-voltage-blocking switch devices required by CSIs than any of the silicon based MOS-gated switches. For example, nearly all GaN power switches are designed as lateral devices built on substrates using silicon, silicon carbide, or sapphire. While the lateral device configuration is generally considered to be an undesirable restriction on GaN switch technology, a lateral MOSFET or HEMT device configuration does not include a body diode, unlike the vertical device configuration. This feature makes the device an intriguing candidate for the reverse-voltage-blocking capability that is required for their use in CSIs.

In addition, the conducting channel of a lateral MOSFET or HEMT is capable of gate-controlled current in both polarities. As a result, lateral MOSFETs can be designed with dual gates in multiple configurations that gives them gate-controlled current capabilities in both current polarities. When combined with the reverse-voltage blocking capabilities described above, this gives the monolithic device the characteristics of a bi-directional (BD) switch (also known as a four-quadrant or ac switch) switch that has been long-sought for a variety of special applications including matrix converters and solid-state circuit breakers.

Some intriguing examples of WBG-based monolithic BD switch devices have been reported in the literature by both academic and industry researchers. One of the most impressive examples is a GaN-based 1,400 V, 100 A monolithic BD switch developed by Panasonic [38] using a lateral device configuration (Figure 4.22(a)). Another example is a SiC-based 1,200 V, 20 A monolithic BD switch developed by Dr B.J. Baliga at NCSU [39] using a vertical device configuration (Figure 4.22(b)). Both of these BD devices can be modeled as two anti-series connected MOSFETs in common-drain configurations. As shown in the simplified cross-section view of the Panasonic device (Figure 4.22(a)), this is appealing because there is a single channel shared by the two merged MOSFETs, resulting in a low on-state voltage.

It is important to note that these monolithic BD switches are versatile devices that are excellent candidates for use in CSIs. Despite the demonstrated technical feasibility of these monolithic BD switches and their promising applications in ac power converters and solid-state ac breakers, none are in commercial production at this time. Although this is disappointing, there is reason for optimism that this will change in the not-too-distant future since multiple device manufacturers are beginning to offer sample BD switch devices to potential customers at the time of this writing.

The second reason why WBG technology is reviving interest in CSIs is their beneficial impact on the size of the CSI's passive component sizes. More specifically, the much higher switching frequencies that are possible using WBG power switches compared to their Si-based counterparts is making it possible to significantly shrink both the mass and the volume of the CSI's dc-link inductor L_{bus} as well as the output filter capacitors C_f. The results of a case study [19] are shown in Figure 4.23 for a 3 kW CSI application that will be discussed later in this chapter. These figures show that for fixed output voltage power (3 kW) with fixed amplitudes of the dc-link ripple current and the output ripple voltage, the required values of L_{bus} and C_f drop inversely with the switching frequency, resulting in monotonic decreases in their masses and volumes as the switching frequency is increased. It

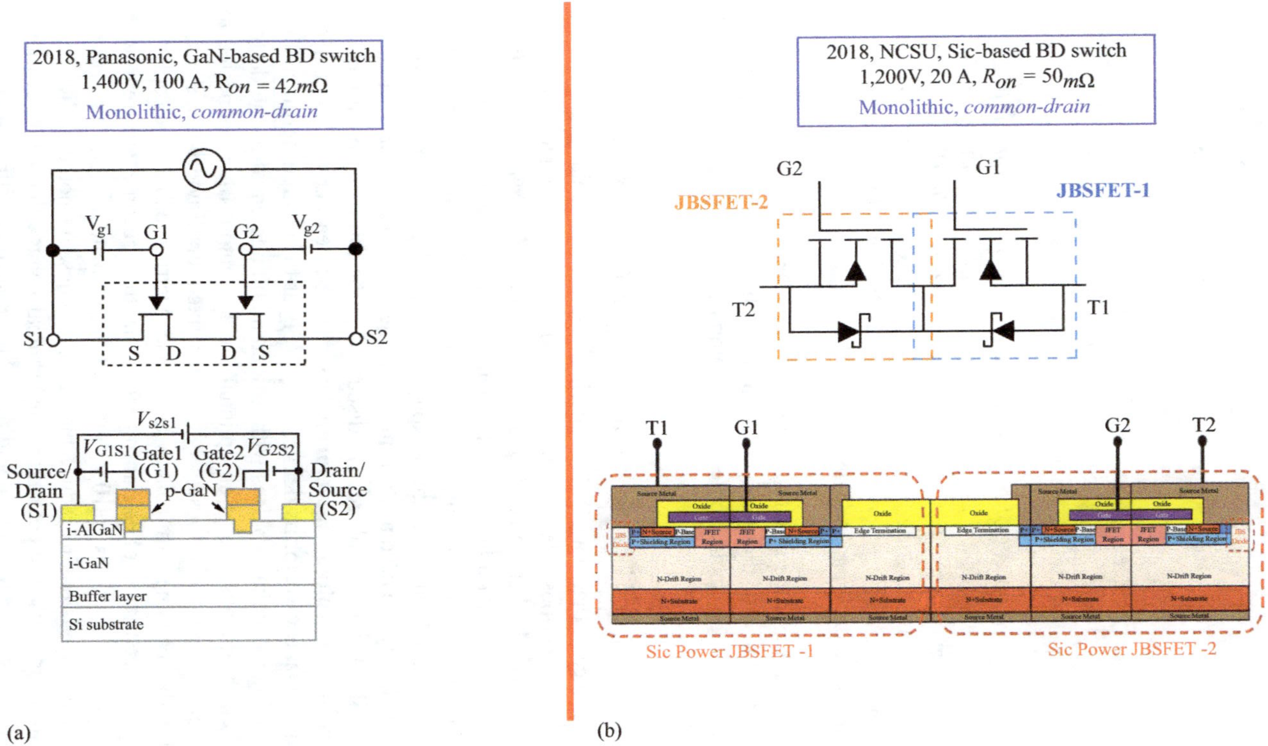

Figure 4.22 Two examples of WBG monolithic bi-directional switch devices using common-drain configurations with two gates that have been successfully designed, built, and tested in prototype quantities. © 2023 Reproduced, with permission, from IEEE [38,39].

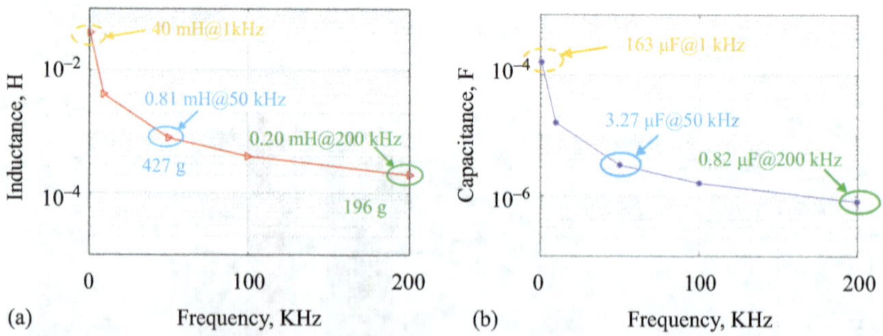

Figure 4.23 Calculated values of (a) dc-link inductance and (b) output filter capacitance for a 3 kW CSI unit showing inverse relationship between required passive component values and the PWM switching frequency, resulting in significant reductions in mass and volume as the switching frequency is raised. © 2023 Reproduced, with permission, from Dr Hang Dai [19].

should be noted that the same is not true for the dc-bus capacitor C_{bus} in most VSIs because the size/volume of these capacitors using either electrolytic or film-type units is typically set by their ripple current amplitude and not by the VSI's switching frequency [40].

4.3.3 Advantages of CSIs over VSIs using WBG switches

Having made the case that new WBG switch technology is making CSIs more feasible than what was possible with the incumbent Si-based switches, the next obvious question is whether CSIs offer any significant performance advantages over the baseline VSIs that dominate nearly all ac applications today. The answer is that they do offer important advantages, including several that address specific disadvantages discussed earlier in this chapter that WBG-based switches produce or aggravate when they are used as drop-in replacements for the silicon switches in conventional VSIs. These advantages are discussed in the following paragraphs.

Sinusoidal output voltage waveforms: As described earlier in this chapter, the line-to-line voltage waveforms delivered by VSIs at their output terminals are pulse width-modulated voltage pulses with a fixed amplitude equal to the dc-link bus voltage V_{bus}. In the absence of an optional 3-phase output filter connected to the inverter's output terminals, these sequences of fixed-amplitude voltage pulses are delivered to the machine terminals by a 3-phase electrical cable. The cable has the electrical characteristics of a transmission line that become apparent with the extremely high dv/dt values (>10 kV/μs) that WBG switches can create in the absence of special provisions to reduce their dv/dt values. As noted earlier, such high dv/dt values can cause the peak machine terminal voltage to double (or more) with cable lengths as short as 1 m. In addition, the repeated charging and discharging of the cable capacitance with each voltage pulse distorts the inverter's output current waveforms and reduces the motor drive efficiency by creating additional losses in the cable which can match or exceed the inverter losses when the cable length increases to 10 m or longer [16].

In contrast, the output waveforms delivered at the output terminals of the CSI are almost perfectly sinusoidal with total harmonic distortion (THD) values that can be less than 2% (Figure 4.24) [16]. The reason for the dramatic improvement compared to the VSI is the capacitive filter required at the CSI's output terminals. The high PWM switching frequency (50 kHz in this case) made possible by the WBG switches plays a major role in suppressing the voltage ripple amplitude in the inverter's output voltage waveforms. The sinusoidal output voltage waveforms also virtually eliminate any pulsed charging/discharging of the cable capacitances regardless of the cable length, eliminating any associated parasitic losses.

Conducted common-mode EMI reduction: As noted earlier in this chapter, the ultra-fast switching of the WBG switches in a conventional VSI significantly increases the amplitude of its CM-EMI frequency spectrum compared to the corresponding frequency spectrum when standard Si-based switches are used. Although techniques are available to reduce these CM-EMI spectral peaks by purposely slowing down the *dv/dt* values during the VSI switch turn-on and turn-off events, this defeats some of the key advantages (e.g., lower switching loss) made available by the much faster switching of the WBG switches. Similarly, the introduction of a larger EMI filter at the VSI's input terminals is undesirable because of the filter's additional mass, volume, and cost.

The combination of L_{bus} and C_f that appears in the CSI topology (Figure 4.20) comprises an *LC* filter that plays a valuable role in suppressing the amplitude of the conducted common-mode (CM) EMI at the CSI's input. Comparative measurements of the measured CM-EMI at the input bus terminals of the baseline VSI and CSI are provided in Figure 4.25(a), showing a much faster roll-off of the CM-EMI frequency spectrum for the CSI in the frequency range from 150 kHz to 2 MHz, resulting in a maximum reduction of 40 dBμV (i.e., 100×) in the CSI's spectrum at

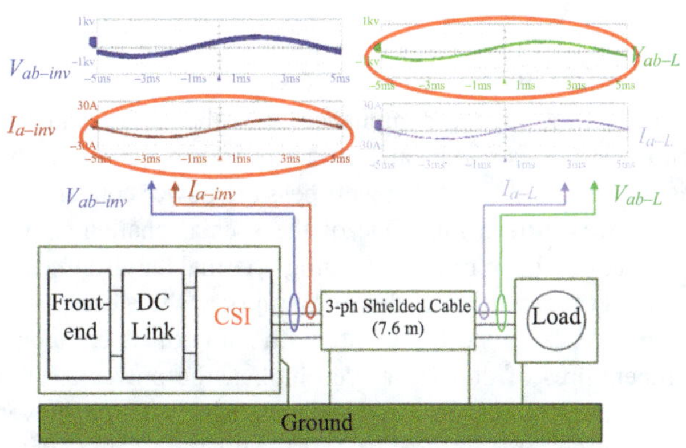

Figure 4.24 Laboratory demonstration of a 3 kW 3-phase CSI exciting a passive 3-phase RL load at the end of a 7.6-m shielded cable, including measured waveforms of the line-to-line voltage and phase current at both ends of the cable

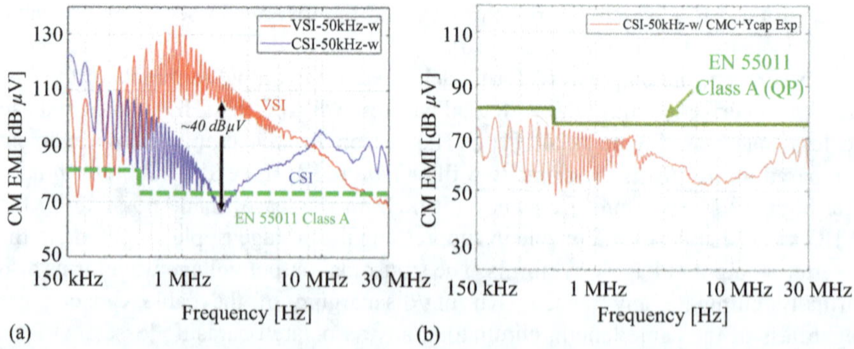

Figure 4.25 Measured common-mode EMI frequency spectra at dc link inverter terminals: (a) comparison of 3 kW VSI and CSI for 50 kHz PWM switching frequency, together with maximum CM EMI limit set by EN 55011 Class A standard; and (b) resulting improvement in CSI's CM EMI spectrum by adding common-mode choke and Y capacitors at CSI dc link terminals. © 2023 Reproduced, with permission, from Dr Hang Dai [19].

2 MHz [19]. The CSI's spectrum amplitude rises beyond 2 MHz due to the impact of parasitic capacitances in the dc-link inductor, but these high-frequency components are relatively easy to suppress.

It can be observed that, despite the advantages of the basic CSI for reducing its CM-EMI frequency spectrum compared to that of the VSI, the CSI's spectrum exceeds the limits of one of the well-known international EMI standards (CISPR EN 55011 Class A(QP)) for industrial equipment over most of the regulated frequency range from 150 kHz to 30 MHz. However, the suppression provided by the CSI makes it comparatively much easier to achieve compliance with the EN 55011 maximum limit by adding a relatively modest common-mode choke and Y-capacitors [41]. This additional filtering results in the updated CSI CM-EMI frequency spectrum shown in Figure 4.25(b) that demonstrates compliance over the entire regulated frequency range. This CSI advantage is particularly attractive in applications such as aerospace that demand tight EMI limits and the lowest possible power converter mass.

High-temperature operation: One of the special challenges posed by integrated motor drives is the more challenging thermal environment imposed by positioning the power electronics in close proximity to the electric machine. While the electric machine is designed to operate with some of its components reaching steady-state temperatures of 200°C or even higher, the power electronics is typically limited to operating in cooler environments. Many silicon power devices are limited to peak junction temperatures of 150°C, although some switches intended for automotive applications now have maximum junction temperatures of 175°C. However, there are many other components in the power converter that are limited to maximum operating temperatures of 125°C or lower. This poses major challenges for the thermal management engineers who have to find engineering solutions that effectively shield the power electronics from the much hotter temperatures reached by the machine's stator windings and other hot spots. A long-term objective is to design power electronics that can operate robustly and reliably in the same thermal environment as the machine in order to minimize the mass, volume, and cost of the special thermal management equipment needed to protect the power electronics from the hotter electric machine.

The introduction of WBG power semiconductors provides the basis for a major step in the desired direction since WBG material (SiC, GaN) maintains its semiconductor properties at much high temperatures than silicon. This is one of the advantages offered by the wide energy bandgap associated with these devices, making it possible for SiC and GaN to continue operating successfully as semiconductor switches at junction temperatures of 300°C or higher. While this is a major improvement over silicon material, much of this advantage is lost unless the packaging of the WBG switches is improved compared to the standard packaging technology used for silicon devices.

Looking beyond the power switches, there are typically several other passive components inside inverters that impose lower maximum temperature limits. In particular, electrolytic and film capacitors that are the standard choices for the dc bus capacitor C_{bus} in VSIs are typically limited to maximum temperatures of 105°C and 125°C, respectively (Figure 4.26). Although some specialty film capacitors are available that can operate at temperatures up to 175°C, their maximum capacitance ratings are very limited, severely restricting their usefulness in VSIs for dc bus capacitors.

In contrast, the dc bus inductors used in CSIs typically have temperature ratings up to 200°C or higher since they are made of the same types of materials used in the machines, i.e., ferromagnetic cores and insulated magnet wire. The

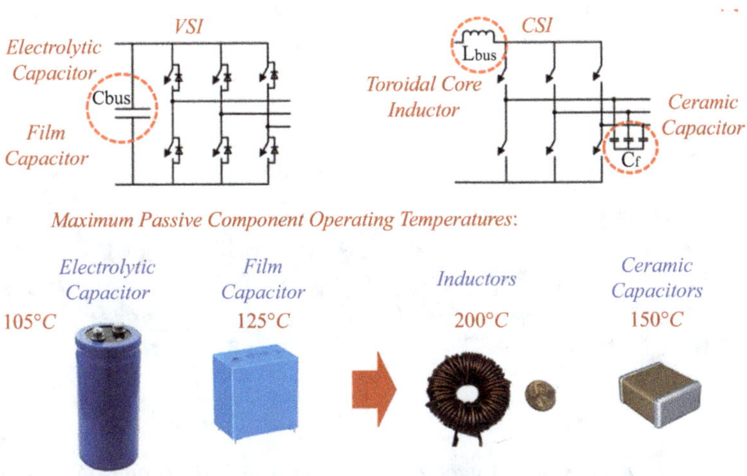

Figure 4.26 Comparison of passive components and their typical maximum temperature limits for VSIs (left) and CSIs (right)

capacitors used for the CSI's output filter capacitors C_f are ac capacitors that typically have values that are far smaller than the dc bus capacitors in VSIs, making ceramic capacitors a good choice for this application. Ceramic capacitors are conveniently available with 150°C ratings, and units using higher-temperature ceramics are available with ratings up to 200°C.

Summarizing, CSIs have a natural advantage over VSIs for high-temperature operation because the passive energy storage components that they require are more conveniently compatible with high-temperature operation than those in VSIs. This advantage makes it easier for CSIs to co-exist in close physical proximity to electrical machines inside the same IMD housing without as much need for specialized thermal management features to protect the power electronics from the machine's higher operating temperature.

Enhanced fault protection: CSIs also offer some appealing advantages in terms of improved fault protection features compared to VSIs that can be valuable in many high-performance applications. Two of them are highlighted in this discussion. The first of these is the value of the dc bus inductor during short-circuit faults, an advantage that was well understood and appreciated during the 1960s and 1970s when CSI popularity was at its highest. More specifically, a shoot-through fault in one of the inverter phase-legs is much more difficult to protect from causing major damage in a VSI compared to a CSI. Since the dc bus in the VSI is designed to behave as a stiff voltage source, a shoot-through fault in a VSI results in very high fast current rise that can destroy the phase-leg switches very quickly unless very fast action is taken to interrupt the fault current.

The amount of time that is available for a healthy gate drive in the faulted phase-leg to respond to turn off the shoot-through current without damaging the power switch devices depends heavily on the switch device type and the dc bus voltage value. While today's mature Si-based IGBTs can survive shoot-through fault currents for >10 μs before failing with 400 Vdc bus voltages, the corresponding survival time for today's GaN HEMT devices is typically less than 0.5 μs (Figure 4.27(a)) for the same 400 Vdc bus voltage, requiring extremely fast

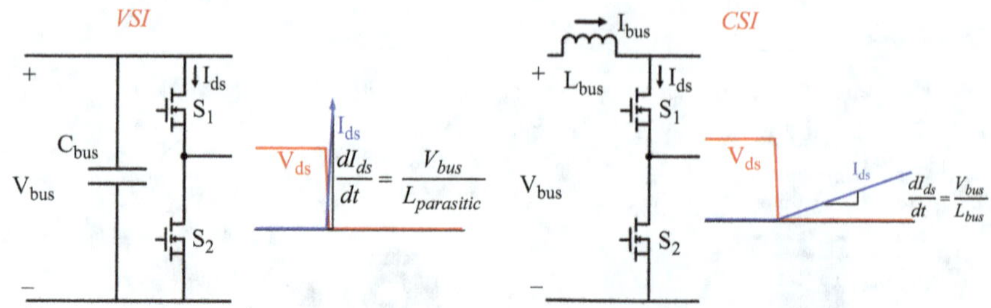

Figure 4.27 Comparison of switch device stresses in VSI (a) and CSI (b) when phase-leg switches S_1 and S_2 are both turned on simultaneously, corresponding to a shoot-through fault condition for VSI

response from the switch gate drives in order to protect the devices from imminent failure [42].

In contrast, the dc bus in the CSI (Figure 4.27(b)) behaves as a voltage source with much higher series impedance because of the dc link inductor that inherently limits the rate of the current rise (i.e., *di/dt*) to the ratio of the dc bus voltage to the inductance (i.e., V_{bus}/L_{bus}) which gives the gate drives much more time, often 10s of milliseconds, to respond by removing the gate drive. In addition, the fault current that must be interrupted is almost always much lower in the CSI than the VSI because of the inductor's effectiveness in limiting the fault current *di/dt* to a much smaller fraction of its corresponding value in the VSI.

In summary, the presence of the dc bus inductor in the CSI is a tremendous gift when developing inverter shoot-through protection, making this type of fault far less dangerous in CSIs compared to conventional VSIs, particularly when WBG switches are being used. Our engineering forbearers who took full advantage of CSIs to help them protect against all-too-frequent inverter shoot-through faults using the earliest days of solid-state thyristor switches in the late 1950s and 1960s would smile and nod in approval. However, it must be acknowledged that the presence of the dc-link inductor in the CSI creates the need for open-circuit fault protection to minimize over-voltage risks [43].

The second advantage of CSIs over VSIs in fault protection specifically applies to applications using permanent magnet (PM) synchronous machines. PM machine are frequently selected for many transportation applications including passenger electric vehicles and new electrified propulsion motor drives for aviation applications where the highest possible power density is required. It has long been known that PM motor drives create some special fault risks because the magnets cannot be conveniently "turned off" when a fault occurs. As a result, if a single switch in a VSI six-switch inverter fails in a short-circuit condition, the PM machine is highly likely to suffer damaging demagnetization of some or all of its rotor magnets even if the gate drives are instantaneously removed from all of the other five healthy switches as soon as the single switch fails [44].

Figure 4.28(a) illustrates this single-switch short-circuit condition following failure of the upper switch in the phase A phase-leg in the upper left corner of the six-switch bridge [44]. This drawing shows that the reverse-conducting diodes associated with the upper switches in the other two healthy phase-legs are responsible for carrying the fault current even if the gate signals are removed from their associated switches. The simulated machine phase currents immediately following the fault application at time $t = 0$ are provided in Figure 4.28(b), showing that the presence of the diodes adds large dc bias currents to the phase currents, aggravating the seriousness of the fault. The effects of this single-switch short-circuit fault can be sufficiently serious to cause additional damage to the inverter and demagnetize the rotor magnets of the PM machine.

Figure 4.28(c) shows why the response of the CSI to the same single-switch fault condition is so much improved compared to the VSI's response [45]. The key reason for the improvement is that none of the six inverter switches in the CSI include reverse-conducting diodes as discussed at the beginning of this chapter, denying the fault current a closed-circuit path to flow provided that gating is

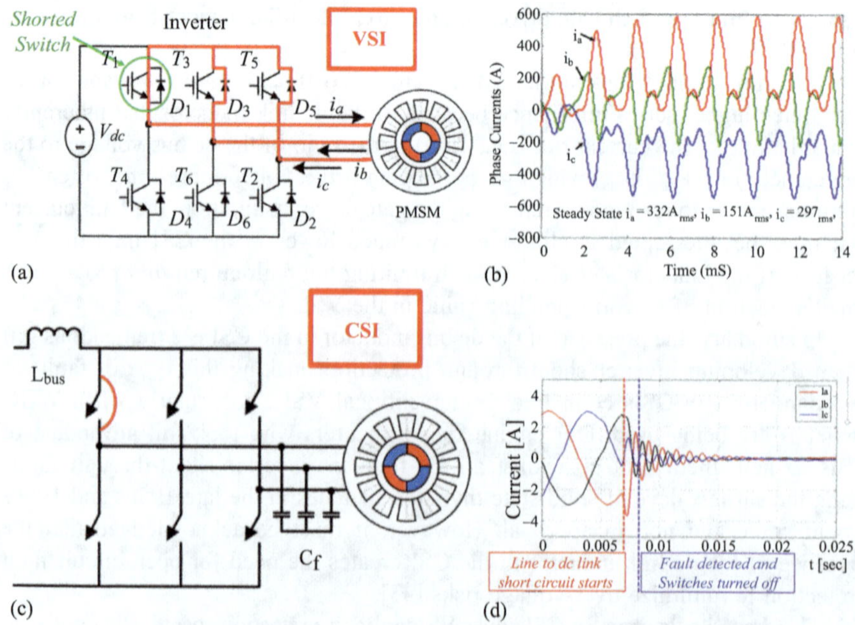

Figure 4.28 Comparison of phase currents, simulated for VSI (a) and measured for CSI. © 2023 Reproduced, with permission, from IEEE [44]; (b) following application of a short-circuit fault to the upper switch in phase a and gate removal from the other 5 healthy switches, highlighting the much more benign fault response of the CSI.

removed promptly from the five healthy switches. Figure 4.28(d) shows measured phase currents for these fault conditions, including delay of the gate removal from the healthy switches by 2 ms (quite long) following the short-circuit failure to represent the time required by the controls to detect the fault. The machine's inductance limits the current rise during the 2 ms delay period, and the phase currents are promptly extinguished following the gate on-state voltage removal because of the absence of the reverse-conducting diodes. None of the switch phase currents reach dangerous levels throughout this complete fault sequence.

The single-switch failure fault discussed here is far from being the only potential fault mode that the PM machine drive can experience, but it certainly rates as a particularly serious one. The fact that CSIs provide an effective means of protecting against a serious fault mode that is nearly impossible to match with two-level VSIs such as the one shown in Figure 4.28(a) is extremely valuable in applications where high reliability and fault tolerance are absolutely critical requirements.

4.3.4 Challenges posed by CSIs using WBG switches

Having presented a series of appealing advantages offered by CSIs, it is important to provide a balanced evaluation of their characteristics by describing some of the special challenges posed by CSIs in comparison to conventional VSIs. Three of them will be discussed in this section:

Need for commercially available monolithic WBG devices with RVB capability: As noted earlier, the CSI requires switches that have reverse-voltage-blocking (RVB) capability, and no monolithic WBG-based devices are presently commercially available with these characteristics. The most promising WBG power switch that is the likeliest to reach commercial production is the monolithic bidirectional switch that offers significant versatility that makes it useful for a significant number of important power electronics applications, including CSIs. In the absence of the commercially available monolithic BD switch, it is possible to implement WBG switches with RVB capability as hybrid switches by connecting two discrete WBG MOSFETs or HEMTs in anti-series combinations to provide this capability [46]. Effectively, one of the two MOSFETs/HEMTs is acting as a synchronous rectifier whenever the combined switch is in its on-state, helping to lower the hybrid switch's total forward voltage drop. Despite this advantage, the hybrid switch cannot be expected to match either the low forward voltage drop of a well designed monolithic WBG-based BD switch or its ultimate cost because the monolithic BD switch is a single WBG chip replacing two discrete WBG switches with approx. twice the total chip area as the monolithic BD switch device. This makes it critical for WBG device manufacturers to initiate commercial production of the WBG-based monolithic BD switch as soon as possible to eliminate this critical barrier to the adoption of CSIs in future integrated motor drive applications.

Need for front-end converter to convert dc supply voltage to regulated current: Another special requirement of the CSI is the need to regulate the current in dc link inductor L_{bus} under all load conditions. The fact that we live in a world dominated by voltage sources (e.g., the utility grid, batteries) rather than current sources imposes special requirements on the CSI's interface with its power source that the VSI avoids. A simple example is shown in Figure 4.29. For simplicity, assume that the power source is a single-phase ac grid connection that behaves as a voltage source. As shown in Figure 4.29(a), the interface between this ac source and a VSI can be as simple as a single-phase bridge rectifier using four uncontrolled diodes. The PWM-based VSI can conveniently regulate its line-to-line output voltage amplitude to any instantaneous value less than the peak amplitude of the ac voltage source (assuming a large bus capacitor C_{bus} connected at the output terminals of the ac rectifier circuit). This reflects the fact that a VSI behaves as a controllable buck converter requiring the output voltage to be less than its input voltage.

In contrast, the CSI's interface with the single-phase ac voltage source in Figure 4.29(b) includes a dc/dc buck converter stage (for motoring operation) at the output of the rectifier circuit in order to regulate the dc bus current fed to the CSI whenever the CSI load voltages are lower than the ac voltage amplitude. This reflects the fact that a CSI behaves as a controllable boost converter (the dual of the VSI) that requires the output voltage to be greater than its input voltage source. While the added dc buck converter in Figure 4.29(b) requires only one additional controlled switch and a diode to implement, these components must be sized to process the full output power, adding losses and cost to the total CSI converter cost.

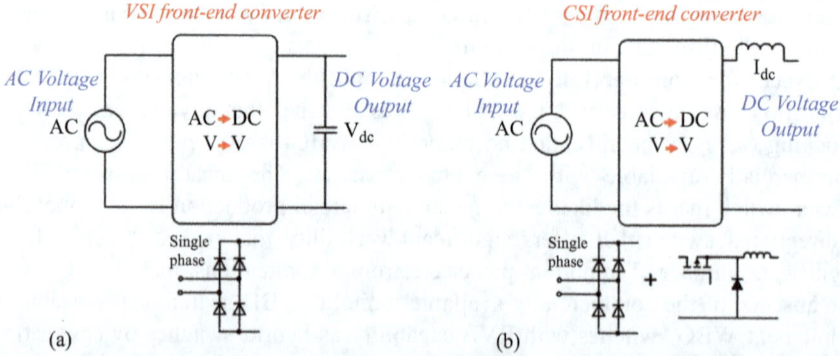

Figure 4.29 Comparison of front-end converter stage requirements for a VSI (a) and CSI (b) assuming a single-phase ac voltage input for both cases. For a PWM-controlled VSI, the front end can be as simple as a 4-diode uncontrolled rectifier, while the CSI needs a controlled dc/dc converter (shown in (b) as a buck converter) in addition to the uncontrolled rectifier

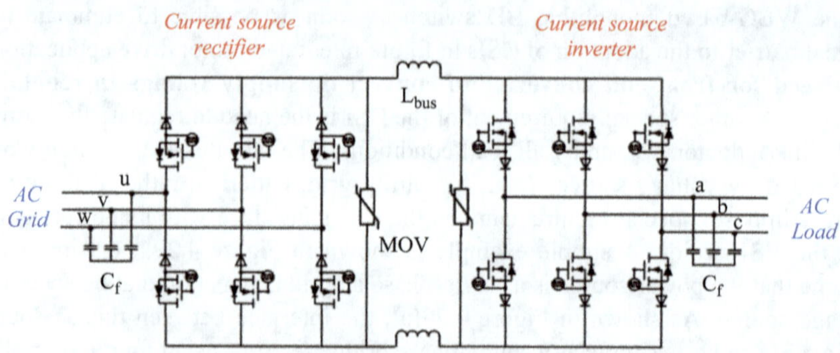

Figure 4.30 Power circuit of a 3-phase back-to-back CSI ac/ac power converter for exciting an adjustable-frequency 3-phase ac load from a fixed-frequency, fixed-amplitude 3-phase ac grid source

However, it should be pointed out that the active front-end converter stage required by the CSI is not always a disadvantage, depending on the application. For example, Figure 4.30 shows the circuit diagram for a three-phase grid voltage source that is connected to a three-phase CSI using a three-phase active current–source rectifier (CSR) [9]. The CSR stage has the identical topology as the CSI but with an inverted power flow from its ac terminals to the dc link during motoring operation, a natural capability of both the CSI and VSI power stages. While the CSR stage's requirement for an additional six RVB switches may seem excessive at first sight, the CSR provides three valuable functions in this "back-to-back" CSR–CSI power converter configuration: (1) dc bus current regulation, as discussed above for the simpler buck converter; (2) regenerative power flow from the CSI back to the 3-phase ac source, which is required in applications such as elevators or hoists; and (3) high power quality and low common-mode EMI at the CSR's ac input terminals because of the sinusoidal ac voltage waveforms delivered at these terminals, matching the high quality of the CSI's sinusoidal output waveforms discussed earlier. In fact, the CSR makes it possible to significantly reduce the requirements for passive *LCL* filters between its input terminals and the ac grid compared to typical filtering requirements for voltage-source rectifier (VSR) power stages that do not naturally deliver sinusoidal output voltage waveforms in the absence of additional filters.

DC bus inductor design, fabrication, and losses: While dc bus capacitors for VSI applications can typically be specified as catalog items offered by capacitor manufacturers in a wide range of voltage, current, and capacitance ratings, dc bus inductors for CSIs generally must be custom-designed for each application requiring careful selections of ferromagnetic core materials, core shape, winding wire size (i.e., cross-sectional area), number of parallel wire strands, etc. [10]. Undoubtedly, the fact that VSIs have been so dominant in inverter applications for the past several decades accounts in part for the relative ease of specifying and acquiring dc bus capacitors compared to their counterpart dc bus inductors in CSIs. The question of whether the convenient availability of wide arrays of alternative ready-built dc bus inductors would ever match today's rich selection of off-the-shelf dc bus capacitor products is an important one that has no obvious answer.

The second issue posed by the CSI's dc bus inductors is their losses which are typically significantly higher than those of the VSI's dc bus capacitors. A major contributing factor to this difference is that all of the dc bus current in a CSI passes through the inductors, contributing resistive I^2R dissipation losses (even in the absence of any PWM effects) which are vanishingly smaller in dc bus capacitors. PWM-induced AC core losses in the ferromagnetic material used by the CSI inductor contribute additional losses. These higher passive component losses in the CSI hurt its efficiency compared to that of the basic VSI. High switching frequencies made possible by WBG switches shrink the required inductance value and the inductor's mass/volume which help to significantly reduce the inductor losses, making the CSI efficiency more competitive with that of VSIs.

The comparison of CSI and VSI efficiencies changes even further in favor of the CSI if the playing field is leveled by requiring the VSI to deliver sinusoidal output voltage waveforms that match the quality of the CSI's sinusoidal output waveforms. Meeting this requirement inevitably requires the VSI topology to change by some combination of introducing multi-level VSI topologies and/or adding passive *LC* filters at the VSI's output terminals. Since the VSI output filters typically require inductors in each of the VSI's output phases, the VSI becomes saddled with some of the same physical constraints and losses associated with the CSI's dc bus inductor. In fact, imposing this output power quality requirement on the VSI goes far toward equalizing the efficiency metrics of the VSI and CSI, particularly when the availability of monolithic BD switches is assumed for the CSI [19].

4.3.5 Case history: 3 kW WBG-based CSI integrated motor drive [19]

4.3.5.1 Overview and specifications

This section provides a summary of a project that combined wide-bandgap power switches, a CSI, and a high-efficiency PM synchronous motor to form a new class of integrated motor drives with outstanding power density and performance characteristics (Figure 4.31). This four-year project, sponsored by the US Advanced Research Projects Agency – Energy (ARPA-E), was focused on developing a 3 kW integrated motor drive appropriate for a variety of applications including pumps, fans, and compressors used in heating, ventilating, and air conditioning (HVAC) equipment for both terrestrial and aerospace systems. A set of demanding performance specifications were set at the beginning of the project that stretched the state-of-the-art in several areas including power density, efficiency, EMI, and waveform power quality.

After considering several alternative configurations, a decision was made to choose an axial-mounted IMD configuration with the power electronics mounted adjacent to one of the machine ends. A commercially produced PM machine was chosen that is designed with a chamber at the non-drive end of the machine. This end chamber is separated from the main motor housing by an aluminum wall that also serves the structural purpose of providing the mounting surface for the non-drive-end bearing housing. A resolver is mounted inside the end chamber on the same aluminum wall and mechanically coupled to the rotor shaft. A decision was made to mount the CSI power electronics inside this end chamber which provides some valuable thermal and electromagnetic insulation from the more demanding environment inside the main motor housing.

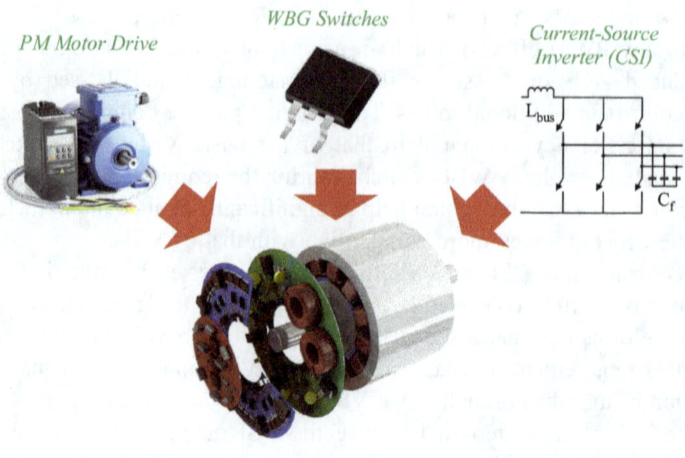

Figure 4.31 Graphic representation highlighting the key technologies that were combined to form a new class of IMD during a 4-year research project. © 2023 PM Motor Drive image, courtesy of The Inverter Drive Supermarket.

An exploded view of the major components in the CSI-IMD configuration is provided in Figure 4.32. There are two annular printed circuit boards (PCBs) that comprise a back-to-back CSR-CSI power converter using the topology shown previously in Figure 4.30. The power switches are mounted along the outer circumference of one of these boards to enhance the cooling. More specifically, air is drawn axially through openings in the end chamber cover using a small fan and then directed radially across the tops of the power device cases and the PCBs before being exhausted through vents on the sides of the end chamber (see Figure 4.32).

Specifications for several of the key electrical input and mechanical output variables for the motor drive are presented in Table 4.4. Since the input power source for the CSI-IMD is three-phase ac power from the utility grid, and demonstration of bidirectional power is desired (i.e., both motoring and braking/generating), the back-to-back CSR–CSI power converter configuration shown in Figure 4.30 was selected for this project.

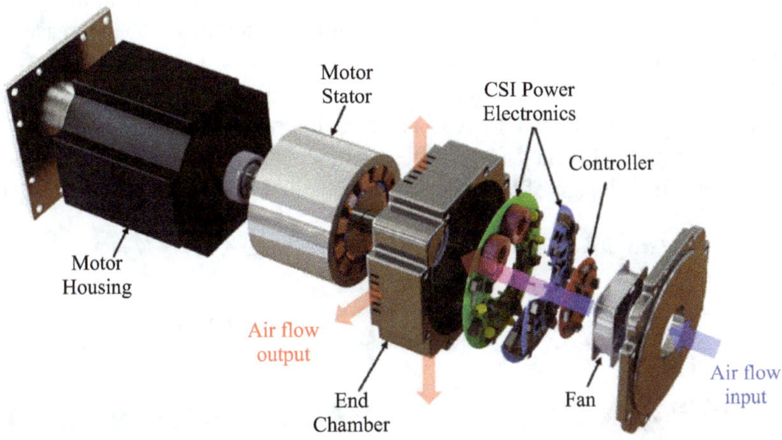

Figure 4.32 Exploded view of prototype 3 kW WBG-based CSI-IMD unit identifying CSI power electronics mounted in an air-ventilated end chamber of the PM machine at the non-drive end of the IMD unit

Table 4.4 Specifications of 3 kW WBG-based CSI IMD [19]

Specifications	**Value**
Rated power	3 kW
Rated torque	16.3 Nm
Rated speed	1,644 rpm
DC link current @ 3 kW	12 A dc
Motor RMS line-to-line voltage @ 3 kW	226 V rms
Motor RMS phase current @ 3 kW	8.4 A rms
PWM switching frequency	125 kHz

4.3.5.2 3 kW prototype CSI-IMD unit

Since the peak line-to-line input voltage is less than 350 V, both SiC and GaN power switches are candidates for this application, and both were aggressively evaluated during the course of this project in an effort to maximize the CSI efficiency. Since no monolithic WBG switches with reverse-voltage-blocking capabilities could be secured for this project, it was necessary to use hybrid switch configurations consisting of two WBG devices connected in series or anti-series for all of the tested CSI configurations. Figure 4.33 shows two of the hybrid switch phase-leg building-block configurations used to construct both the CSR and CSI power stages [47]. The version on the left side of the figure uses a series-connected MOSFET (or HEMT) and SiC Schottky diode for each RVB switch that offers gating simplicity at the price of higher losses. After using this hybrid switch configuration during the first half of the 3-year project, the alternative configuration shown on the right side of Figure 4.33 was adopted during the second half of the project consisting of two anti-series MOSFETs (or HEMTs) in a common-source configuration with two gate drives per hybrid switch. As indicated in the figure, the change to the second hybrid switch configuration resulted in a 0.5% increase in the CSI efficiency at 3 kW output power from 97.5% to 98%.

The design of the dc link inductor is critical to the CSI's performance metrics because its mass/volume and losses play a major role in determining the IMD's power density and efficiency [10]. After significant analysis and multiple design iterations, the final design of the identical dual inductors selected for the CSI-IMD

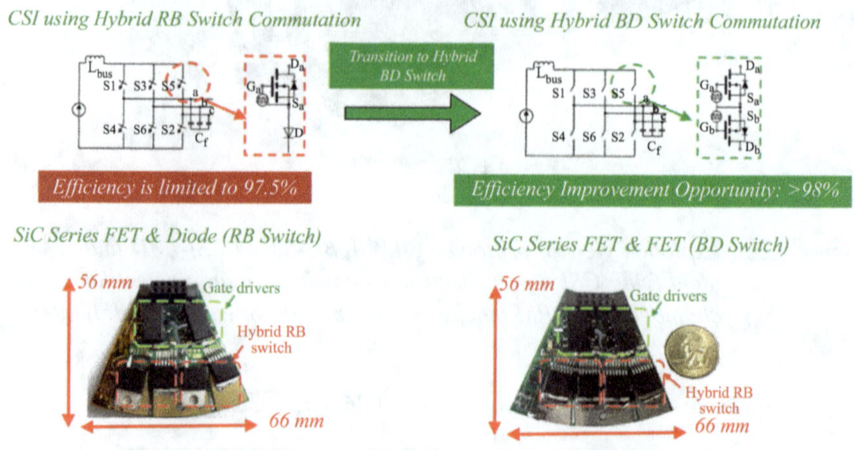

Figure 4.33 Illustration showing two of the hybrid switch configurations used during the CSI-IMD project to implement bidirectional switches using two series-connected discrete WBG-devices. The version on the left uses a SiC MOSFET in series with a Schottky diode, and the version on the right uses two SiC MOSFETs connected in anti-series. The circuit configurations of both hybrid switches and their compact implementations for the IMD unit are shown. © 2023 Reproduced, with permission, from IEEE [47].

prototype unit (one inductor in each dc bus as shown in Figure 4.30) uses a toroidal core filled with a high-performance iron-nickel (FeNi) powder iron alloy (Figure 4.34). One of the valuable advantages of this particular alloy is that it retains a high percentage of its original permeability value (with pure ac excitation) even when the total inductor current includes a large dc current component.

Figure 4.35 provides a view of the final version of the prototype CSI-IMD unit, showing that the complete power electronics, including the front-end active

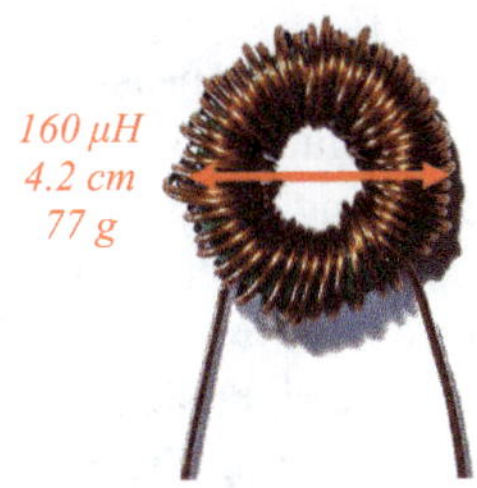

Figure 4.34 View of one of the two identical dc link inductors used in the 3 kW prototype CSI-IMD unit consisting of a high-performance powder iron core with a toroidal winding capable of operating at frequencies up to 125 kHz

Figure 4.35 View of prototype WBG-based CSI-IMD unit with non-drive end cover removed to view 3-phase ac/ac power electronics. Half of the visible power board is dedicated to current-source rectifier stage and the second half to the CSI. The two dc link inductors are visible, eliminating the need for dc link capacitors

rectifier stage, fits compactly inside the housing of the 3 kW PM machine. The power electronics consists of two printed circuit boards (PCBs) mounted adjacent to each other; the outer of the two PCBs (visible in Figure 4.35) includes all of the SiC FET switches mounted along the outer periphery of the card to enhance convection cooling using forced air delivered by a small 3 W fan mounted in the machine's end cover (not visible). The smaller red PCB mounted above the two power cards contains a DSP controller that is sufficiently powerful to deliver three-phase PWM gate drive commands to both the CSR and CSI power stages at PWM switching frequencies up to 125 kHz.

Figure 4.36 shows examples of key measured performance metrics for the 3 kW prototype CSI-IMD unit. Figure 4.36(a) provides measured three-phase voltage and current waveforms during rated 3 kW operation, demonstrating the near perfect sinusoidal waveshapes delivered by the CSI [48]. The measured total harmonic distortion (THD) of the current waveforms is less than 1.5% with a PWM switching frequency of 100 kHz. The right side of Figure 4.36(b) shows the measured efficiencies vs. output power for both the Year 2 and Year 3 (final) versions of the CSI (inverter only), demonstrating that the CSI efficiency for the Year 3 unit exceeds 98% for load levels from 50% load (1.5 kW) to full load (3 kW).

Table 4.5 provides a summary comparison of the CSI-IMD's measured and target values of the key performance metrics that were set at the beginning of the 4-year project. All key performance metrics were met or exceeded. For at least three of these metrics, the measured values exceeded the targets by large margins.

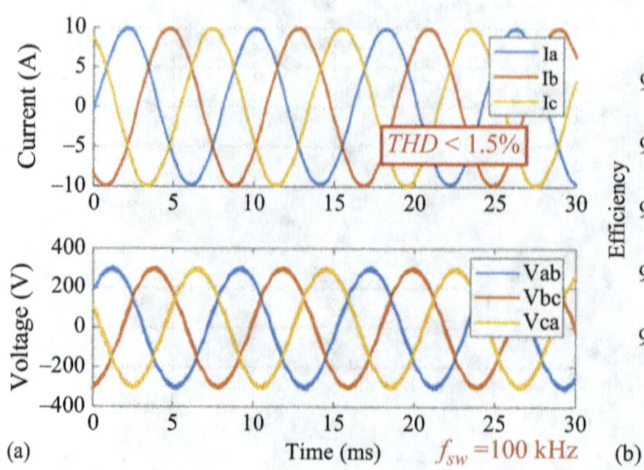

Figure 4.36 *(a) Measured IMD motor phase current (upper) and line-to-line voltage waveforms for 100 kHz switching frequency, demonstrating current total harmonic distortion (THD) less than 1.5% [48]; and (b) measured CSI efficiency of Year 2 benchtop CSI-IMD unit and Year 3 prototype CSI-IMD unit (see Figure 4.35) vs. output power, with Year 3 CSI exceeding 98% efficiency for all loads greater than 50% rated. © 2023 Reproduced, with permission, from IEEE [48].*

Table 4.5 Performance metrics of prototype 3 kW WBG-based CSI-IMD unit

Metrics	Targets	Measured with motor load
CSI efficiency	>98% @ 3 kW >96% @ 1.5 kW	98.1% @ 3 kW 98.0% @ 1.5 kW
Phase current THD @ 8.2 Arms	<8%	<1.5%
CSI conducted CM EMI	<100 dBμV 150 kHz–30 MHz	78 dBμV (max.)
CSI specific power density (w/o controller)	>8 kW/kg	>8.7 kW/kg
CSI volumetric power density (w/o controller)	>5 kW/liter	9.14 kW/liter

For example, the measured common-mode EMI profile was sufficiently suppressed in the prototype CSI-IMD unit to meet the requirements of a well-established CISPR EMI standard for industrial equipment, EN 55011 Class A (QP).

4.3.6 CSI-IMD technology scalability

The scalability of the WBG-based CSI-IMD technology to higher power levels is an important topic of continuing research and development activities. For example, a project is currently under way to develop a high-performance electric vehicle traction drive with a peak output power of 100 kW based on wide-bandgap power electronics using the same CSI-IMD concepts discussed in the preceding section [49]. In addition to the challenges associated with increasing the CSI peak power rating by more than 33 times, the EV traction drive requires the maximum motor speed to increase by nearly an order of magnitude to 20,000 rev/min, in addition to providing a constant-power speed ratio (CPSR) of 3:1, which is a typical requirement for passenger EV traction drives. A volumetric power density target of 33 kW/L for the combined inverter plus motor requires the CSI-IMD technology to push well beyond the present limits of today's Si-based EV motor drive technology using VSIs to achieve success.

Looking ahead even further into the future, a study is also under way to investigate the scalability of WBG-based CSI-IMD technology up to power levels of 2 MW for aerospace propulsion applications. In view of the ultra-demanding requirements for reliability in aerospace equipment, this application is providing a perfect opportunity to explore the modularization of CSI integrated motor drives. The includes highlighting some of the natural advantages that CSIs offer for enhanced fault tolerance with permanent magnet synchronous machines compared to VSI-based motor drives. By simultaneously pushing the limits of motor drive power density and reliability, this long-term research initiative is providing exciting opportunities to develop advanced CSI integrated motor drives for challenging aerospace applications that will meet or exceed the performance metrics achieved by any other known motor drive technology.

4.4 Closing comments

The development of wide bandgap power semiconductor represents a disruptive technology revolution in the field of power electronics that is still unfolding at the time of this writing. Just as the last major technology disruption in the form of silicon MOS-gated power semiconductor devices (i.e., IGBTs, power MOSFETs) required more than a decade during the 1980s–1990s to become dominant in the power electronics marketplace, WBG power semiconductor devices are still in the midst of emerging as the dominant power device technology of the modern era. Despite the incomplete maturity of WBG-based power device technology today, the promise of SiC and GaN power devices is already apparent in the early market entries and laboratory prototype power converters that are convincing demonstrations of the impressive performance improvements offered by WBG technology.

Integrated motor drives are positioned to be one of the major beneficiaries of several key advantages that WBG power semiconductor devices offer including higher switching frequencies attributable to their low switching losses, lower forward on-state voltage drops, higher blocking voltage ratings, and higher operating temperatures. These advantages are tailor-made for IMD applications, enabling the achievement of substantially higher power densities (both mass and volume), higher efficiency, and improved compatibility with more hostile high-temperature operating conditions. The Tesla 3 integrated motor traction drive that was first produced in 2017 represents a technology milestone and wake-up call for competitors who were not convinced of the readiness of WBG technology for production electric vehicle applications when it first appeared.

The appearance of a disruptive new technology such as WBG power semiconductor devices gives researchers and entrepreneurs an opportunity to aggressively question whether new WBG power devices should be treated solely as drop-in replacements for silicon IGBTs and MOSFETs in VSIs. The result of this willingness to challenge standard conventions is leading to exciting new developments including promising WBG-based bidirectional (BD) power switches that can block voltage and conduct gate-controllable current in both polarities. These new BD switches are reviving interest in well-known but highly under-utilized power converter topologies such as the classic matrix converter that requires only 9 of these BD switches to achieve bidirectional power flow between two ac sources/loads operating at different frequencies, voltages, and power factors.

The opportunities opened by WBG technology to develop reverse-voltage-blocking power switches is also reviving interest in high-frequency CSIs that offer a variety of special advantages for IMD applications. These valuable advantages include the elimination of temperature-limited bus capacitors, and the delivery of sinusoidal output voltage and current waveforms that suppress dangerous voltage overshoots at the load terminals as well as common-mode EMI. Prototype demonstrations of WBG-based IMDs have confirmed the intriguing opportunities offered by this technology. These special features include reduced vulnerability to dangerous short-circuit faults in permanent magnet machine drives that is particularly valuable in applications demanding the highest possible power density and reliability such as IMDs in aerospace actuators and propulsion equipment.

In summary, there are many exciting developments awaiting us as IMDs take full advantage of evolving WBG technology during the years ahead.

References

[1] B.J. Baliga, "Power semiconductor device figure of merit for high frequency applications," *IEEE Electron Device Letters*, vol. 10, p. 455, 1989.

[2] B. Ozpineci and L. Tolbert, "Smaller, faster, tougher," *IEEE Spectrum*, vol. 48, no. 10, pp. 45–66, 2011, doi:10.1109/MSPEC.2011.6027247.

[3] B.J. Baliga, *Wide Bandgap Semiconductor Power Devices: Materials, Physics, Design, and Applications*. UK: Elsevier Science, 2018.

[4] D. Han, "Conducted common-mode electromagnetic interference in wide bandgap semiconductor devices based DC-fed motor drives: challenges and solutions", PhD thesis, Dept. of Electrical and Computer Engineering, University of Wisconsin – Madison, 2017.

[5] J. Millán, P. Godignon, X. Perpiñà, A. Pérez-Tomás, and J. Rebollo, "A survey of wide bandgap power semiconductor devices," *IEEE Transactions on Power Electronics*, vol. 29, no. 5, pp. 2155–2163, 2014, doi:10.1109/TPEL.2013.2268900.

[6] E.A. Jones, F. Wang, and B. Ozpineci, "Application-based review of GaN HFETs", in *Proceedings of IEEE Workshop Wide Bandgap Power Devices Application* (*WiPDA*), October 2014, pp. 24–29.

[7] K. Shenai, "High-density power conversion and wide-bandgap semiconductor power electronics switching devices," *Proceedings of the IEEE*, vol. 107, no. 12, pp. 2308–2326, 2019, doi:10.1109/JPROC.2019.2948554.

[8] K. Olejniczak, T. Flint, D. Simco, *et al.*, "A compact 110 kVA, 140°C ambient, 105°C liquid cooled, all-SiC inverter for electric vehicle traction drives," in *2017 IEEE Applied Power Electronics Conference and Exposition* (*APEC*), 2017, pp. 735–742, doi:10.1109/APEC.2017.7930776.

[9] R.A. Torres, H. Dai, W. Lee, T.M. Jahns, and B. Sarlioglu, "Development of current-source-inverter-based integrated motor drives using wide-bandgap power switches," in *2019 IEEE 15th Brazilian Power Electronics Conference and 5th IEEE Southern Power Electronics Conference* (*COBEP/SPEC*), 2019, pp. 1–6, doi:10.1109/COBEP/SPEC44138.2019.9065675.

[10] R.A. Torres, H. Dai, T.M. Jahns, and B. Sarlioglu, "Design of high-performance toroidal DC-link inductor for current-source inverters," in *2019 IEEE Applied Power Electronics Conference and Exposition* (*APEC*), 2019, pp. 2694–2701, doi:10.1109/APEC.2019.8722096.

[11] R.A. Torres, H. Dai, W. Lee, K. Saviers, T.M. Jahns, and B. Sarlioglu, "Investigation of cooling techniques and enclosure types for integrated motor drives," *2022 IEEE Energy Conversion Congress and Exposition (ECCE)*, Detroit, MI, USA, 2022, pp. 1–8, doi:10.1109/ECCE50734.2022.9947618.

[12] F. Qi, M. Wang, and L. Xu, "Investigation and review of challenges in a high-temperature 30-kVA three-phase inverter using SiC MOSFETs," *IEEE Transactions on Industry Applications*, vol. 54, no. 3, pp. 2483–2491, 2018, doi:10.1109/TIA.2018.2796059.

[13] K. Yamaguchi, K. Katsura, and T. Jikumaru, "Motor loss and temperature reduction with high switching frequency SiC-based inverters," in *2017 IEEE 5th Workshop on Wide Bandgap Power Devices and Applications (WiPDA)*, 2017, pp. 127–131, doi:10.1109/WiPDA.2017.8170534.

[14] Y. Abdullah, X. Li, K. Wang, J. Wang, L. Liu, and S. Bala, "High temperature design of a GaN based modular integrated drive with natural cooling using metal clad PCBs," in *2019 IEEE Energy Conversion Congress and Exposition (ECCE)*, 2019, pp. 4012–4017, doi:10.1109/ECCE.2019.8912701

[15] P. Yi, P.K.S. Murthy, and L. Wei, "Performance evaluation of SiC MOSFETs with long power cable and induction motor," in *2016 IEEE Energy Conversion Congress and Exposition (ECCE)*, 2016, pp. 1–7, doi:10.1109/ECCE.2016.7855344.

[16] H. Dai, R.A. Torres, W. Lee, T.M. Jahns, and B. Sarlioglu, "High-frequency evaluation of two-level voltage-source and current-source inverters with different output cables," in *2020 IEEE Energy Conversion Congress and Exposition (ECCE)*, 2020, pp. 2403–2410, doi:10.1109/ECCE44975.2020.9235681.

[17] H. Zeng, J. Swanke, T.M. Jahns, and B. Sarlioglu, "High-frequency modeling and inter-turn voltage distribution analysis of a modular electric machine for electric aircraft propulsion," in *2022 AIAA/IEEE Electric Aircraft Technologies Symposium* (*EATS*), 2022, pp. 1–8.

[18] D. Han, S. Li, Y. Wu, W. Choi, and B. Sarlioglu, "Comparative analysis on conducted CM EMI emission of motor drives: WBG versus Si devices," *IEEE Transactions on Industrial Electronics*, vol. 64, no. 10, pp. 8353–8363, 2017, doi:10.1109/TIE.2017.2681968.

[19] H. Dai, "Investigation of current source inverter using wide bandgap for integrated motor drive applications", PhD thesis, Dept. of Electrical and Computer Engineering, University of Wisconsin – Madison, 2021.

[20] B. Zhang and S. Wang, "A survey of EMI research in power electronics systems with wide-bandgap semiconductor devices," *IEEE Journal of Emerging and Selected Topics in Power Electronics*, vol. 8, no. 1, pp. 626–643, 2020, doi:10.1109/JESTPE.2019.2953730.

[21] A. von Jouanne, R. Collin, M. Stephens, *et al.*, "Motor bearing current characterization in SiC-based variable frequency drive applications," in *2020 IEEE Energy Conversion Congress and Exposition (ECCE)*, 2020, pp. 2718–2725, doi:10.1109/ECCE44975.2020.9235925.

[22] Creamer Media, "Bearing protection needed for optimum performance." Available: https://www.engineeringnews.co.za/article/bearing-protection-needed-for-optimum-performance-2013-01-10 [Accessed: 02-May-2023].

[23] "Discovering and Preventing Bearing Current Damage." Available: https://est-aegis.info/2018/02/discovering-and-preventing-bearing-current-damage [Accessed: 11-Mar-2023].

[24] D. Han, C.T. Morris, W. Lee, and B. Sarlioglu, "A case study on common mode electromagnetic interference characteristics of GaN HEMT and Si MOSFET power converters for EV/HEVs," *IEEE Transactions on Transportation Electrification*, vol. 3, no. 1, pp. 168–179, 2017, doi:10.1109/TTE.2016.2622005.

[25] S. Liu, H. Lin, and T. Wang, "Comparative study of three different passive snubber circuits for SiC power MOSFETs," in *2019 IEEE Applied Power Electronics Conference and Exposition (APEC)* , Anaheim, CA, USA, 2019, pp. 354-358, doi: 10.1109/APEC.2019.8722302.

[26] C. Vadstrup, X. Wang, and F. Blaabjerg, "LC filter design for wide band gap device based adjustable speed drives," in *2014 International Power Electronics and Application Conference and Exposition*, 2014, pp. 1291–1296, doi:10.1109/PEAC.2014.7038049.

[27] M. Pastura, S. Nuzzo, M. Kohler, and D. Barater, "Dv/Dt filtering techniques for electric drives: review and challenges," in *IECON2019 – 45th Annual Conference of the IEEE Industrial Electronics Society*, 2019, pp. 7088–7093, doi:10.1109/IECON.2019.8926663.

[28] B. Narayanasamy and F. Luo, "A survey of active EMI filters for conducted EMI noise reduction in power electronic converters," *IEEE Transactions on Electromagnetic Compatibility*, vol. 61, no. 6, pp. 2040–2049, 2019, doi:10.1109/TEMC.2019.2953055.

[29] D. Han, C.T. Morris, and B. Sarlioglu, "Common-mode voltage cancellation in PWM motor drives with balanced inverter topology," *IEEE Transactions on Industrial Electronics*, vol. 64, no. 4, pp. 2683–2688, 2017, doi:10.1109/TIE.2016.2633234.

[30] A.M. Hava and E. Ün, "Performance analysis of reduced common-mode voltage PWM methods and comparison with standard PWM methods for three-phase voltage-source inverters," *IEEE Transactions on Power Electronics*, vol. 24, no. 1, pp. 241–252, 2009, doi:10.1109/TPEL.2008.2005719.

[31] "Tesla Model 3: Inverter Teardown," Nagoya University. Available: https://www.youtube.com/watch?v=fj4KBVgJsGA [Accessed: 11-Mar-2022].

[32] "Teardown expert Sandy Munro Compares Tesla, Nissan, Jaguar Inverters." Available: https://chargedevs.com/newswire/teardown-expert-sandy-munro-compares-tesla-nissan-jaguar-inverters/ [Accessed: 11-Mar-2022].

[33] "Performance Analysis of the Tesla Model 3 Electric Motor using MotorXP-PM." Available: https://motorxp.com/wp-content/uploads/mxp_analysis_TeslaModel3.pdf [Accessed: 11-Mar-2022].

[34] "Compare Side-by-Side: 2019 Tesla 3." Available: https://www.fueleconomy.gov/feg/Find.do?action=sbs&id=41415&id=41416&id=41188 [Accessed: 11-Mar-2022].

[35] J. Swanke, H. Zeng, D. Bobba, T.M. Jahns, and B. Sarlioglu, "Design and testing of a modular high-speed permanent-magnet machine for aerospace propulsion," in *2021 IEEE International Electric Machines & Drives Conference* (*IEMDC*), 2021, pp. 1–8, doi: 10.1109/IEMDC47953.2021.9449536.

[36] Y. Cong, N. Adina, Z. Wei, *et al.*, "Submodule design of a 2 kV 1 MW integrated modular motor drive for aviation applications," in *2021 IEEE 8th Workshop on Wide Bandgap Power Devices and Applications* (*WiPDA*), 2021, pp. 345–350, doi:10.1109/WiPDA49284.2021.9645119.

[37] M. Ohadi, "The game changing potential of co-design in development of light and ultra-efficient all-electric powertrain for aviation," Presented at the 2020 Annual Meeting of the NSF Center for Power Optimization of Opto-Thermal Systems (POETS), University of Illinois, Urbana-Champaign, 19 May 2020.

[38] Y. Kudoh, K. Mizutani, N. Otsuka, *et al.*, "Single to two-phase matrix converter using GaN-based monolithic bidirectional switch for driving symmetrical two-phase motor," in *2014 IEEE Energy Conversion Congress and Exposition (ECCE)*, 2014, pp. 3186–3191, doi:10.1109/ECCE.2014.6953833.

[39] K. Han, A. Agarwal, A. Kanale, *et al.*, "Monolithic 4-terminal 1.2 kV/20 A 4H-SiC bi-directional field effect transistor (BiDFET) with integrated JBS diodes," in *2020 32nd International Symposium on Power Semiconductor Devices and ICs* (*ISPSD*), 2020, pp. 242–245, doi:10.1109/ISPSD46842.2020.9170064.

[40] J. Wang, Y. Li, Y. Han, "Integrated modular motor drive design with GaN power FETs," *IEEE Transactions on Industry Applications,* vol. 51, no. 4, 2015, pp. 3198–3207, doi:10.1109/TIA.2015.2413380.

[41] H. Dai, R.A. Torres, T.M. Jahns, and B. Sarlioglu, "Analysis and suppression of conducted common-mode EMI in WBG-based current-source converter systems," *IEEE Transactions on Transportation Electrification*, vol. 8, no. 2, pp. 2133–2148, doi:10.1109/TTE.2022.3144426.

[42] X. Lyu, H. Li, Y. Abdullah, *et al.*, "A Reliable ultra-fast three step short circuit protection method for E-mode GaN HEMTs," in *2019 IEEE Applied Power Electronics Conference and Exposition* (*APEC*), 2019, pp. 437–440, doi:10.1109/APEC.2019.8722299.

[43] M.S. Hamad, A.K. Abdelsalam, B.W. Williams, "Over-voltage protection of single phase grid connected current source inverters using a simplified passive network," in *IET Conference on Renewable Power Generation (RPG 2011),* Edinburgh, 2011, pp. 1-5, doi: 10.1049/cp.2011.0159.

[44] G. Choi and T.M. Jahns, "PM synchronous machine drive response to asymmetrical short-circuit faults," *IEEE Transactions on Industry Applications*, vol. 52, no. 3, pp. 2176–2185, 2016, doi:10.1109/TIA.2016.2518992.

[45] Y. Zhang, W. Zhang, and T.M. Jahns, "Investigation of single-switch short-circuit fault characteristics of a PM machine drive with a current source

inverter," in *2015 IEEE International Electric Machines & Drives Conference (IEMDC)*, 2015, pp. 967–973, doi: 10.1109/IEMDC.2015.7409178.

[46] H. Dai, R.A. Torres, T.M. Jahns, and B. Sarlioglu, "Characterization and implementation of hybrid reverse-voltage-blocking and bidirectional switches using WBG devices in emerging motor drive applications," in *2019 IEEE Applied Power Electronics Conference and Exposition (APEC)*, 2019, pp. 297–304, doi:10.1109/APEC.2019.8721910.

[47] H. Dai, R.A. Torres, W. Lee, T.M. Jahns, and B. Sarlioglu, "Integrated motor drive using soft-switching current-source inverters with SiC- and GaN-based bidirectional switches," in *2020 IEEE Energy Conversion Congress and Exposition (ECCE)*, 2020, pp. 2372–2378, doi:10.1109/ECCE44975.2020.9236394.

[48] T.M. Jahns and B. Sarlioglu, "The incredible shrinking motor drive: accelerating the transition to integrated motor drives," *IEEE Power Electronics Magazine*, vol. 7, no. 3, pp. 18–27, 2020, doi:10.1109/MPEL.2020.3011275.

[49] F. Chen, H. Ding, S. Lee, W. Feng, T.M. Jahns, and B. Sarlioglu, "Current source inverter based large constant power speed ratio SPM machine drive for traction applications," in *2020 IEEE Transportation Electrification Conference & Expo (ITEC)*, 2020, pp. 216–221, doi: 10.1109/ITEC48692.2020.9161471.

Chapter 5

Thermal management of integrated motor drives

Rafal Wrobel[1], Liliana De Lillo[2] and Xiang Shen[3]

5.1 Introduction

When reviewing technology related with integrated motor drives (IMDs), it is natural to look first at the most recent developments in high-specific-output and high efficiency electrified propulsion, on-board power generation and actuation. All of these have been recently receiving increased attention in the context of global efforts towards the net-zero emission and more sustainable society. Particularly, transportation (future mobility) has been undergoing dynamic transformative changes, with numerous technological challenges to be addressed [1,2]. Here, both system integration and thermal management have been identified as essential to development of the next-generation IMDs. Interestingly, when exploring the motor drive technology, it is evident that the concept of IMDs is not new. IMDs have been with us for much longer than one might think. The early examples of integrating electronics with electrical machines include both automotive and aerospace applications introduced in the 1960s [3,4]. Here, car alternators and aircraft generators were the first to combine both electrical machine and power electronics (three-phase full-wave discrete diode rectifiers) in a single package to provide a specific application driven, technologically competitive products. Figure 5.1 presents the early design solutions including the rotating rectifier assembly of an actively air-cooled brushless AC generator [3] and alternator assembly [5] for aerospace and automotive applications, respectively.

Unsurprisingly, these early examples were faced with similar challenges to the modern-day IMDs. In the context of thermal management, both dedicated and shared heat removal from the machine body and power electronics were used, e.g. dedicated heat sinking of power electronics for the car alternators and shared air or oil cooling for the aircraft generators. Interestingly, both application examples use the early integration architecture to this day as the preferred solution, of course with more advanced (power dense) machine designs and improved power electronics ratings.

[1]School of Engineering, Newcastle University, UK
[2]Department of Electrical and Electronic Engineering, University of Nottingham, UK
[3]Department of Mechanical and Construction Engineering, Northumbria University, UK

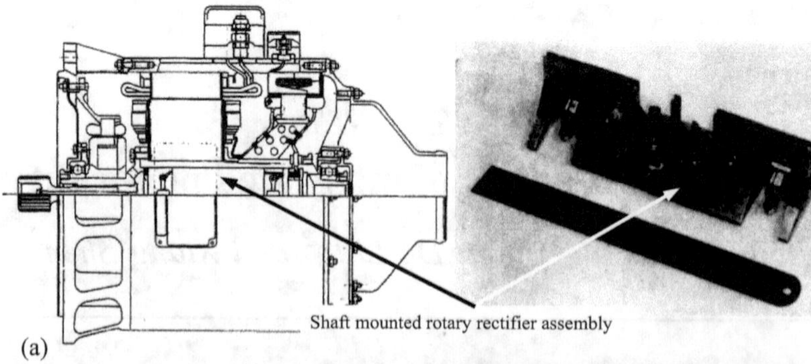

(a)

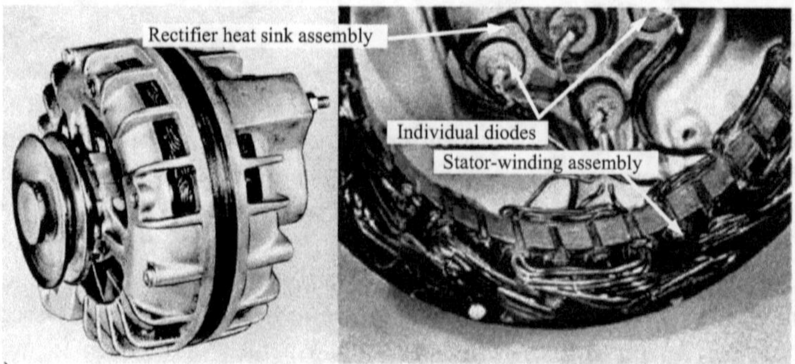

(b)

Figure 5.1 Early examples of integrating power electronic components with electrical machines, (a) actively air-cooled brushless AC generator with shaft mounted rotary rectifier for aerospace applications (50 kVA, 6,000 rpm) [3]; (b) passively air-cooled DC open-frame alternator with heat sink mouthed rectifier for vehicle applications (420 W, 1,250 rpm) [5].

Although diode rectifiers do not constitute a motor drive, the initial developments discussed here are considered to be a steppingstone to the modern day IMDs, where a more comprehensive approach to the system integration is employed, e.g. incorporating a complete inverter with auxiliary and control circuitry. It is important to note that the motor drive integration is just one of the work streams related with IMDs. Subsystems like, mechanical suspension, gearing (transmission), clutching and braking are frequently integral part of developing new integrated systems for propulsion or actuation [4].

This chapter is focused predominantly on high-specific-output and high efficiency applications, but one must remember that the IMDs have been commonly used in multitude of everyday consumer goods. Applications like consumer's electronics, health and beauty goods, home appliances, garden and power tools, personal mobility vehicles and many others. Figure 5.2 presents examples of IMDs for such applications, an electric toothbrush and an electric-bicycle drive, both

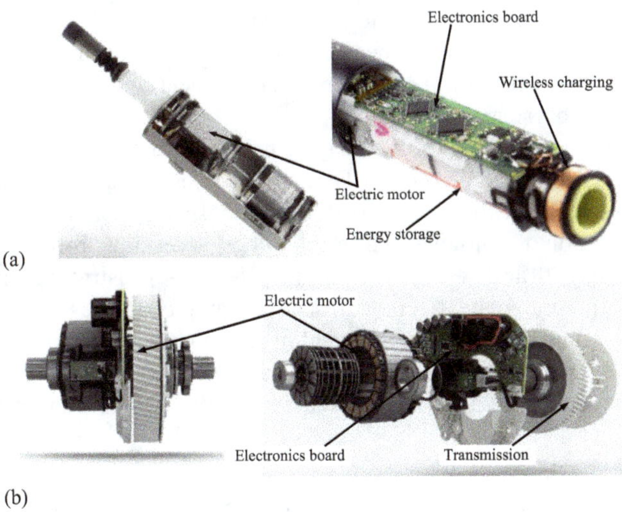

Figure 5.2 Examples of consumer goods application of IMDs, (a) electric toothbrush by Oral-B – [6], (b) electric-bicycle mid-drive by Bosh (250 W, 85 Nm) – [7] (note that only IMDs without the product enclosure are shown)

revealing high level of system integration [6,7]. Here, both examples use natural convection to assure reliable operation of the IMDs, but devices like vacuum cleaners or power drills frequently use integrated active air-cooling. Note that the top-level system architecture shown in Figure 5.2(a) is not dissimilar to what the modern electrified transportation is aimed at, with subsystems like wireless charging, energy storage, drive, control and communication electronics and electric motor packaged in a single unit for convenience and comfort. It is important to note that these relatively low-power applications are also faced with challenges associated with the system integration and thermal management. Balancing the overall system functionality with more detailed design targets for the individual subsystems is indeed essential for successful deployment of the IMDs across all power levels.

Some of the potential benefits associated with adopting IMDs (at a consumer level) have already been briefly mentioned. A more detailed list includes the reduced mass, reduced volume, improved efficiency, lower cost, and improved functionality and robustness [8]. However, integrating together two very different subsystems like motor and drive has several challenges related with physical implementation and compatibility, i.e., electrical and electromagnetic, mechanical, and thermal design factors. In particular, the thermal management aspects have been very important for successful deployment of high-specific-output and high efficiency IMDs [8–10]. It is worth to mention that the continuous evolution of power devices from Silicon (Si) to wide-band gap Silicon Carbide (SiC) and Gallium Nitride (GaN) offers new routes for motor drive integration, with increased operating temperature, thermal conductivity and efficiency, on the one hand, and increased switching speed/frequency and reduced footprint, on the other hand [11]. All these impose interesting engineering challenges associated with the close subsystem integration.

The following sections of this chapter provide a detailed technology overview of thermal management of IMDs across variety of applications. Starting with a generic classification of alternative thermal management techniques and system integration topologies. The technology overview follows then with several commercial and research examples of IMDs illustrating the key concepts for an integrated heat removal. The chapter continues with alternative technology avenues for advanced thermal management of integrated motors and drives. The authors' observations and forecasts of the further evolution to thermal management of IMDs are discussed in the concluding section of the chapter.

5.2 Thermal management of electrical machines

5.2.1 Thermal management of electrical machines – system classification

In general, thermal management of electrical machines and power electronics is governed by three fundamental heat transfer mechanisms: conduction, convection, and radiation. The heat transfer mechanisms occur simultaneously with different intensity, subject to a specific machine (thermal management system) design. There are two main functional groups of thermal management of IMDs, passive and active. The passive heat removal relies on free convection and radiation primarily from the machine outer periphery, e.g., a naturally ventilated motor housing with fins. In contrast, active heat removal makes use of forced convention, where a cooling fluid (gas or liquid) is introduced to remove heat from selected parts of the machine assembly, e.g. a motor housing with liquid jacket, a stator assembly with forced air ventilated channels. It is important to note that ability to effectively transport heat from across heat sources to the heat removing features is essential in both functional groups. Here, conduction is frequently the dominant heat transfer mechanism. A more detailed classification of the passive and active thermal management includes indirect and direct cooling methods. The indirect approach makes use of heat removing elements, which are in contact with active parts (heat sources) of the IMD assembly, e.g., an oil cooled rotor via a hollow shaft, an air cooled stator via housing. The heat removing techniques, where active components are in direct contact with cooling media, are categorised as direct cooling, e.g., end-windings spray oil cooling, oil flooded stator-winding assembly [12–18]. Figure 5.3 shows schematically classification of the discussed thermal management techniques listing numerous examples. Note that cryogenically cooled IMDs including superconducting electrical machines are also considered here. Although their commercial implementation has been receiving increased interest, in the context of next-generation electrified propulsion and power generation for aircraft

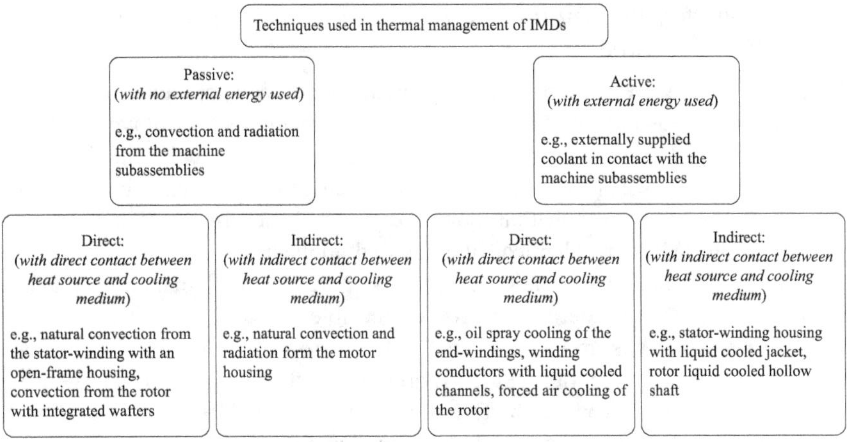

Figure 5.3 Classification of alternative techniques used in thermal management of IMDs

applications, examples of such IMD implementations are mainly limited to the individual motor and drive subsystems.

5.2.2 Mechanical and thermal integration of IMDs – system architectures

In principle, thermal management of electrical machines for IMDs is analogous to that used in DMDs. However, there are small differences related to packaging of the individual motor-drive subsystems, which usually share a portion of the overall mechanical assembly for IMDs. There are two generic approaches to the mechanical integration of IMDs including axial and radial arrangement of the motor-drive subassemblies [8–10]. For the axial packaging, both subsystems are arranged along the motor axial length, with the motor end-windings or endplate being the shared plane. The radial layout of IMDs makes use of the motor outer surface in radial direction inwards or outwards, subject to a specific machine topology, e.g., a radial-flux machine with outer rotor for the radial inwards interfacing plane and a radial-flux machine with inner rotor for the radial outward shared surface. Both packaging approaches enable for shared or dedicated cooling of the individual motor-drive subassemblies. The shared thermal management is representative of a series cooling system, where a given heat removal medium is used across both motor and drive subsystems. In contrast, the parallel thermal management approach employs dedicated cooling of the motor-drive subsystems individually [9]. For highly integrated IMDs system, a clear distinction between the mechanical and thermal packaging types listed here might be in some way difficult to make, i.e., where hybrid design solutions are preferred or more effective. It is important to note that an appropriate balancing of the two mechanical and thermal designs, i.e., subsystem proximity and separation, is very important to achieving a well-balanced system combing the attributes of IMDs discussed earlier.

5.2.3 Conventional (non-cryogenic) thermal management of IMDs – examples

There is a large variety of research and commercial developments available to choose from to illustrate the thermal management techniques employed in IMDs. Consequently, only a number of recent examples, which are representative of the specific system classification or architectures discussed earlier, have been selected here. It is important to note that although the generic heat removal techniques used across multiple IMDs are identical, the detailed application-driven design choices are usually different. That detailed design and practical implementation makes a specific mechanical and thermal integration feasible.

Figure 5.4 presents an example of IMD with forced air indirect cooling and axial arrangement of the motor (PMSM) and drive subassemblies. The system has been developed for in-wheel vehicle traction applications [19]. Both initial hardware prototype and complete drive train implementation (with two-stage 18:1 reduction gearbox) are shown in Figure 5.4(a) and (b), respectively. When reviewing the IMD mechanical and thermal packaging, a number of observations can be made. Here, the motor and the drive are separated by an end-plate, on which a set of custom ultra-small power electronics modules (SiC, 1,200 V, 150 A, 250°C) is directly mounted. The end plate acts as a heat spreader into the IMD's outer periphery with fins. The forced air over the IMD housing is provided by the

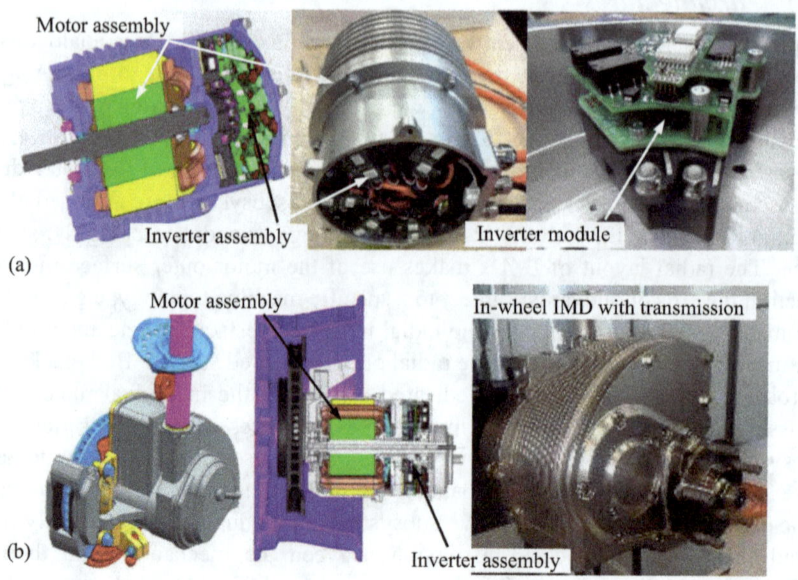

Figure 5.4 Example of forced air cooled IMD for in-wheel traction applications – axial arrangement of the motor and inverter (20 kW, 20,000 rpm), (a) prototype of IMD with a single inverter module shown and (b) complete IMD-based traction system [19]

vehicle motion with airflow in the range of 5–6 m/s. The drive subassembly is arranged in a multi-layered fashion, with temperature sensitive components away from the main heat source, i.e., end-winding end-plate assembly. Replacing the forced air cooling with a liquid cooled alternative could further improve effectiveness of heat removal from the in-wheel IMD example.

The next example, Figure 5.5, has also been developed for in-wheel traction applications [20]. However, a direct-drive IMD configuration is shown here. The outer rotor (hub) PMSM has a liquid cooled heat removal assembly packaged within the stator inner bore. Both motor and axially mounted drive share the same indirect active cooling system (with 50/50 WEG coolant, 13l/min at 50°C). The IMD's generated heat is removed via the radial and orthogonal thermal paths. To aid a good heat transfer from the machine body into the liquid jacket, an appropriate impregnation (high thermal conductivity encapsulation) of the stator-winding assembly is used. A similar strategy is employed when packaging the integrated electronics [20–22]. As the hub motor is exposed to multitude of environmental factors and operational stress, encapsulation of the individual IMD's subassemblies also provides a protection against harsh environment, i.e., ingress of any unwanted elements. It is important to note that the IMD includes a sealing system, which alongside the component encapsulation offers a robust design solution.

An example of indirect actively liquid cooled IMD with radial integration of the motor-drive subassemblies is shown in Figure 5.6. Similarly, as in the previous examples, the IMD system has been developed for vehicle propulsion applications (front axle). Here, the motor-drive also includes mechanical transmission (a two-stage reduction gearbox), which has been packaged in one compact unit. A series liquid cooling system has been employed (with WEG coolant, flow in the range of

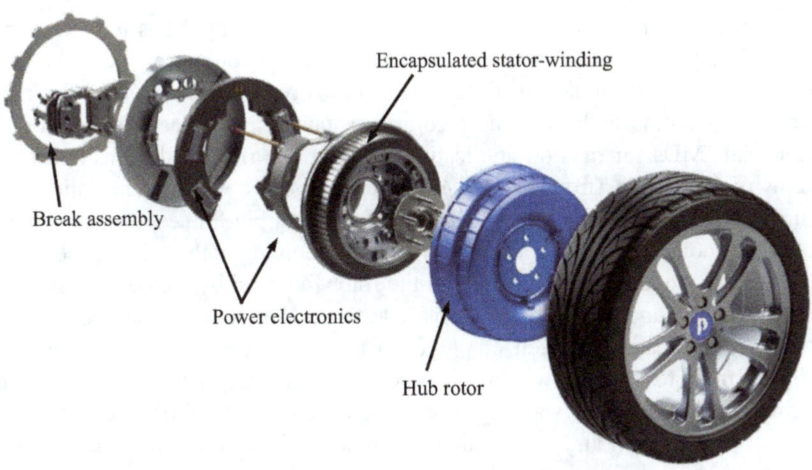

Figure 5.5 Example of actively liquid cooled IMD for in-wheel traction applications – axial arrangement of the motor and inverter by Protean Electric (60 kW, 1,600 rpm) [20]

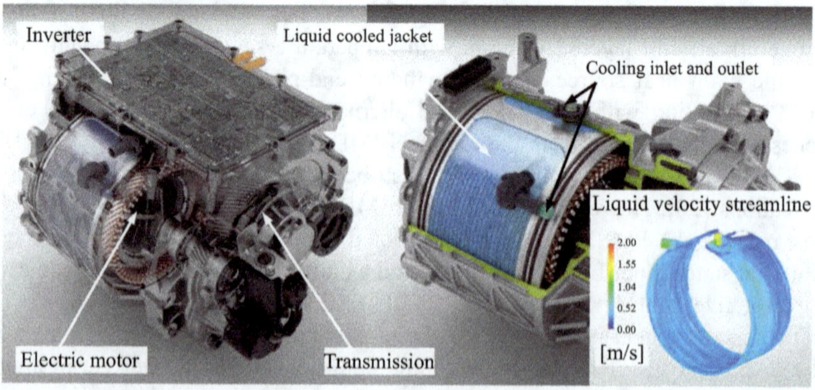

Figure 5.6 Example of forced liquid cooled IMD for traction applications by BorgWarner (125 kW, 10,600 rpm) – radial outwards arrangement of the motor and inverter [20]

8l/min to 15l/min at 65°C). Note that both motor (PMSM) and drive have their individual enclosures, which are then mounted together. The way both sub-assemblies are fitted suggests that thermal management provided within the individual units is the main heat removal path, i.e., a limited share between two cooling subsystems. Interestingly, the liquid jacket design adopted here is not integral to the motor housing and instead a grooved sleeving mounted onto the stator-winding assembly is used. This then forms sealed cooling channels when fully assembled within the motor enclosure. An example of coolant velocity streamline for such liquid jacket is shown in Figure 5.6.

The next example of advanced thermal management of IMDs is presented in Figure 5.7 [23–25]. Here, a series indirect active liquid cooling of the IMD's sub-assemblies has been employed. Both motor and drive are integrated in the radial plane, which seems to be the most common approach used across multitude of commercial IMDs for traction applications. Figure 5.7(a) presents constituent subsystems of the single (front axle) IMD system together with a schematic of the cooling path. The WEG coolant enters a set of compact pin-fin heat exchangers used to cool the motor-drive electronics first. Then the cooling path includes the hollow shaft assembly to aid heat removal from the rotor followed by the bearing endplate, stator jacket and another bearing end-plate at the outlet. Note that an IM with aluminium squirrel-cage has been employed here. Consequently, a more aggressive cooling of the rotor and end-windings has been implemented. The rotor squirrel-cage end-rings with integrated wafters have been introduced to further enhance heat removal from the rotor and end-windings by mixing air within the motor end region cavity (with forced liquid cooled endplates). Further to this, the actively cooled bearing endplate at the outlet acts as an oil cooler for the gearbox assembly.

In principle, the twin (rear axle) IMD system shares the same thermal management architecture as the single IMD, Figure 5.7(b). Here, however, the two

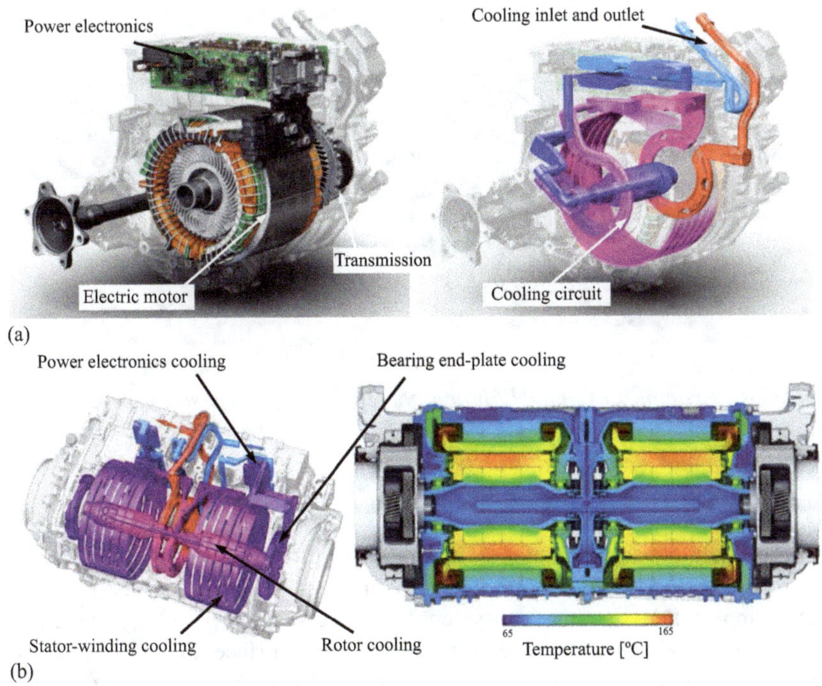

Figure 5.7 Examples of actively liquid cooled IMD for traction applications by Audi – radial outwards arrangement of the motor and inverter, (a) single IMD concept together with heat removal path (125 kW, 13,500 rpm), (b) twin coaxial IMD together with heat removal path and example of temperature distribution across individual motor

parallel cooling paths for the individual IMDs are merged before entering the rotor hollow shaft assembly to compensate for any difference in the coolant temperate, i.e., due to torque vectoring [23]. Also, the order in which the bearing endplates are cooled is different due to the coaxial IMD arrangement. The coolant separates into two parallel paths before entering the bearing endplate at the outlet. Such arrangement allowed for a compact water jacket between the two IMDs, with low flow losses and uniform flow distribution across the individual motor-drives. Figure 5.7(b) also shows temperature predictions within the twin-axle IMD assembly reviling uniform temperature distribution for the individual IMDs (continuous rated power at 13,500 rpm operating point is considered here).

The examples discussed so far include IMDs exclusively for on-road, consumer vehicle applications. However, as discussed earlier subject of IMDs is not limited to the automotive industry and there is a multitude of interesting developments related to other sectors of transportation. The next example, Figure 5.8, presents IMD developed with focus on, but not limited to, aerospace propulsion.

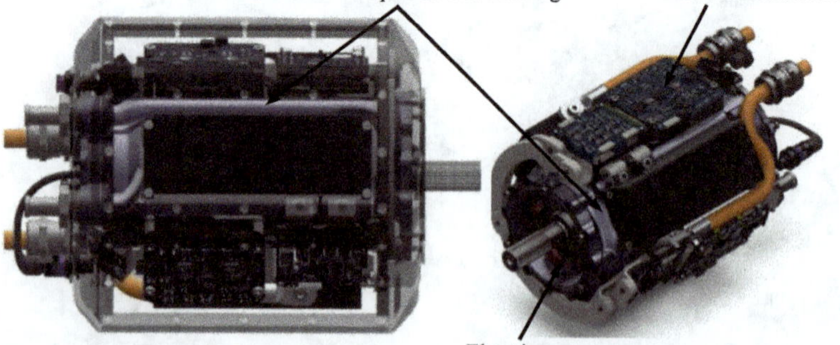

Figure 5.8 Examples of actively liquid cooled IMD for aerospace propulsion by H3X (200 kW, 20,000 rpm) – radial outwards arrangement of the motor and modular inverter [26]

This interesting concept is representative of the indirect liquid cooling with radial outward packaging of the motor and drive subassemblies. In contrast to the previous examples, here the drive subsystem fully utilises outer periphery of the motor enclosure, i.e. available electronics-motor interfacing surface [26]. In general, such approach has been considered previously [10], however, in this application an AM housing (heat exchanger) with integrated cooling channels is used to ensure a close system integration. The AM housing provides a compact and lightweight unit, where both mechanical and thermal design aspects are tailored for both motor and drive needs. This PMSM-based IMD offers exceptionally high power density, 11 kW/kg and 13 kW/kg continuously for geared (4:1 planetary gear box) and ungeared version, respectively. Here, also WEG coolant is employed with relatively high flow rate between 15l/min and 30l/min and temperature in the range of −40°C to 60°C. It is important to note that the use of AM in construction of electrical machines, power electronics and drives (PEMD) has been receiving increasing interests, with numerous examples of compact and highly effective heat removal concepts discussed in the literature [27,28].

The last example in this section presents a MW range output power IMD for marine propulsion, which has been developed specifically for submarine applications. Although this is not the mainstream of electrified technology for transportation, the motor and drive integration for large machines is very interesting indeed. The IMD application calls for a high level of redundancy, consequently a 12-phase winding with IGBT-based modular drive configuration is implemented here. The motor is capable of operating with just a single drive unit, of course at reduced output power. Interestingly, the bell-shaped inner PM rotor assembly allows for the drive modules to be packaged radially inwards. Both the winding and drive modules are actively cooled with fresh water, which is circulated by pumps and cooled by seawater in heat exchangers. To provide a degree of redundancy, there are two independent cooling circuits. The heat exchangers and pumps are located on the top

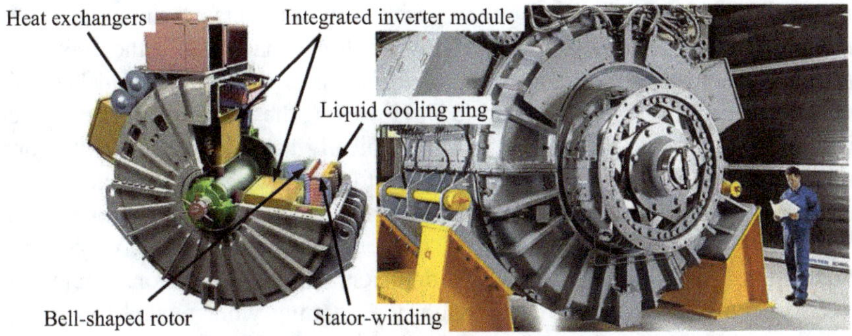

Figure 5.9 Example of actively liquid cooled IMDs for marine applications by Siemens AG (2 MW) – radial inwards arrangement of the motor and modular inverter [29]

of the IMD unit, Figure 5.9. Here the heat exchangers are designed as tubular coolers made of titanium alloy. Further to these the complete IMD includes anticondensation heaters, speed and monitoring computer (fault management) together with a set of sensors for fault detection [29].

When analysing the examples of IMDs presented in this section, it is evident that the use of a specific thermal management approach is application driven. For applications like on-road consumer vehicles, where relatively a large number of IMD examples already exist, some generic observations related to the system packaging and thermal management can be made. Here, IMDs with integrated transmission (reduction gearbox) seems to be preferred solution. Interestingly, mechanical integration of such IMDs appears to be very similar, where the motor and drive subsystems are in close proximity. However, as both subsystems have their own dedicated heat removal features/paths, e.g., liquid cooled housing for the motor and heat exchanger for the drive, in principle they could operate independently as discrete units. Such a 'weak' thermal coupling between IMD subassemblies enables a simpler and more cost effective approach when designing and manufacturing IMDs, see Figures 5.6 and 5.7. On the other hand, a 'strong' thermal link, where the motor and drive share a single heat removal unit, offers greater reduction in volume and weight, but at increased cost and complexity, see Figures 5.4, 5.5 and 5.8. The high level of integration required for IMDs with the 'strong' thermal link calls for less conventional solutions, e.g. application of novel design, new materials and manufacturing/integration techniques.

5.2.4 Cryogenic thermal management of IMDs – examples

The use of cryogenic cooling in electrical machines and power electronics has been gaining more attention in recent years predominantly in the context of the future on-board power generation and propulsion for the next generation of low-/zero emission aircrafts [30]. In general, the concept of cryogenically cooled (superconducting) machines is not new, with early attempts of employing superconductors [here LTS, −269°C (~4 K)] in electrical machines in the 1960s. It is important to note that cryogenically cooled machines with conventional winding materials have also been considered [31]. However, the mainstream research and development activities are focused on cryogenically cooled machines with superconducting windings and/or magnets, subject to the specific machine architecture and application. The latest examples of superconducting machines employ HTS, −196°C (~77 K), and include a broad range of applications, e.g., magnetically levitated (MAGLEV) trains, marine propulsion, power generation, aircraft and automotive propulsion [30–33]. Although, the ongoing work related to the cryogenically cooled DMD subsystems is rich with examples [31–35], the integration of cryogenically cooled electrical machines and power electronics has had less attention. Here, the available examples are limited to the most recent large development programs [36].

Conceptually, the latest developments of superconducting machines for aircraft propulsion favour fully superconducting architectures, where HTS are employed for both stator windings (with ac excitation) and rotor stacks or bulks working as magnets [36–38]. It is important to note that one of many challenges associated with successful implementation of fully superconducting electrical machines is associated with the winding ac power losses, frequently leading to quenching and winding failure. The overall machine design is dictated to a large extend by the integrated thermal management systems, i.e., cryostat and cryocooler, where appropriate. Figure 5.10 presents a concept of cryogenically cooled rotor with HTS (stacks) magnets developed in the EU Horizon 2020 project, Advanced Superconducting Motor Experimental Demonstrator (ASuMED) [36–39]. Here, the rotor cryostat uses helium in a gaseous form. To provide a low heat leakage, the rotor cryostat requires a rotating seal to separate the cooling fluid and vacuum insulation. A set of, ferrofluidic seals have been employed in this design. The stator cryostat is based on a capillary system, which is integrated directly into the stator wining and cooled with liquid hydrogen. Such dual cryostat integrated thermal management approach allows for the stator and rotor subassemblies to be treated individually to provide the most effective solution, i.e., highest cooling capability. Some of the motor metrics include 99% efficiency and 20 kW/kg power density.

The next examples, Figure 5.11(a), shows a concept of cryogenically cooled IMD as a part of the US Department of Energy, Advanced Research Projects Agency-Energy (ARPA-E) programmes: Range Extenders for Electric Aviation with Low Carbon and High Efficiency (REEACH), and Aviation-class Synergistically Cooled Electric-motors with iNtegrated Drives (ASCEND) [40]. Here, a fully integrated all-electric powertrain system at a power density of $\geq$ 12 kW/kg with an efficiency at $\geq$ 93%, is considered. The complete IMD includes, a fully superconducting electrical machine cooled at −253°C (~40 K), GaN-based power inverter cooled at −153°C (~120 K) and magnetocaloric adaptive cryocooler with liquefied natural gas (here, Bio-LNG) at −161°C (~112 K) used as a heat sink. Figure 5.11(b) presents a schematic representation of the complete IMD power train. As the project is relatively new, available information on the IMD cooling and integration is rather

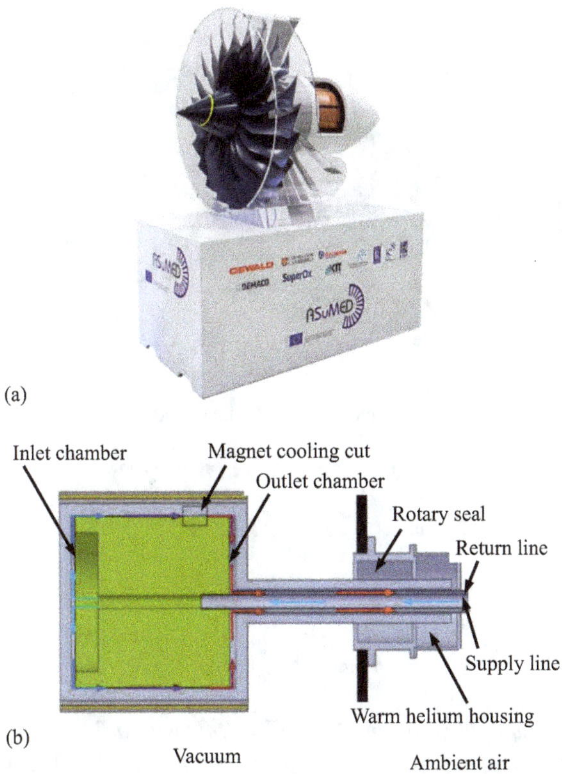

Figure 5.10 Example of fully superconducting electrical machine for aerospace propulsion ASuMED project (1 MW, 6,000 rpm), (a) engine integrated mockup motor and (b) schematic of integrated rotor cooling [36]

limited [40]. The initial drawings suggest the axial integration of the motor and drive, with a cold plate arrangement for the power electronics.

It is evident that the cryogenically cooled IMDs offer a step change in power density for all-electric power trains. However, there are multitude of issues related with practical (and reliable) implementation of such propulsion systems, especially when fully superconducting machine designs are considered. Providing a compact highly integrated and effective cooling system alongside well managed winding ac power losses are some of the fundamental key challenges. Here, as mentioned earlier, the design choices associated with the machine packaging to a large extend are dictated by the superconductor materials used and associated thermal management system. The latest technology examples presented in this section look very interesting indeed, but available information is rather limited to make any generic observations regarding efficacy of the proposed design solutions.

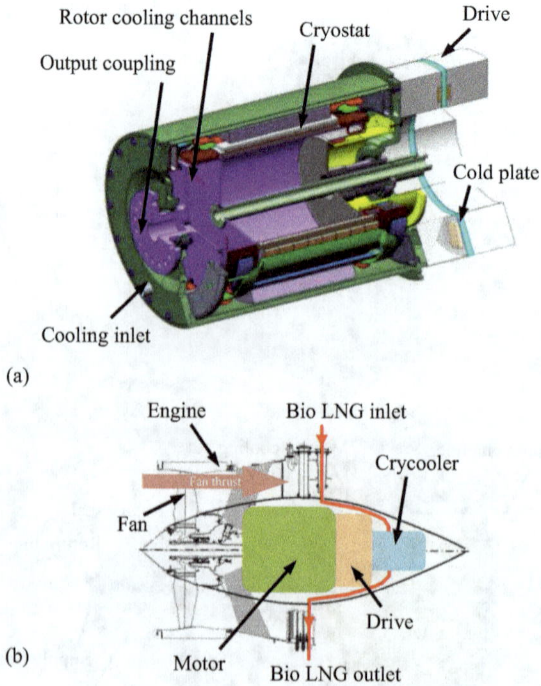

Figure 5.11 Example of cryogenically cooled IMD with a fully superconducting electrical machine for aerospace propulsion SOARING project (2.5 MW, 5,000 rpm), (a) schematic representation of the integrated motor and drive and (b) engine integration of the IMD [40]

5.3 Thermal management of power converters for IMD

Power electronic converters are typically more sensitive to high operating temperatures than the motors that they drive. This is due to the nature of the materials and interconnect systems employed. In comparison, motors are simpler in terms of the number of different materials and thermal interfaces used. Typically, motors would use magnetically permeable irons or steels for the stators, electrically conductive copper or aluminium for the conductors and the casings would normally be constructed with cast iron or aluminium. There is also a need for high performance insulating materials and often permanent magnets are also used. All of these materials have a relatively high temperature capability compared to power electronic parts. The mass of the conductors and stator housings also offer a thermal buffer for highly transient performance demands, even in motors where temperature sensitive magnetic materials are employed.

Conversely power converters utilise semiconductor materials whose junction temperature is often limited to 150°C which are mounted onto cooling systems with soft solders which again have a temperature limitation. The physically small size of the semiconductor devices together with the losses associated with their operation often generates high heat flux levels and therefore the thermal management of the power converters are to some extent a more significant challenge. An excursion beyond the thermal limit of a motor for a short period can often be tolerated with minimal damage but a power converter will simply fail.

The thermal management of power converters in general is a challenge which increases in difficulty as the physical size reduces since the surface area available to remove heat is reduced. This implies that the thermal management of a compact integrate drive, where there may also be a heating influence of a motor in close proximity, will be an even greater challenge and as such, must be integral to the design process from the outset.

Close proximity integration demands power devices and modules capable of operating in harsh conditions, with elevated temperatures and resistance to the damaging effects of thermal cycling. Cooling the integrated power converter requires extensive thermal analysis of the converter and machine both individually and as a unit. Adopting separate cooling systems for the converter and machine potentially increases the volume, cost and maintenance demands of the integrated motor drive while utilising a unified cooling strategy can yield size and weight reductions but increases design complexity.

5.3.1 Packaging of power electronics module

The power converter in an IMD is the most temperature sensitive component in the system. There are many parts of a power converter which through their operation exhibit electrical losses which need to be characterised accurately [41]. The majority of these losses however are created within the power semiconductor devices and there are multiple factors which affect these losses [42] such as switching strategy and frequency, semiconductor device type and material together with operating temperature. Electrical losses need to be managed in order to ensure that maximum temperatures are not exceeded.

Due to the often harsh thermal requirements there has been a growing demand for an efficient, high temperature operable converter which has led to advancement in research into wide-band gap semiconductor devices such as silicon carbide (SiC) and gallium nitride (GaN), both having the potential to operate at increased temperatures when compared to silicon-based semiconductors [43,44]. Environments which require an extreme range of temperature operation, from very low to high, such as aircraft actuation [45] or propulsion, automotive traction or environmentally exposed industrial processes and renewable energies [46], may benefit from increased semiconductor performance capabilities. The ability for a power module to operate at a high temperature could result in a more compact thermal management system for the IMD and associated reduction in costs of the cooling system. A higher power converter temperature however may require increased capability of the control and gate drive area if these are not thermally isolated from each other. As such, system wide benefits of higher temperatures may be hard to realise. When considered in isolation, wide band gap semiconductor devices like SiC and GaN can theoretically operate at temperatures above 500°C while Si-based devices are restricted to junction temperatures lower than 175°C [47,48]. The packaging of the power module primarily restricts its performance and maximum operating temperature [49,50]. The module undergoes high thermal and thermo-mechanical stresses when operated over a high temperature range, leading to failures in the system. Degradation of polymeric materials used in the module packaging, differences in coefficient of thermal expansion (CTE) leading to mechanical wear as well as the creation of intermetallic compounds, which may weaken the joints, also occur at high temperatures [47,51,52]. The present trend for high temperature capability is to develop power modules which utilise materials which remain stable at higher temperatures such as an increase usage of ceramic composites, direct bonded copper laminates and low temperature co-fired ceramics (LTCC) to create more complex functional structures while attempting to match the CTEs to improve reliability [53,54].

The main features of a power module which affect the thermal performance and reliability of the converter are the ways in which the functional parts are bonded together and how the electrical connections are formed such as: interconnects, die attachment, substrate choice and encapsulation [49]. The key to understanding the requirements of a thermal management system is the understanding of the structures available to the designer to create the power sections and the ways in which these are integrated. In general, there is a need to simultaneously consider the need to control the electrical conduction requirements, the conduction of thermal energy and the need for electrical isolation. In order to achieve this, a typical power module contains many different structural layers with different electrical and thermal properties. These requirements often conflict as some materials are sub optimal in fulfilling the required functions, for example, many electrical insulators are poor thermal conductors compared to the electrical conductors. It is important to note that each functional layer has a different thermal resistance which impedes the eventual passage of thermal energy to the ambient.

Figure 5.12 shows a typical stack up of the different layers needed to create a functional power converter. These parts are required to conduct the electrical energy required to drive the electrical machine together with the thermal energy which is not required but needs to be managed. As well as each layer representing a thermal resistance, the interfaces between functional parts represent a potential mechanical failure zone and advanced cooling techniques and high reliability systems attempt to reduce the number of thermal interfaces in the design.

5.3.2 Electrical interconnects

Conventional power modules use aluminium, copper or gold wire bonds as the main current path for interconnection between the top side of the devices and the substrate, as shown in Figure 5.1. The wire bond is often considered to be the weakest and most delicate part of the power module [55]. At temperatures above 200°C, the reliability and mechanical stability of aluminium bonds is drastically reduced with wire bond lift-off or wire bond breakages the most common failures.

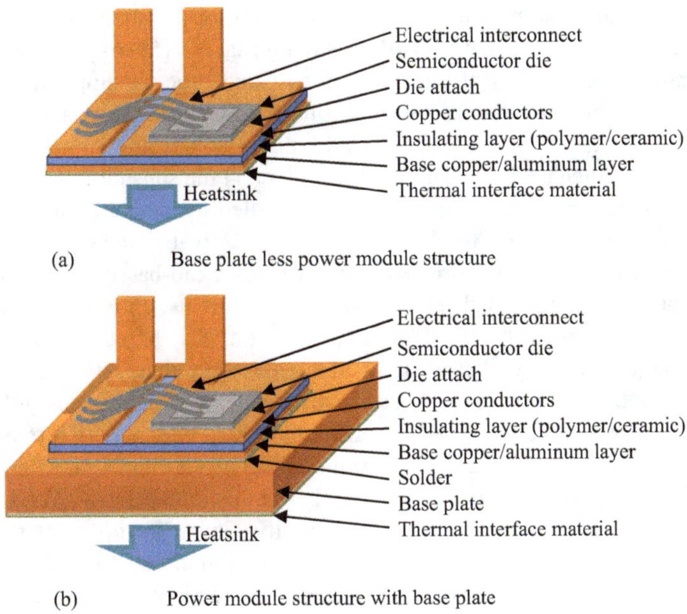

Figure 5.12 Typical stack up of a power converter section showing the different functional layers and thermal interfaces. (a) Baseplate-less power module and (b) power module with baseplate.

These failures usually occur at the heel of the wire bond where the highest stress occurs. Alternative interconnect options that offer high-temperature operation and reliability over 200°C will improve the lifetime of power electronic modules in IMDs. An investigation into large diameter platinum wires for high-temperature application has shown that despite exhibiting exceptional mechanical strength at high temperatures, CTE mismatch with the semiconductor and low electrical conductivity are some of its disadvantages [56]. Silver alloy wire bonds have also been investigated [57] with improvements shown in reliability but cost may be an issue compared to aluminium. Copper wire bonds offer an improvement over aluminium wire bonds in terms of current carrying capability and robustness [58] and with reduced cost with respect to silver. Higher energy is needed however for the ultrasonic welding processes and as such, care is needed when bonding to the semiconductors. Copper wire bonds are often used for substrate to external interconnections and aluminium bonds are used for the semiconductor interconnect.

5.3.3 Die attachment

Potentially, the die attachment is one of the most critical interface layers when considering the thermal management of the power module within the IMD. The die attachment is the connecting interface between the substrate and the semiconductor die as shown in Figure 5.1 providing a mechanically stable and thermally efficient connection to the substrate material. Materials used for die attachment range from high lead and lead free solders to silver sintered layers [59–61]. Solders are the most common die attachment due to their cost and process simplicity but often have a low working temperature and ruggedness [62]. Here, solder alloy pre-forms or pastes containing a mixture of flux and balls of solder are applied to the substrate. The die is then positioned above the solder and the substrates are then passed through an oven where the temperature is controlled to facilitate the melting and cooling of the solder layer. The flux is temperature activated and is used to remove oxide layers which may have formed on the surfaces. Lead-based solders have been largely replaced due to global legislation to reduce its use in industrial processes and products. The majority of solder recipes are tin (SN)-based solder alloys and some are characterised by working temperatures of more than 280°C. However, its relatively low thermal conductivity can restrict efficient heat transfer in high performance power modules. Zinc alloy solders have also been reported to operate at temperatures above 350°C [63]. Silver Sintering offers an improved die attachment option regarding enhanced ruggedness and heat transfer capability. Working temperatures up to 650°C and thermal conductivity almost an order of magnitude higher than solder have been reported [64]. In a similar way to soldering, nano-silver paste is screen printed to the areas beneath the semiconductor die. These then may or may not be dried in an oven (depending on the type of material used) and then, following the die placement, local temperature and pressure are applied for a set time to create a near solid silver bond-line. One disadvantage however is that a high pressure (around 30 MPa) is required to achieve a sintered bond leading to potential for cracks or possible destruction of the relatively fragile ceramic substrate if the process is not closely controlled. A further disadvantage is the process time required for each high pressure sintered bond. This does not allow for an increasing in production speeds without an increased number of bonding machines. Conversely increasing flow rates in solder reflow ovens is relatively straightforward. An alternative to high pressure sintered joints are the silver particle pressureless bonding systems and while these systems compete readily with solder systems, they are generally less robust than the high pressure systems [65]. Due to the relatively high cost of the raw materials in the silver nano particle sintering systems, recent studies have attempted to determine the optimum bond line thickness while maintaining a reasonable strength [66].

5.3.4 Encapsulant

The encapsulant serves multiple functional requirements. The first is to act as a protective barrier for the power devices and components from the environment where pollutants such as dust or moisture should be eliminated or where physical abrasive damage is a possibility. It is often also necessary in systems where the dielectric strength of air is insufficient to ensure that the working voltages as well as transient test voltages for isolation and certification purposes have to be maintained. A less important functional requirement but potentially advantageous for IMD's is that some encapsulants are thermally conductive and can act as heat spreaders although these are often still generally worse than the substrate materials used [67,68]. In applications where the power electronics are not thermally decoupled from the motor, such as in close proximity drive integration (for instance, back iron mount or winding mounted integration), the PE encapsulant may also be subjected to high ambient and operating temperatures. It is therefore important that the chosen encapsulant is able to withstand the desired temperatures, harsh environment, and voltage breakdown while maintaining acceptable thermal requirements. Typical encapsulation materials are silicone gel, epoxy resins, injection moulded thermoset plastics, cement-based [69] preparations and perylene with the latter showing high working temperatures more than 200°C although its cost often makes its use prohibitive. Several encapsulants claim to have working temperatures more than 250°C [70]; however, reliability studies claim that many encapsulants slowly degrade before reaching 250°C [51].

5.3.5 Substrates

The substrate creates a mechanically stable platform and forms the basis of the electrically conductive connections between the source or emitter connections of vertical semiconductor devices which are directly bonded to the substrates using the die attachment techniques described above, and other connections which are required which may or may not be connected to the top of the devices using the interconnect methods. The substrates often also provide electrical isolation to the base of the module as in the baseplate style of power module shown in Figure 5.1(b) or to the bottom surface of the module and hence coolant system as in the baseplate-less style of module shown in Figure 5.1(a).

The substrate typically consists of a thin, electrically insulating material sandwiched between a layer of patterned copper on the top surface and a metallic thermally conductive layer on the bottom. One of the most cost-effective substrate materials is the Insulated Metal Substrate (IMS) [71,72] which consists of a copper, or more typically aluminium base layer, a polymetric electrical insulating layer and then a thin layer of copper (typically $<100\ \mu$m but higher thicknesses are available) which is patterned using standard PCB production techniques. The main benefit of IMS is the cost but suffers from thermal performance due to the thermal resistance of the polymers used for the insulating layer. Higher thermal performance can be obtained by using ceramic to form the insulating central layer. Similar to IMS, ceramic substrates have an electrically conductive top and bottom layer, sandwiching in this case a ceramic middle layer. Typically, only copper is used for the top and bottom layers. There are generally two classes of ceramic substrate used for power applications [73] and these are based on the manufacturing process used to create the layered patterned structured. Direct bonded copper (DBC) substrates, as the name implies are created using continuous layers of copper foil and green (uncured) state ceramic which are fired at high temperatures, as such the copper is directly bonded to the substrate creating both a good thermal and mechanical bond. The patterning of the copper layers is then done using wet etch PCB processes. Active metal brazed (AMB) substrates utilise intermediate titanium layers in between the ceramic and the copper layers to enable a good mechanical bond with certain ceramics which cannot be directly bonded to copper alone. Although this method introduces additional complexity to the process as the copper foil patterns need to be pre-formed together with the braze layers and then fired in a high temperature oven with the ceramic layers, in general is more robust [74] in terms of thermal cycling with respect to DBC substrates.

5.3.6 Power electronic cooling methodologies

As described above, it is crucial that the passage of thermal energy from critical semiconductor parts to ambient is managed effectively. While high-temperature solutions with potentially highly capable exotic semiconductor devices are often sought, presently available mass production techniques for power modules and functional interface material limitations mean that the operating temperatures of the devices is limited and as such, effective cooling is indispensable. Also, while efficiency and volume of the IMD plays a large part in the design process, the energy used in driving the thermal management system must also be taken into consideration. Very high heat transfer coefficients are possible [75] for the power converter sections by using high velocity fluid cooling systems for example, allowing significant size reductions but this is often achieved at the cost of excessive energy used for pumps and heat exchangers. A trade off can be found by including the cooling system power requirements as 'energy loss' when looking at the entire system efficiency. Each cooling methodology has advantages and disadvantages in terms of, volume, raw material used, cooling ability, energy used to drive the thermal management system and finally reliability.

The thermal energy can be dissipated directly to the ambient in forced air cooled or passively cooled systems. However passively cooled systems are rarely used due to the amount of energy which needs to be removed in high power systems. Where liquids are available as the cooling media, significant space and size reductions can be achieved using closed loop thermal management systems for the power stage and motor. It should be noted that there still needs to be a heat exchanger in order to dissipate the waste heat to the ambient which is often not taken into account in power density calculations but can have a significant beneficial impact when the system is considered, especially if multiple IMD's are included with a single thermal management system. Examples of cooling techniques [76,77] which are fundamental to understand the mechanisms involved in an IMD cooling management system are included in the following sections. Starting from forced-air cooling and subsequently moving to cold-plate, pin-fin, turbulator or jet impingement when using liquid working fluids.

5.3.7 Forced air cooled

Forced air cooling can be utilised where the power electronic devices can be spread over a reasonably large area and where a sufficient quantity of air is available [78]. Care must be taken to ensure that the critical elements of the system are able to be effectively cooled under all environmental conditions.

One key advantage of forced air-cooled methodologies is that the thermal energy is delivered directly to the ambient air and no further heat transfer systems are required, this is in direct contrast to liquid-based cooling systems where secondary heat exchangers are required for the system in order to dissipate the energy. Additional challenges when considering forced-air cooling may be dictated by the needs of the specific application [79]. Here an example of the importance of finding the optimum size/weight of the heatsink while considering also the fan characteristics is presented.

5.3.8 Cold plate

A cold plate, one example shown in Figure 5.13, is often considered the simplest form of liquid cooled heat-sink. There are various ways in which these can be constructed, and their performance generally improves with more complex manufacture. The most common form is where channels are bored into a solid aluminium base plate to accept a copper pipe which contains the working fluid. The surfaces are then ground flat such that power modules can then be mounted.

As with standard air cooled heatsinks, a layer of thermal grease is required between the power module and the cold plate, introducing a further thermal resistance into the system. Although the construction method is relatively simple, the use of long straight tubes creates low amounts of turbulence within the fluid flow. Further, stratified boundary layers can form at the edges of the pipe, limiting the potential for heat transfer compared to other methods [80]. Structures which introduce features to increase the turbulence and mixing of the working fluid can increase the heat transfer potential but this is often done with the requirement for an increased pressure drop and hence pumping power [81]. Internal structures are often difficult to realise and proprietary brazing or cold-welding techniques are needed to join multiple layers together to form a liquid tight structure. Cold-plates

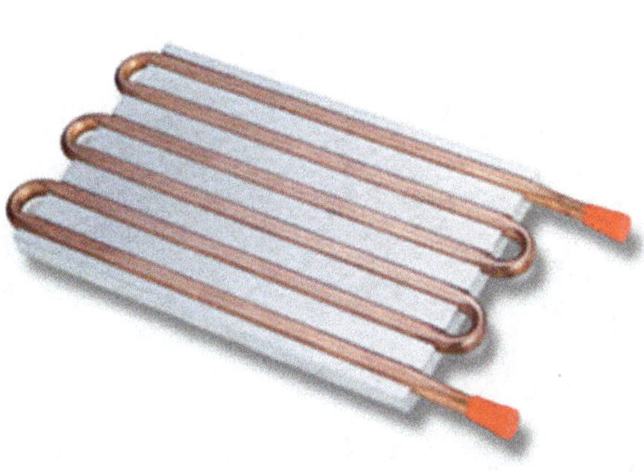

Figure 5.13 A typical cold-plate arrangement showing a machined aluminium base with copper coolant inserts

are almost entirely manufactured from metals and raw material usage is high compared to other liquid cooled technologies.

5.3.9 Pin-Fin

A higher performance ratio between pumping power used and thermal transfer can be achieved using pin-fin coolers [82]. An example of a power module including Pin/Fin structures is shown in Figure 5.14. These coolers allow the working fluid to directly cool the base plate of a power module, removing the need for a thermal interface material used when mounting a module onto a cold-plate. The surface area of the baseplate of the power module is increased with the inclusion of protrusions shaped as either pins or fins, these features also increase the turbulence and mixing within the working fluid, increasing thermal transfer. This modified power module base-plate must be used together with a liquid guide structure. Care must be taken when designing the protrusions as there are many factors which influence the heat transfer, such as, dimensions, shape, height, fluid flowrate, thermal conductivity of the baseplate. One drawback is that each baseplate needs to be sealed against the fluid guide using O-rings, a potential failure point. While the pin-fin baseplate is always metallic (copper or AlSiC) unlike the cold-plate, the majority of the liquid guide part can be formed using low weight polymer materials if needed reducing raw material usage.

5.3.10 Jet impingement

Jet impingement cooling, as the name suggests, is designed to directly cool particular heat sources within the power circuit [83–85]. Very high heat transfer coefficients can be achieved using this method with reasonable pumping powers when compared to either cold-plate or pin-fin arrangements. Jet impingement systems are typically designed to serially cool the different heat sources (normally active semiconductor devices) by spraying the coolant onto the substrate directly

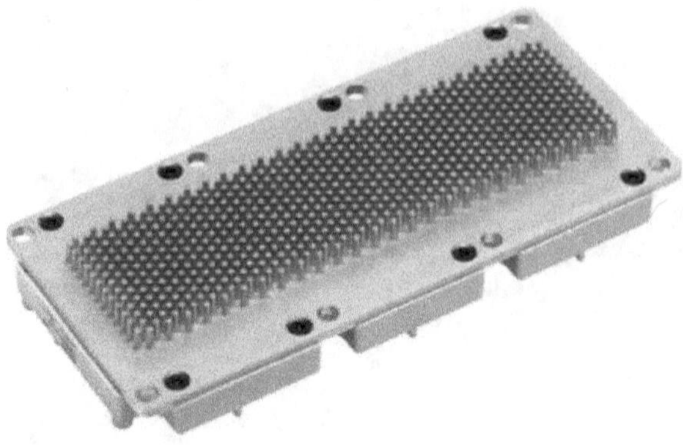

Figure 5.14 A HybridPACK2TM power module package from Infineon which includes copper pin structures on the baseplate in the liquid flow areas

underneath the devices to be cooled. The coolant follows a serpentine flow path to sequentially cool each device which can make the design of such systems quite challenging. Since the main purpose of a jet impingement cooler is to direct the liquid, less expensive plastic parts can be used instead of typical metallic parts needed for cold plates or pin-fin systems but due to the complexity of the fluid flow paths, 3D-printed coolers are often needed. Another aspect which may be a disadvantage in some systems, although the volume required is significantly reduced compared to other methods, is that due to the very significant reduction in expensive metallic parts there is a dramatic reduction in the thermal mass of the cooler system so there is little or no transient overload capability the ability to operate with an inhibited/damaged coolant flow system and this should be taken into account when calculating the desired reliability of the system. Figure 5.15 shows a representation of one cell of a typical jet impingement system. A cut-away view highlights the direction of the fluid flow within the structure.

5.3.11 Turbulator

Turbulators [86] are another example of high-performance cooling system which, like the jet impingement systems, can be physically created with relatively inexpensive plastic structures but are simpler to manufacture than jet impingement systems. These systems are positioned directly beneath the power substrate onto which the power devices are mounted. Similar to a jet-impingement system, the purpose of the structure is to guide the liquid coolant medium to the areas of heat generation in a parallel/serial manner. A baffle or disturbing structure close to each of the heat sources creates local turbulence to increase the local heat transfer coefficient directly beneath the power device and disturb any stagnant coolant areas at the point where it is needed. Due to the complex flow that occurs, computational fluid dynamics modelling methods are necessary in order to develop an effective cooler. These structures, once designed, are simpler to manufacture and classical

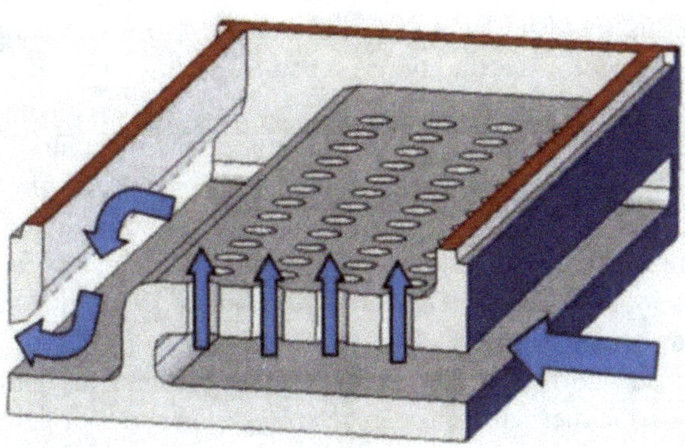

Figure 5.15 A cross section of a 3D printed structure to realise a jet impingement system where different substrate areas, corresponding to heat sources are sprayed in series

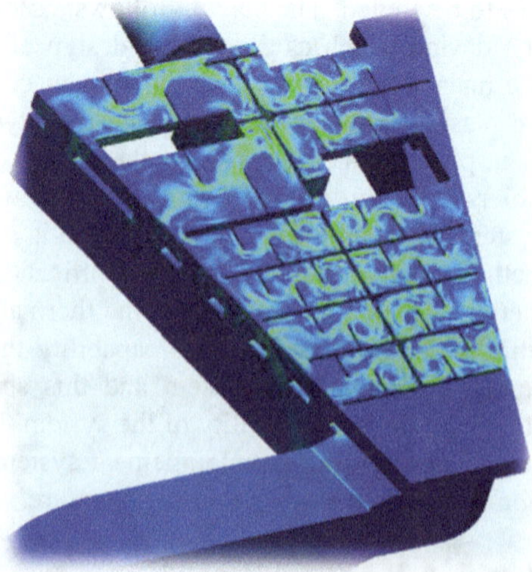

Figure 5.16 A computational fluid dynamics map of the fluid velocities in a turbulator structure for an IMD (courtesy of Electrical Cooling Solutions Ltd)

injection moulding techniques can be used. Figure 5.16 shows an example CFD flow map of a turbulator design for an integrated drive power converter section.

5.3.12 IMD cooling arrangements

In integrated motor drives, the converter is mounted on the motor casing or in the machine enclosure, combining both thermal and mechanical requirements of these components into a single design problem. By considering the same housing for power converter and motor, the physical size of the more drive is reduced and so is the raw material usage of its manufacturing process. However, adopting a unified cooling system instead of the traditional separated options means that the thermal demands of both converter and machine need to be considered simultaneously.

Since the thermal demands of power electronics and machines vary (due to the large heat flux of the power electronics component compared with the machine), a unified cooling system that efficiently accommodates both power converter and motor adds complexity to the system [87]. The use of a combination of active and passive cooling systems is often adopted in modern IMDs due to the rigorous thermal demands of the system.

In most industrial drives and IMDs in literature, when a unified active cooling system is adopted, the thermal paths can be arranged in series or parallel depending on the physical requirements of the system. In a parallel path, both heat sources, the power converter and the motor are treated independently from each other while in a serial cooling system usually the power converter is cooled first and then the motor. An example of a hybrid parallel/serial arrangement is shown in Figure 5.17. Here,

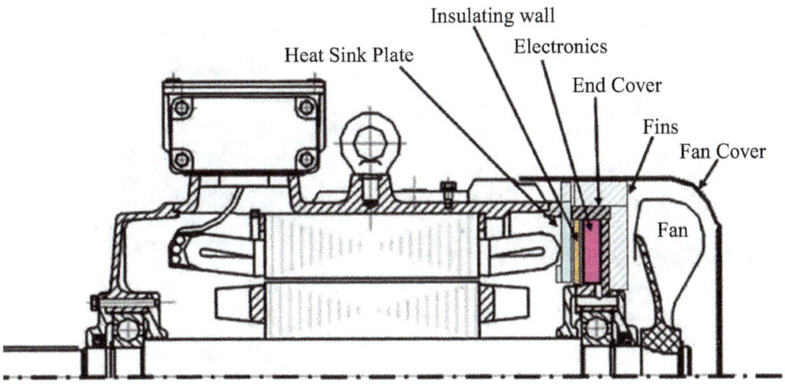

Figure 5.17 A cross section of a forced air cooled industrial integrated drive solution [88]

the motor fan directly forces air onto the power electronics contained within the end plate of the motor. Depending on the fan and cowling design, the pre-warmed air can then be directed towards the motor stator (serial), or, a separate path for fresh cool air can be directed towards the motor (parallel).

Specifically, when the power electronics and motor are integrated to create an IMD where the heat sources are in close proximity, research has been done into structures which thermally integrate the power electronic module with the motor back iron or windings, for example [89,90]. Here the heat flow paths to the stator thermal interfaces should be carefully designed to maintain acceptable temperatures on the semiconductor devices.

Thermal isolation is often the preferred solution due to the different thermal requirements of the components. Thermal isolation involves the sectioning of the power module from the motor heat sources in a thermally controlled compartment. The closer the converter is to heat sources however, like the stator windings and back iron, the harder it is to isolate. Once thermally isolated, the designer is then free to choose either a parallel arrangement or serial.

In general, a parallel structure allows the designer to separate the thermal management problem into separate developments with potentially differing coolant systems with different temperatures and flow rates, but this may increase the total system cost. A serial system, on the other hand, an example of which is shown in Figure 5.18, uses a single coolant flow, simplifying the coolant system requirements but increasing the difficulty of the IMD design as both power converter and motor thermal management needs to be addressed simultaneously.

A serial, unified path for the machine and converter offers a common path for heat flow and can simplify system requirements. The thermal path is from the coolant to the PE then machine (or machine before PE). Emphasis is often given however, to cooling the power electronic components first as the machine is usually viewed as more thermally 'robust' and can handle elevated temperatures for short

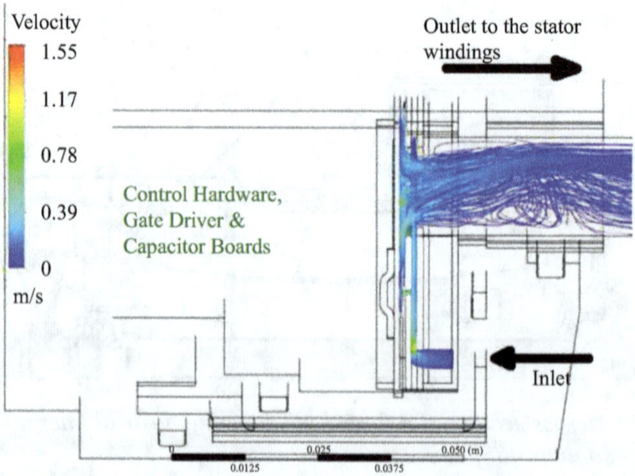

Figure 5.18 A CFD map of a serially cooled IMD where the inlet first cools the power substrates and then is guided towards the stator of the motor [91]

times while semiconductors have a much shorter thermal time constant and rarely tolerate any form of overtemperature beyond their maximum ratings.

The varying thermal demands of both PE and machine and the potential for large temperature gradient in the thermal path together with the physical constraints of the actual integration process itself makes its implementation challenging [91].

In [92,93], the machine housing is used as a common thermal node for the cooling of both motor and power converter. Cavities machined in the motor housing allow for a forced-cooled water system to be used, however the drawback with this arrangement is the large pressure drop in the coolers and large temperature gradient that exists in the common thermal path due to the difference in thermal demands of heat sources.

In a passive cooling system, no external active cooling system is utilised, and heat dissipation is achieved through a combination of natural convection, use of a connected thermal mass, naturally occurring ventilation and thermal isolation. Thermal management using a passive cooling system can be very challenging, but the difficulties can be alleviated by, reducing the overall power ratings and hence losses incurred in the system, utilising high operating temperature components, and increasing efficiency.

5.3.13 Summary

In general, the cooling techniques which can be employed all have different characteristics, and each should be evaluated with respect to the performance of both the motor and power electronics parts at the IMD design time [94]. The desire for a compact solution in order to minimise the raw materials used during manufacture is often at odds with the need to remove heat from critical parts. Care needs to be taken in ensuring that the most critical parts are cooled before the less sensitive parts in a series cooled solution or that sufficient cooling media is available for the sensitive parts in parallel systems [8,45,95].

The choice of cooling method is should also be based on the desired pumping power or energy used to move the coolant medium (gas or liquid) versus the heat transfer coefficient that is required for the application. In general, heat transfer coefficients increase with fluid velocity/flow rate, heat capacity and fluid thermal conductivity but the energy used by the thermal management system also increases with fluid velocity and viscosity. Each cooling methodology described above exhibits different thermal transfer coefficients compared to the amount of energy required to achieve similar performance. For highly compact designs however where the heat flux densities are high, there may be little choice in which type of system can be used. Figure 5.19 shows typical heat transfer coefficients which can be achieved for different cooling methods.

The manufacturing complexity and the ability of the cooling system to deliver reliable cooling arrangements for the more thermal-sensitive power electronics are key challenges for the design engineer [94,96]. Higher manufacturing complexity leads to higher manufacturing cost but lower through life costs due to efficiency and weight improvements. Also, the transient thermal impedance of the chosen cooling system has an impact on reliability, where rapidly changing motor currents introduce a very dynamic heating effect in the semiconductor devices. The ability of the power converter to effectively ride through these highly transient events is also dependent on the chosen cooling technology and is generally related to the thermal mass within the system. For example, a cold plate or pin-fin thermal management system will in general have a higher thermal mass than a turbulator or jet impingement system, as such, will be more robust to short term overloads but on the other hand their usage corresponds to an increased usage of raw materials. Thermal management systems which exhibit higher coefficients of thermal transfer

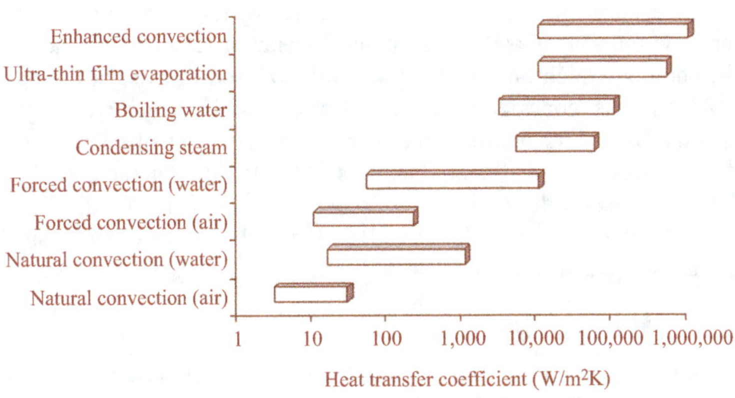

Figure 5.19 Typical heat transfer coefficients for different types of cooling system

such as jet impingement or turbulator will need to be rated for the peak energy dissipation but in general will have a reduced bill of materials.

The thermal management of either a power converter [97] or motor is in itself challenging. The difficulty increases when these items are combined to form an IMD and great care must be given to finding a suitable solution which meets the differing needs of the parts of the system however, the benefits in terms of reductions in raw material usage and energy consumption outweigh the effort required to overcome the challenges.

5.4 Outlook

Modern and future power electronics integration will require IMDs with higher power density and higher efficiency, associated with limited installation and operation space. Effective thermal management is paramount to successful implementation of IMDs, which is a significant thermal challenge in a field where thermal design has often been of secondary importance to electromagnetic performance. The current mechanical design of the machine configuration normally reserves very limited space embedded in or on the housing for the converter installation, with very little consideration of the genuine needs of the IMD design, which introduces significant challenges for the thermal development of the IMDs [11,98].

The converter losses make major contributions to the high heat flux density in current IMDs. Unlike electrical machines, the heat flux density for power semiconductor devices of the converter (typical range: 50–500 W/cm^2) could be up to 500 times that for magnetic components (typical range: 0.1–3 W/cm^2) and even more if compared with capacitors (typical range <0.1 W/cm^2) as a result of the size reduction [99]. When reviewing the existing literature devoted to the future (less conventional) methods for thermal management of electrical machines and drives, it is evident that there is no specific single design avenue or technology being favoured at the moment.

Such a high heat flux density is simply caused by the more and more rigorous operation environment of IMDs, i.e., higher operating temperature and narrower space for cooling, which unfortunately, is still expected to exist in the upcoming future. Nevertheless, considering the trends in electronic device with high power and heat flux density, to optimize the thermal management of adjacent coupling integration, researchers and manufacturers are focusing on advanced cooling technologies, along with traditional cooling methods [11,100,101] and novel materials with higher thermal tolerance [102–105], which can hopefully shed some lights on future thermal management of IMDs.

5.4.1 Liquid metal cooling

Liquid metal is a general term for a type of metal or metal alloy that is liquid at a relatively low temperature. This type of metal not only has the extremely intense thermal and electrical conductivity of metal elements, but also has the fluidity of liquid elements. The thermal conductivity and the specific heat capacity of liquid metal are significantly greater than traditional thermal conductivity agents such as methanol and water.

In fact, liquid metal has been used as a coolant in nuclear reactions since 1963. The first nuclear-powered submarine USS Nautilus used lead-bismuth alloy as the coolant. However, the heat transfer performance was far from ideal due to the intensive corrosiveness on structural materials [106]. In 2005, Li *et al.* found a computer chip cooling method using low melting point metal or its alloy as coolant, and, for the first time, they applied liquid metal cooling to electronics. Later in 2010, ASME's Journal of Electronic Packaging awarded the best paper of the year to the liquid metal cooling technology paper 'Design of Practical Liquid Metal Cooling Device for Heat Dissipation of High Performance CPUs' [107]. As a novel heat transfer material, liquid metal has initiated revolutions in thermal management areas including thermal resistance processing, phase change heat transfer design, and electronic packaging processing.

Typical low melting point metals are summarised in Table 5.1. Low melting alloys, often referred as fusible alloys, such as gallium-containing alloys [111], alkali-containing alloys [112] and Lipowitz's alloy [113], can also be used as coolants due to their better thermal conductivity and stability.

As a solid–liquid phase change material, liquid metal is an ideal heat transfer medium for a new generation of heat dissipation technology due to its characteristics such as great thermal conductivity, similar flow pattern compared to ordinary liquids, single-phase flow, weak corrosiveness, etc. The advantages of the liquid metal cooling technology are summarised below:

- Wide application on most high power-dense electronic components.
- Integration of heat conduction and heat convection, providing expanded heat dissipation area.
- Lower operating noise.

Table 5.1 Properties of typical metals with low melting point [108–110]

Liquid metals	**Density (kg/m³)**	**Melting point (°C)**	**Thermal conductivity (W/m°C)**	**Specific heat (kJ/kg°C)**
Lithium	515	186.0	41.3	4.389
Potassium	664	63.2	54.0	0.780
Tin	730	232.0	15.1	0.221
Sodium	927	97.83	86.9	1.380
Bismuth	979	271.4	8.1	0.122
Rubidium	1,470	38.85	29.3	0.363
Cesium	1,796	28.65	17.4	0.236
Gallium	5,907	29.8	29.4	0.370
Indium	7,030	156.8	36.4	0.230
Mercury	13,546	−38.87	8.34	0.139

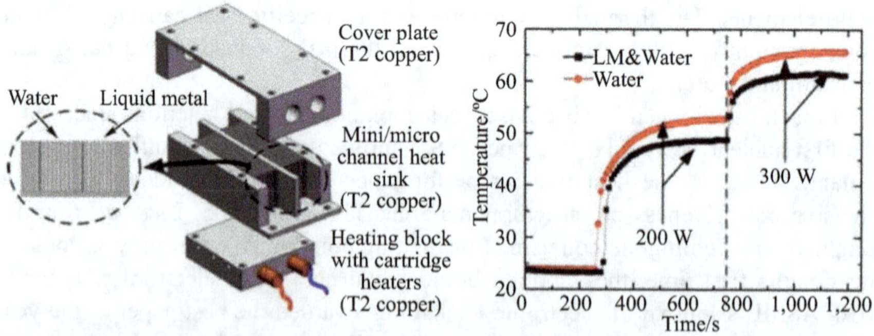

Figure 5.20 An example of hybrid liquid metal/water mini-channel heat sink cooling and the thermal performance comparison, adapted from [114]

- Compact volume with improved cooling performance, compared to traditional cooling methods.
- High stability and reliability due to no actuating mechanism in the cooling system.

Figure 5.20 presents an example of a hybrid mini-channel cooling with liquid metal and water. The water cooling and liquid metal cooling utilise neighbouring mini channels in the heat sink to provide approximately 7–8% improved cooling performance, compared to the conventional water cooling at 200 W and 300 W power. As liquid metal convection cooling is easy to use and stop, it responds effectively to changes in the heat load of the heat source, preventing it from overheating and hence can also be used as a backup/buffering solution when other cooling methods fail.

5.4.2 *Phase change materials (PCMs) cooling*

PCMs offer an interesting design avenue when considering winding designs with an extended overload capability. The advantages of PCMs are quite straightforward because of the simple structure with no moving parts: reliable and silent operations, high heat transfer efficiency, longer service life due to lower temperatures, etc.

One of the most typical PCMs is heat pipes (HPs), consisting of a tube/pipe shell (envelope), a wick and a working medium (working fluid). It has an evaporation section and a condensation section, as shown in Figure 5.21. A HP usually adopts a metal tube with closed ends, and the inner wall is attached to the wick. The tube is vacuumized and filled with a proper amount of working fluid before sealing. The thermodynamic cycle in heat pipes starts with heat entering the pipe and small amount of working fluid in the evaporation section evaporating into a vapour.

The heat is then carried by the vapour to the cooler part of the pipe (the condensation section). Subsequently the vapour condenses back to the working fluid and hence releases the carried heat. The condensed fluid is then retracted to the inner wick structure through capillary action and complete a thermodynamic cycle. A common way to enhance heat transfer efficiency is to embed heat pipes into the base of a heat sink [115], allowing heat pipes in direct contact with the heat source and thereby eliminating the gap between the heat source and the embedded heat pipes, therefore improving the efficiency of heat transfer to the surrounding materials. A conventional wicked heat pipe is presented in Figure 5.21.

In recent years, heat pipes have been widely used in the cooling of power electronic equipment due to their excellent thermal conductivity (the thermal conductivity can approach 250 times that of copper [117]) and excellent temperature uniformity. In view of the compact size of the IMDs, the following micro heat pipes with typical characteristics are paid more attention to and hence are suggested in this subsection.

Micro/ultra-thin flat heat pipes (MFHPs, UFHPs) are ideal for controlling the temperature of hot areas through the highly efficient heat conduction between its flat smooth metal surface and concentrated heat sources. The heat dissipation area is effectively increased due to the M/UFHP's unique flat geometry, as shown in Figure 5.22. The thermal conductivity of the M/UFHP is restricted by the thermal resistance of the heat pipe wall. Hence, to improve the working performance of the M/UFHP, a key point

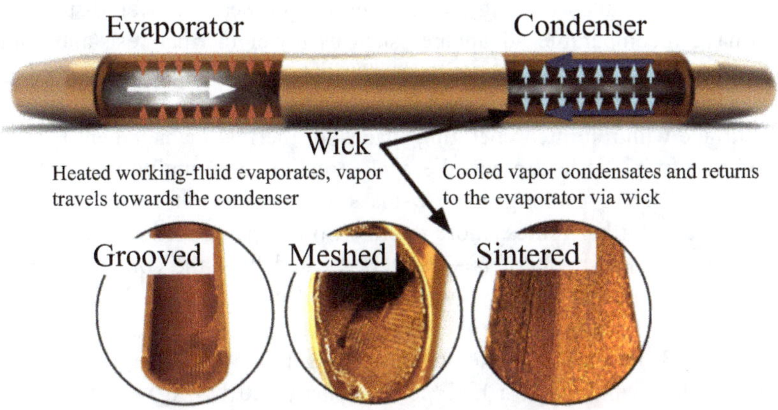

Figure 5.21 The thermodynamic cycle in a typical wicked heat pipe [116]

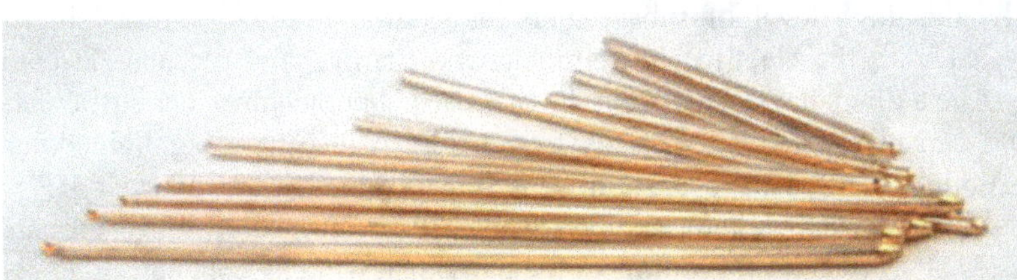

Figure 5.22 A popular variety of M/UFHPs is copper-water flat heat pipes whose flat height can be lower than 2 mm and maximum heat flux density can be more than 300 W/cm²

is to improve the thermal conductivity of the heat pipe wall. It is also worth to mention that some inherent characteristics of M/UFHPs, such as a small external surface area and single shape, will limit the applications in smaller electric devices.

Micro/ultra-thin vapour chambers (MVCs, UVCs) are essentially another form of M/UFHPs whose applications can be limited at certain occasions by the strip shape and the small external contacting area as a result of fabricating from traditional cylindrical heat pipes. M/UVCs can be easily made to thinner than 1 mm and the maximum heat transfer capacity is proportional to the thickness [118]. The effective thermal conductivity of contemporary M/UVCs can be more than 1,000 W/(m·K) [119,120], and can even reach 12,000 W/(m·K) in an extreme case of a 0.4-mm thick UVC [121], which can conduct approximately 40 times more heat than copper. It makes M/UVCs ideal solution components for IMDs as they can be designed to various geometries even with bends, through holes or structural flanges, making M/UVCs good choices to eliminate the hot spots formation, leading to improved temperature uniformity to prevent unnecessary thermal stresses and strains.

Micro/ultra-thin loop heat pipes (MLHPs, ULHPs) are popularly applied in the key industries such as aerospace (e.g., cooling solutions in more electric aircraft (MEA)), telecommunications (e.g., telecom microprocessor cooling), defence (e.g., high-efficiency avionics cooling), etc. Over longer heat transfer distances, M/ULHPs have incomparable advantages such as lower thermal resistance, longer product lifetime, better customization possibility (design flexibility, bendability and routability), making them particularly suitable to attach closely to the heat source surface within limited operating spaces for thermal management [115,122].

Different from flat heat pipes, the wick structure is normally only in the evaporator of M/ULHPs, resulting in a low-mass solution that can be utilized in more complicated and further routes (more than 20 m) compared to the traditional HPs. Furthermore, as the evaporator is separately located from the condenser, entrainment effects can be maximumly suppressed, and consequently the heat transfer efficiency can be significantly improved [123,124].

Figure 5.23 presents examples of integrating HPs within the individual conductors/electrical winding of a high-power-density (20 kW/kg) electric machine concept for the future aircraft propulsion [125]. Here the authors combined AM aluminium coils with copper–water HPs. The proposed HP-based thermal management system provides the main heat removal path via an appropriate HE. Clearly, this high level of subassembly integration imposes several challenges associated with the electrical, electromagnetic thermal and mechanical compatibility. The authors have shown that the proposed approach provides sufficient heat removal capability. The initial results from hardware tests have highlighted that the compatibility matters have a significant impact on efficacy of such HP-based thermal management system. Further to the HP-based heat removal, the authors evaluated also direct liquid cooling of the individual winding conductors [128]. The proposed systems enabled to achieve the conductor current densities, 14 A/mm^2 and 20 A/mm^2 for the direct liquid cooling and HP-based concepts, respectively. It is worth to note that the HP-based thermal management of electrical machines has had considerable attention in recent years as it enables targeted heat removal from

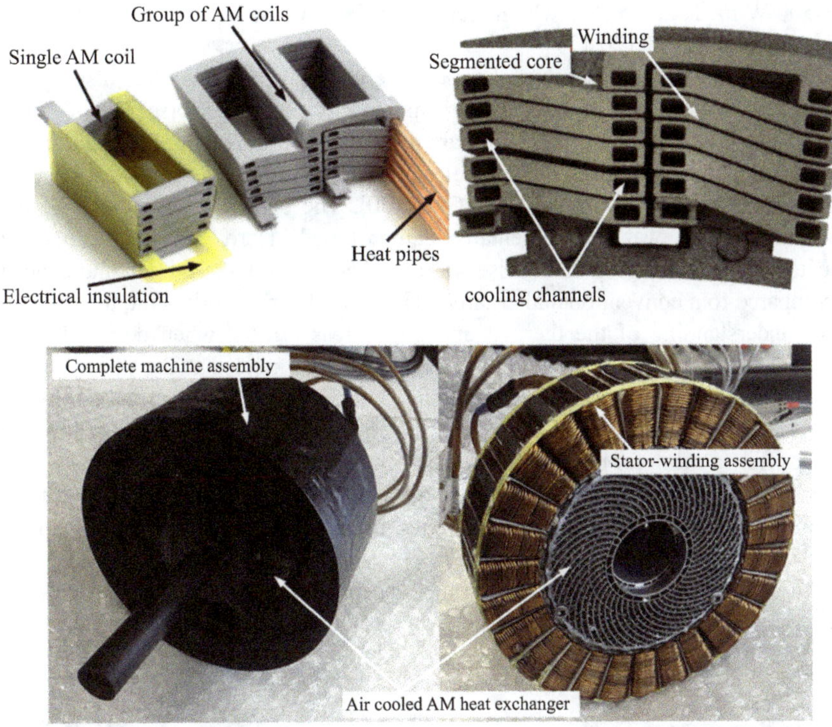

Figure 5.23 Examples of targeted heat removal: (top) additive manufacturing (AM) winding with integrated heat pipes/actively cooled channels (250 kW, 5,000 rpm) [125]; (bottom) stator-winding assembly with integrated AM air-cooled heat exchanger (600 W, 1,000 rpm) [126,127]

intricate composite motor subassemblies. In contrast, thermal management of power electronics employing the HP technology is well established and commonly used across a broad range of applications [129].

A concept of air-cooled ultra-lightweight integrated HE for extracting heat from the complete stator-winding assembly is shown in Figure 5.4, bottom. Here, the fully assembled HE occupies the inner bore of the stator core pack [126,127]. Note that the HE assembly also provides a mechanical support for the rotor. Such an arrangement provides a thermal path for an effective heat extraction. It is important to note that an outer-rotor machine topology, which was developed for all-electric propulsion system of high-altitude UAVs, is considered here. The HE offers extremely high working surface area due to the application of design features possible only when using AM, i.e. lattice structure. Balancing the heat removal and pressure drop across from the HE while maintaining mechanically sound light-weight structure has been found essential for this application. Here, the aircraft generated airflow is harnessed for the motor cooling at the aircraft take-off. Further to this, application of aluminium alloy (AlSi10Mg) enables achieving the required performance of HE, with an equivalent thermal conductance of stator to HE equal to ~500 W/m^2 K at 1.5 m/s, which increases to ~1,400 W/m^2 at 7.5 m/s for the air flow intake at 20°C [126,127].

Along with heat pipes, other phase-change-materials-based cooling also present outstanding prospects to be applied for improving cooling performance of IMDs. Examples of cooling with other PCMs are presented in Figure 5.5, showing a family of AM conductors with different channel profiles to accommodate for PCM. Here the authors used paraffin (RT70HT) as PCM [130]. The benefit of improved thermal transient respond was experimentally demonstrated. Here, approximately 60% reduction of winding temperature rise was achieved for PCM-based winding concept as compared to a conventional alternative. However, further work is required to gain better understanding of the thermal and power loss effects, when developing the complete winding/coil demonstrator. It is important to note that the phase change mechanism has been used here to store the generated heat (latent heat) during the PCM phase transition from solid to liquid. In general, the phase change, e.g., liquid to vapour offers a mean for highly effective heat transport, which is orders of magnitude better than that for commonly used solid metallic materials. HPs and vapour chambers are some of the examples of employing the phase change in thermal management of electronics and electrical machines [129].

Figure 5.24(b) presents an interesting concept of an AM concentrated coil with integrated cooling channels, Here, the coil geometry was designed to reduce the

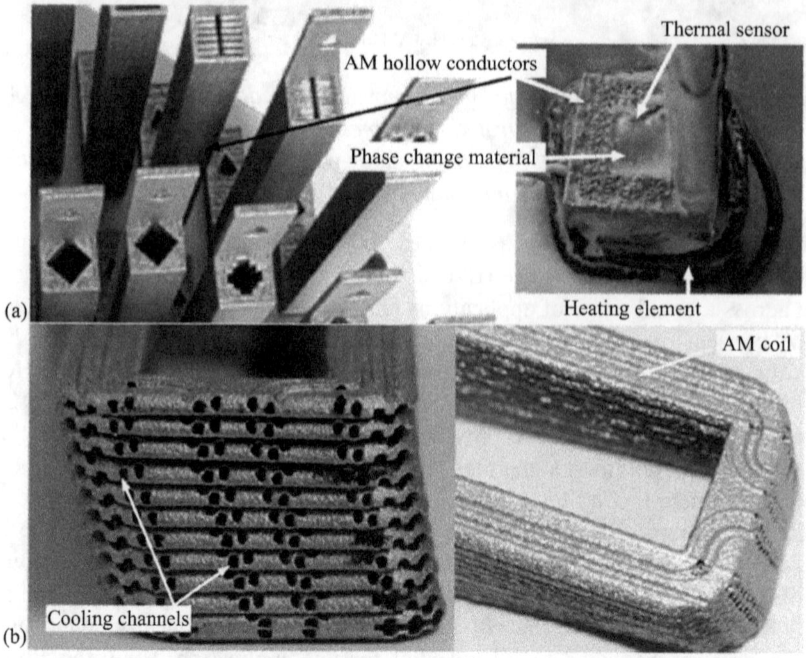

Figure 5.24 Examples of targeted heat removal, (a) a family of AM hollow conductors with phase material [130]; (b) AM coil with integrated channels [131]

winding ac power losses as well as to provide a high heat removal capability [131]. The authors used aluminium alloy (AlSi10Mg) in construction of the coil. Interestingly, the cooling channels are designed to reduce the overall winding power loss generated at ac operation. It was experimentally demonstrated that the proposed approach allows for direct heat extraction form the winding, with conductor current density up to 100 A/mm^2 (WEG liquid coolant with 1.9 l/min at 30°C). This is well beyond the existing figure of merit for the windings with direct liquid cooling. Again, it is important to mention that a more conventional technology like directly cooled hollow conductors also offers a considerable improvement [132].

5.4.3 Cryogenic cooling

As described in Section 6.2, cryogenic cooling usually requires the machine operating in a cryogenic liquid at low temperature to activate remarkable cooling function for the active parts, resulting in a significantly improved torque output, reduced machine size, lower losses and hence higher power density and efficiency, compared to the machines with conventional cooling designs [133–135]. Cryogenic cooling is commonly associated with superconducting machines. However, it is important to note that such a cooling approach is not exclusive to superconductive windings. Some attempts were made to use cryogenic-based thermal management for machines with conventional windings.

Examples of typical cryogenic cooling designs in electrical machines are presented in Figure 5.25. Figure 5.25(a) shows an example of an SRM, which was cryogenically cooled using liquid nitrogen [138]. The authors managed to increase the winding current density significantly from 11 A/mm^2 for conventional cooling to 100 A/mm^2 for a cryogenically cooled machine, with the 50% duty-cycle of the excitation. Note that such an increase of current density results from the reduced temperature of the winding material, i.e., reduced electrical resistivity, that is orders of magnitude lower (a factor of 100 or better) than at room temperature. Also, to further enhance heat removal from the winding body, the split-end coils/winding was employed.

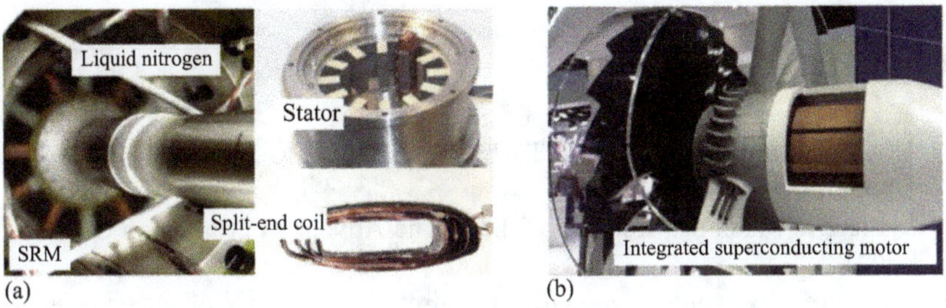

Figure 5.25 Examples of typical cryogenic cooling machines: (a) SRM demonstrator with conventional Cu winding – NASA [136]; (b) integrated superconducting motor for aerospace propulsion – ASuMED project [137]

The superconducting machines for propulsion/on-board power generation have been undergoing dynamic advances [139]. Figure 5.25(b) presents an example of the latest development, an integrated superconducting machine for aerospace applications. This ambitious motor development is funded by the EU (advanced superconducting motor experimental demonstrator (ASuMED) project) [137]. Some of the motor performance targets include 99% efficiency, 1 MW output power, 6,000 rpm rated speed, 20 kW/kg power density, with windings constructed using second-generation (2G) high-temperature superconductor (HTS) tape (dc field magnets and ac windings). Some of the more detailed metrics include magnetic and electric loadings larger than 2.5 *T* and 450 kA/m, respectively. All these are well beyond what is achievable when using conventional machine technology and in line with the existing technology forecasts [140,141].

Clearly the selected examples presented in this section do not encompass all possible routes for improvement associated with thermal management of IMDs. Other cutting-edge cooling technologies such as microchannel cooling [142,143], thermoelectric cooling [144,145], spray cooling [146–148] and synthetic jet cooling [149,150] also present outstanding prospects to be applied for improving cooling performance of IMDs. There is also a large variety of activities devoted to development of new materials, e.g., multifunctional composites, nanofluids and electrical insulation systems and others, which might allow for further improvements [151–153]. In principle, the array of available techniques for a focused heat removal is relatively well understood and includes the fundamental heat and mass transfer mechanisms. However, when considering the 'design for application', aspects of a practical implementation make the design choices less obvious, where frequently conflicting design targets may exist. When comparing the available performance measures e.g., the winding conductor current density, it is evident that significant improvements are achievable even when employing more conventional manufacturing process and materials, but with close integration between the heat source and heat the sink in mind. One should remember that thermal management of IMDs is one of the key aspects of a much broader, multidisciplinary design process, where balancing of the generated power losses (heat) and rate of heat removal is always an important aspect of the design considerations.

References

[1] "Accelerating Ambition – Technology Strategy 2019," UK Aerospace Technology Institute, 2019.

[2] "Electric Machines Roadmap 2020," The Advanced Propulsion Centre UK, 2021.

[3] A. Ford, "Brushless generators for aircraft—a review of current developments," *Proceedings of the IEE-Part A: Power Engineering*, vol. 109, no. 47, pp. 437–452, 1962.

[4] "Electric Machines – Industry Challenges 2020 – 2035+," Advanced Propulsion Centre UK, 2021.

[5] "Service Reference Book Alternator (AC Generator) Service," Chrysler Corporation, 1960.

[6] T. Isakeit. "A Teardown Peek Inside Oral B's Seriously Connected iO Toothbrush." https://www.ifixit.com/News/43726/a-teardown-peek-inside-oral-bs-seriously-connected-io-toothbrush.

[7] R. B. GmbH. "Bosch eBike Drive Units." https://www.bosch-ebike.com/en/products/drive-unit.

[8] T. M. Jahns and H. Dai, "The past, present, and future of power electronics integration technology in motor drives," *CPSS Transactions on Power Electronics and Applications*, vol. 2, no. 3, pp. 197–216, 2017.

[9] R. Abebe, G. Vakil, G. L. Calzo, *et al.*, "Integrated motor drives: state of the art and future trends," *IET Electric Power Applications*, vol. 10, no. 8, pp. 757–771, 2016.

[10] T. M. Jahns and B. Sarlioglu, "The incredible shrinking motor drive: accelerating the transition to integrated motor drives," *IEEE Power Electronics Magazine*, vol. 7, no. 3, pp. 18–27, 2020.

[11] R. Wrobel, "A technology overview of thermal management of integrated motor drives–electrical machines," *Thermal Science and Engineering Progress*, vol. 29, p. 101222, 2022.

[12] Y. C. Chong, D. Staton, Y. Gai, H. Adam, and M. Popescu, "Review of advanced cooling systems of modern electric machines for eMobility application," in *2021 IEEE Workshop on Electrical Machines Design, Control and Diagnosis (WEMDCD)*, 2021, New York, NY: IEEE, pp. 149–154.

[13] M. Popescu, D. A. Staton, A. Boglietti, A. Cavagnino, D. Hawkins, and J. Goss, "Modern heat extraction systems for power traction machines—a review," *IEEE Transactions on Industry Applications*, vol. 52, no. 3, pp. 2167–2175, 2016.

[14] Y. Yang, B. Bilgin, M. Kasprzak, *et al.*, "Thermal management of electric machines," *IET Electrical Systems in Transportation*, vol. 7, no. 2, pp. 104–116, 2017.

[15] S. Nategh, *Thermal Analysis and Management of High-Performance Electrical Machines*, KTH Royal Institute of Technology, 2013.

[16] G. Moreno, S. Narumanchi, K. Bennion, S. Waye, and D. DeVoto, "Gaining traction: thermal management and reliability of automotive electric traction-drive systems," *IEEE Electrification Magazine*, vol. 2, no. 2, pp. 42–49, 2014.

[17] A. Tikadar, D. Johnston, N. Kumar, Y. Joshi, and S. Kumar, "Comparison of electro-thermal performance of advanced cooling techniques for electric vehicle motors," *Applied Thermal Engineering*, vol. 183, p. 116182, 2021.

[18] P.-O. Gronwald and T. A. Kern, "Traction motor cooling systems: a literature review and comparative study," *IEEE Transactions on Transportation Electrification*, vol. 7, no. 4, pp. 2892–2913, 2021.

[19] K. Akatsu and S. Tanimoto, "Air-cooled multi-phase dual-winding in-wheel motor integrated with ultra small SiC module," in *2020 IEEE Energy*

Conversion Congress and Exposition (ECCE), 2020, New York, NY: IEEE, pp. 41–46.
[20] P. Electric, ProteanDrive®-Pd18-Product-Data-Sheet. Available: https://www.proteanelectric.com/f/2020/02/ProteanDrive®-Pd18-Product-Data-Sheet-2020.pdf
[21] D. Kostic Perovic, "Making the impossible, possible–overcoming the design challenges of in wheel motors," *World Electric Vehicle Journal*, vol. 5, no. 2, pp. 514–519, 2012.
[22] C. J. Ifedi, B. Mecrow, J. Widmer, G. Atkinson, S. Brockway, and D. Kostic-Perovic, "A high torque density, direct drive in-wheel motor for electric vehicles," *6th IET International Conference on Power Electronics, Machines and Drives (PEMD 2012)*, p. 24–24, 2012.
[23] J. Doerr, G. Fröhlich, A. Stroh, and M. Baur, "The electric drivetrain with three-motor layout of the Audi E-tron S," *MTZ Worldwide*, vol. 81, no. 7, pp. 16–25, 2020.
[24] J. Doerr, N. Ardey, G. Mendl, G. Fröhlich, R. Straßer, and T. Laudenbach, "The new full electric drivetrain of the Audi e-tron," in *Der Antrieb von morgen 2019*, New York, NY: Springer, 2019, pp. 13–37.
[25] Audi e-tron Electric Motors & Setup. Available: https://electrichasgoneaudi.net/models/e-tron/drivetrain/motor/
[26] H3X. "Integrated Motor Drives for Electric Aircraft." https://www.h3x.tech/.
[27] R. Wrobel and B. Mecrow, "A comprehensive review of additive manufacturing in construction of electrical machines," *IEEE Transactions on Energy Conversion*, vol. 35, no. 2, pp. 1054–1064, 2020.
[28] P. S. Ghahfarokhi, A. Podgornovs, A. Kallaste, *et al.*, "Opportunities and challenges of utilizing additive manufacturing approaches in thermal management of electrical machines," *IEEE Access*, vol. 9, pp. 36368–36381, 2021.
[29] Siemens radial inwards arrangement of the motor and modular inverter.
[30] E. National Academies of Sciences and Medicine, *Commercial Aircraft Propulsion and Energy Systems Research: Reducing Global Carbon Emissions*, London: National Academies Press, 2016.
[31] G. V. Brown, "Cryogenic electric motor tested," in *Research and Technology* 2003, 2004.
[32] K. S. Haran, S. Kalsi, T. Arndt, *et al.*, "High power density superconducting rotating machines—development status and technology roadmap," *Superconductor Science and Technology*, vol. 30, no. 12, p. 123002, 2017.
[33] T. Shinzato, S. Arakawa, H. Oyama, H. Saka, and T. Hayasaki, "Development of high-temperature superconducting motor for automobiles," *SEI Technical Review*, vol. 75, pp. 62–65, 2012.
[34] T. J. Haugan, "Development of superconducting and cryogenic power systems and impact for aircraft propulsion," in *Presentation, Energy Materials and Applications, Orlando, FL*, vol. 25, 2013.

[35] M. Armstrong, "Superconducting turboelectric distributed aircraft propulsion," in *Paper presented Cryogenic Engineering Conference/International Cryogenic Materials Conference*, 2015.

[36] F. Grilli, *et al.*, "Superconducting motors for aircraft propulsion: the Advanced Superconducting Motor Experimental Demonstrator project," *Journal of Physics: Conference Series*, vol. 1590, no. 1, p. 012051, 2020.

[37] E. Pardo, F. Grilli, Y. Liu, S. Wolftädler, and T. Reis, "AC loss modeling in superconducting coils and motors with parallel tapes as conductor," *IEEE Transactions on Applied Superconductivity*, vol. 29, no. 5, pp. 1–5, 2019.

[38] E. Pardo, F. Grilli, and T. Reis, "AC loss modeling of the stator of a 1 MW REBCO superconducting motor for aviation," in *7th International Workshop on Numerical Modelling of High Temperature Superconductors*, 2021.

[39] "ASuMED – Deliverable System Topology Report," 2017.

[40] P. Kshirsagar, Superconducting Motor and Cryo-Cooled Inverter Engine: SOARING. Available: https://www.arpa-e.energy.gov/sites/default/files/2021-03/08_Parag_ARPA-E_ASCEND_SOARING_kickoff_Jan28_public.pdf

[41] C. Xiao, G. Chen, and W. G. Odendaal, "Overview of power loss measurement techniques in power electronics systems," *IEEE Transactions on Industry Applications*, vol. 43, no. 3, pp. 657–664, 2007.

[42] L. Zhang, X. Yuan, X. Wu, C. Shi, J. Zhang, and Y. Zhang, "Performance evaluation of high-power SiC MOSFET modules in comparison to Si IGBT modules," *IEEE Transactions on Power Electronics*, vol. 34, no. 2, pp. 1181–1196, 2019, doi: 10.1109/tpel.2018.2834345.

[43] P. G. Neudeck, R. S. Okojie, and C. Liang-Yu, "High-temperature electronics – a role for wide bandgap semiconductors?," *Proceedings of the IEEE*, vol. 90, no. 6, pp. 1065–1076, 2002, doi: 10.1109/jproc.2002.1021571.

[44] A. Hassan, Y. Savaria, and M. Sawan, "GaN integration technology, an ideal candidate for high-temperature applications: a review," *IEEE Access*, vol. 6, pp. 78790–78802, 2018, doi: 10.1109/access.2018.2885285.

[45] W. Lee, S. Li, D. Han, B. Sarlioglu, T. A. Minav, and M. Pietola, "A review of integrated motor drive and wide-bandgap power electronics for high-performance electro-hydrostatic actuators," *IEEE Transactions on Transportation Electrification*, vol. 4, no. 3, pp. 684–693, 2018, doi: 10.1109/tte.2018.2853994.

[46] M. Parvez, A. T. Pereira, N. Ertugrul, N. H. E. Weste, D. Abbott, and S. F. Al-Sarawi, "Wide bandgap DC–DC converter topologies for power applications," *Proceedings of the IEEE*, vol. 109, no. 7, pp. 1253–1275, 2021, doi: 10.1109/jproc.2021.3072170.

[47] R. Khazaka, L. Mendizabal, D. Henry, and R. Hanna, "Survey of high-temperature reliability of power electronics packaging components," *IEEE Transactions on Power Electronics*, vol. 30, no. 5, pp. 2456–2464, 2015, doi: 10.1109/tpel.2014.2357836.

[48] T. Funaki, J. C. Balda, J. Junghans, *et al.*, "Power conversion with SiC devices at extremely high ambient temperatures," *IEEE Transactions on*

Power Electronics, vol. 22, no. 4, pp. 1321–1329, 2007, doi: 10.1109/tpel.2007.900561.

[49] F. Hou, W. Wang, L. Cao, *et al.*, "Review of packaging schemes for power module," *IEEE Journal of Emerging and Selected Topics in Power Electronics*, vol. 8, no. 1, pp. 223–238, 2020, doi: 10.1109/jestpe.2019.2947645.

[50] R. W. Johnson, C. Wang, Y. Liu, and J. D. Scofield, "Power device packaging technologies for extreme environments," *IEEE Transactions on Electronics Packaging Manufacturing*, vol. 30, no. 3, pp. 182–193, 2007, doi: 10.1109/tepm.2007.899158.

[51] Y. Yao, Z. Chen, G. Lu, D. Boroyevich, and K. D. T. Ngo, "Characterization of encapsulants for high-voltage high-temperature power electronic packaging," *IEEE Transactions on Components, Packaging and Manufacturing Technology*, vol. 2, no. 4, pp. 539–547, 2012, doi: 10.1109/tcpmt.2011.2173344.

[52] Y. Yao, G. Lu, D. Boroyevich, and K. D. T. Ngo, "Survey of high-temperature polymeric encapsulants for power electronics packaging," *IEEE Transactions on Components, Packaging and Manufacturing Technology*, vol. 5, no. 2, pp. 168–181, 2015, doi: 10.1109/tcpmt.2014.2337300.

[53] K. Klein, O. Rämer, E. Hoene, *et al.*, "Low inductive full ceramic SiC power module for high-temperature automotive applications," in *PCIM Europe Digital Days 2020; International Exhibition and Conference for Power Electronics, Intelligent Motion, Renewable Energy and Energy Management*, 7–8 July 2020, pp. 1–8.

[54] C. M. DiMarino, B. Mouawad, C. M. Johnson, D. Boroyevich, and R. Burgos, "10-kV SiC MOSFET power module with reduced common-mode noise and electric field," *IEEE Transactions on Power Electronics*, vol. 35, no. 6, pp. 6050–6060, 2020, doi: 10.1109/tpel.2019.2952633.

[55] P. A. Agyakwa, L. Yang, M. R. Corfield, and C. M. Johnson, "A non-destructive study of crack development during thermal cycling of Al wire bonds using x-ray computed tomography," in *CIPS 2014; 8th International Conference on Integrated Power Electronics Systems*, 25–27 Feb. 2014, pp. 1–5.

[56] C. J. D. and M. H. A., *Extreme Environment Electronics*, Boca Roton, FL: CRC Press, 2012.

[57] C. H. Cheng, H. L. Hsiao, S. I. Chu, Y. Y. Shieh, C. Y. Sun, and C. Peng, "Low cost silver alloy wire bonding with excellent reliability performance," in *2013 IEEE 63rd Electronic Components and Technology Conference*, 28–31 May 2013, pp. 1569–1573, doi: 10.1109/ectc.2013.6575781.

[58] F. Hille, R. Roth, C. Schäffer, *et al.*, "Reliability aspects of copper metallization and interconnect technology for power devices," *Microelectronics Reliability*, vol. 64, pp. 393–402, 2016.

[59] M. Knoerr, S. Kraft, and A. Schletz, "Reliability assessment of sintered nano-silver die attachment for power semiconductors," *Presented at the 12th Electronics Packaging Technology Conference*, 2010.

[60] C. Buttay, A. Masson, and J. Li, "Die attach of power devices using silver sintering bonding process optimization and characterization," *Presented at the The IMAPS High Temperature Electronics Network*, 2011.

[61] V. R. Manikam and K. Y. Cheong, "Die attach materials for high temperature applications: a review," *IEEE Transactions on Components, Packaging and Manufacturing Technology*, vol. 1, no. 4, pp. 457–478, 2011, doi: 10.1109/tcpmt.2010.2100432.

[62] Y. Huang, Y. Luo, F. Xiao, and B. Liu, "Failure mechanism of die-attach solder joints in IGBT modules under pulse high-current power cycling," *IEEE Journal of Emerging and Selected Topics in Power Electronics*, vol. 7, no. 1, pp. 99–107, 2019, doi: 10.1109/jestpe.2018.2871474.

[63] A. Haque, Y. S. Won, A. S. M. A. Haseeb, and H. H. Masjuki, "Investigations on Zn-Al-Ge alloys as high temperature die attach material," Presented at the 3rd Electronics System Integration Technology Conference ESTC, 2010.

[64] L. Guo-Quan, J. N. Calata, Z. Zhiye, and J. G. Bai, "A lead-free, low-temperature sintering die-attach technique for high-performance and high-temperature packaging," in *Proceedings of the Sixth IEEE CPMT Conference on High Density Microsystem Design and Packaging and Component Failure Analysis (HDP '04)*, 3–3 July 2004, pp. 42–46, doi: 10.1109/hpd.2004.1346671.

[65] H. Miyoshi, K. Endoh, and S. Kurita, "Application of silver nano particle to pressureless bonding onto a copper surface – consideration of substitute material for lead solder," *Presented at the 8th International Conference on Integrated Power Electronics Systems*, 2014.

[66] Z. Deng, G. Zou, Q. Jia, *et al.*, "Effect of Ag sintered bondline thickness on high-temperature reliability of SiC power devices," *IEEE Transactions on Components, Packaging and Manufacturing Technology*, vol. 11, no. 11, pp. 1889–1895, 2021, doi: 10.1109/tcpmt.2021.3110997.

[67] M. Locatelli, R. Khazaka, S. Diaham, *et al.*, "Evaluation of encapsulation materials for high-temperature power device packaging," *IEEE Transactions on Power Electronics*, vol. 29, no. 5, pp. 2281–2288, 2014, doi: 10.1109/tpel.2013.2279997.

[68] S. Tanimoto and K. Matsui, "High junction temperature and low parasitic inductance power module technology for compact power conversion systems," *IEEE Transactions on Electron Devices*, vol. 62, no. 2, pp. 258–269, 2015, doi: 10.1109/ted.2014.2359978.

[69] B. Boettge, F. Naumann, S. Behrendt, *et al.*, "Material characterization of advanced cement-based encapsulation systems for efficient power electronics with increased power density," in *2018 IEEE 68th Electronic Components and Technology Conference (ECTC)*, 29 May–1 June 2018, pp. 1258–1269, doi: 10.1109/ectc.2018.00194.

[70] C. P. Wong, J. M. Segelken, and J. W. Balde, "Understanding the use of silicone gels for nonhermetic plastic packaging," *IEEE Transactions on*

Components, Hybrids, and Manufacturing Technology, vol. 12, no. 4, pp. 421–425, 1989, doi: 10.1109/33.48998.

[71] A. Lindemann and G. Strauch, "Properties of direct aluminum bonded substrates for power semiconductor components," *IEEE Transactions on Power Electronics*, vol. 22, no. 2, pp. 384–391, 2007, doi: 10.1109/tpel.2006.889898.

[72] G. Emre, O. Burak, and C. Shajjad, "Design, analysis, comparison and experimental validation of insulated metal substrates for high power wide-bandgap power modules," *Journal of Electronic Packaging Journal of Electronic Packaging*, vol. 142, no. 4, p. 041107, 2020, doi: 10.1115/1.4047409.

[73] K. Wang, Z. Qi, F. Li, L. Wang, and Xu Yang, "Review of state-of-the-art integration technologies in power electronic systems," *CPSS Transactions on Power Electronics and Applications*, vol. 2, no. 4, pp. 292–305, 2017.

[74] H. Miyazaki, Y. Zhou, S. Iwakiri, *et al.*, "Improved resistance to thermal fatigue of active metal brazing substrates for silicon carbide power modules using tough silicon nitrides with high thermal conductivity," *Ceramics International*, vol. 44, no. 8, pp. 8870–8876, 2018.

[75] C. M. Johnson, A. Castellazzi, R. Skuriat, P. Evans, J. Li, and P. Agyakwa, "Integrated high power modules," Presented at the 7th International Conference on Integrated Power Electronics Systems, 2012.

[76] S.S. Kang, "Advanced cooling for power electronics," Presented at the 7th International Conference on Integrated Power Electronics Systems, 2012.

[77] J. Schulz-Harder, "Review on highly integrated solutions for power electronic devices," Presented at the 5th International Conference on Integrated Power Electronics Systems, 2008.

[78] D. Christen, M. Stojadinovic, and J. Biela, "Energy efficient heat sink design: natural versus forced convection cooling," *IEEE Transactions on Power Electronics*, vol. 32, no. 11, pp. 8693–8704, 2017, doi: 10.1109/tpel.2016.2640454.

[79] A. Castelan, B. Cougo, J. Brandelero, D. Flumian, and T. Meynard, "Optimization of forced-air cooling system for accurate design of power converters," in *2015 IEEE 24th International Symposium on Industrial Electronics (ISIE)*, 3–5 June 2015, pp. 367–372, doi: 10.1109/isie.2015.7281496.

[80] R. Skuriat and C. M. Johnson, "Thermal performance of baseplate and direct substrate cooled power modules," in *2008 4th IET Conference on Power Electronics, Machines and Drives*, 2–4 April 2008, pp. 548–552, doi: 10.1049/cp:20080581.

[81] Y.P. Zhang, X.L. Yu, Q.K. Feng and R.T. Zhang, "Thermal performance study of integrated cold plate with power module," *Applied Thermal Engineering*, vol. 29, no. 17–18, pp. 3568–3573, 2009.

[82] O. Hohlfeld and A. Herbrandt, "Direct cooled modules – integrated heat sinks," *Presented at the 7th International Conference on Integrated Power Electronics Systems*, 2012.

[83] R. Skuriat and C. M. Johnson, "Direct substrate cooling of power electronics," in *2009 13th European Conference on Power Electronics and Applications*, 8–10 Sept. 2009, pp. 1–10.

[84] R. Whitt, D. Huitink, A. Emon, A. Deshpande, and F. Luo, "Thermal and electrical performance in high-voltage power modules with nonmetallic additively manufactured impingement coolers," *IEEE Transactions on Power Electronics*, vol. 36, no. 3, pp. 3192–3199, 2021, doi: 10.1109/tpel.2020.3015226.

[85] S. Jones-Jackson, R. Rodriguez, and A. Emadi, "Jet impingement cooling in power electronics for electrified automotive transportation: current status and future trends," *IEEE Transactions on Power Electronics*, vol. 36, no. 9, pp. 10420–10435, 2021, doi: 10.1109/tpel.2021.3059558.

[86] K. Olesen, R. Bredtmann, and R. Eisele, "'ShowerPower' new cooling concept for automotive applications," *Presented at the Automotive Power Electronics*, 2006.

[87] S. Chowdhury, E. Gurpinar, G. J. Su, T. Raminosoa, T. A. Burress, and B. Ozpineci, "Enabling technologies for compact integrated electric drives for automotive traction applications," in *2019 IEEE Transportation Electrification Conference and Expo (ITEC)*, 19–21 June 2019, pp. 1–8, doi: 10.1109/itec.2019.8790594.

[88] S. Pickering, P. Wheeler, F. Thovex, and K. Bradley, "Thermal design of an integrated motor drive," in *IECON 2006 – 32nd Annual Conference on IEEE Industrial Electronics*, 6–10 Nov. 2006, pp. 4794–4799, doi: 10.1109/iecon.2006.348109.

[89] A. Robert, *End Winding Mount Integrated Motor Drive*, UK: University of Nottingham, 2018.

[90] R. Abebe, G. Vakil, G. L. Calzo, T. Cox, C. Gerada, and M. Johnson, "FEA based thermal analysis of various topologies for Integrated Motor Drives (IMD)," in *IECON 2015 – 41st Annual Conference of the IEEE Industrial Electronics Society*, 9–12 Nov. 2015, pp. 001976–001981, doi: 10.1109/iecon.2015.7392390.

[91] L. d. Lillo, B. Ahmadi, L. Empringham, M. Johnson, J. Espina, and R. Abebe, "Next generation integrated drive, NGID: a novel approach to thermal and electrical integration of high power density drives in automotive applications," in *2018 IEEE Energy Conversion Congress and Exposition (ECCE)*, 23–27 Sept. 2018, pp. 1228–1232, doi: 10.1109/ecce.2018.8558001.

[92] A. Tenconi, F. Profumo, S. E. Bauer, and M. D. Hennen, "Temperatures evaluation in an integrated motor drive for traction applications," *IEEE Transactions on Industrial Electronics*, vol. 55, no. 10, pp. 3619–3626, 2008, doi: 10.1109/tie.2008.2003099.

[93] F. Farina, D. Rossi, A. Tenconi, F. Profumo, and S. E. Bauer, "Thermal design of integrated motor drives for traction applications," in *2005 European Conference on Power Electronics and Applications*, 11–14 Sept. 2005, p. 10, doi: 10.1109/epe.2005.219711.

[94] R. A. Torres, H. Dai, T. M. Jahns, B. Sarlioglu, and W. Lee, "Thermal analysis of housing-cooled integrated motor drives," in *2021 IEEE Transportation Electrification Conference & Expo (ITEC)*, 21–25 June 2021, pp. 1–6, doi: 10.1109/itec51675.2021.9490174.

[95] A. H. R. Mohamed, H. Vansompel, and P. Sergeant, "Power density boosting techniques for reconfigurable integrated modular motor drives," *IEEE Transactions on Energy Conversion*, vol. 37, pp. 1–1, 2022, doi: 10.1109/tec.2022.3146780.

[96] U. Choi, S. Jørgensen, and F. Blaabjerg, "Impact of cooling system capacity on lifetime of power module in adjustable speed drives," *IEEE Journal of Emerging and Selected Topics in Power Electronics*, vol. 7, no. 3, pp. 1768–1776, 2019, doi: 10.1109/jestpe.2019.2912387.

[97] E. Laloya, O. Lucia, H. Sarnago, and J. M. Burdío, "Heat management in power converters: from state of the art to future ultrahigh efficiency systems," *IEEE Transactions on Power Electronics*, vol. 31, no. 11, pp. 7896–7908, 2016, doi: 10.1109/tpel.2015.2513433.

[98] R. Abebe, G. Vakil, G. Lo Calzo, *et al.*, "Integrated motor drives: state of the art and future trends," *IET Electric Power Applications*, vol. 10, no. 8, pp. 757–771, 2016.

[99] M. Gerber and M. Marz, "System integration in automotive power systems," in *2005 European Conference on Power Electronics and Applications*, 2005, New York, NY: IEEE, pp. 10 p. 10.

[100] X. Shen, B. Mecrow, X. Deng, C. Donaghy-Spargo, R. Whalley, and N. Chakraborty, "Direct air cooling of high-power permanent magnet machines," in *2019 IEEE Energy Conversion Congress and Exposition (ECCE)*, 2019, New York, NY: IEEE, pp. 5637–5644.

[101] Q. Li, X. Shen, B. Mecrow, and X. Deng, "Hybrid cooling design of permanent magnet synchronous machines," in *The 10th International Conference on Power Electronics, Machines and Drives* (*PEMD 2020*), 2021.

[102] J. Mao, G. Chen, and Z. Ren, "Thermoelectric cooling materials," *Nature Materials*, vol. 20, no. 4, pp. 454–461, 2021.

[103] M. Harada, D. Morioka, and M. Ochi, "Thermal and mechanical properties of tetra-functional mesogenic type epoxy resin cured with aromatic amine," *Journal of Applied Polymer Science*, vol. 135, no. 16, p. 46181, 2018.

[104] Y. Yao, X. Zeng, K. Guo, R. Sun, and J.-b. Xu, "The effect of interfacial state on the thermal conductivity of functionalized Al2O3 filled glass fibers reinforced polymer composites," *Composites Part A: Applied Science and Manufacturing*, vol. 69, pp. 49–55, 2015.

[105] T. Zhang, J. Sun, L. Ren, *et al.*, "Nacre-inspired polymer composites with high thermal conductivity and enhanced mechanical strength," *Composites Part A: Applied Science and Manufacturing*, vol. 121, pp. 92–99, 2019.

[106] A. Zrodnikov, A. Efanov, Y. I. Orlov, P. Martynov, V. Troyanov, and A. Rusanov, "Heavy liquid metal coolant–lead–bismuth and lead–technology," *Atomic Energy*, vol. 97, no. 2, pp. 534–537, 2004.

[107] Y. Deng and J. Liu, "Design of practical liquid metal cooling device for heat dissipation of high performance CPUs," *Journal of Electronic Packaging*, vol. 132, p. 031009, no. 3, 2010.

[108] N. H. March, *Liquid Metals: Concepts and Theory*, Cambridge: Cambridge University Press, 2005.

[109] T. E. Faber, *Introduction to the Theory of Liquid Metals*, Cambridge: Cambridge University Press, 2010.

[110] H. Ge, H. Li, S. Mei, and J. Liu, "Low melting point liquid metal as a new class of phase change material: an emerging frontier in energy area," *Renewable and Sustainable Energy Reviews*, vol. 21, pp. 331–346, 2013.

[111] Y. Gao and J. Liu, "Gallium-based thermal interface material with high compliance and wettability," *Applied Physics A*, vol. 107, no. 3, pp. 701–708, 2012.

[112] W. Van der Lugt and W. Geertsma, "Liquid alloys with strong chemical interactions," *Journal of non-crystalline solids*, vol. 61, pp. 187–200, 1984.

[113] T. Daeneke, K. Khoshmanesh, N. Mahmood, *et al.*, "Liquid metals: fundamentals and applications in chemistry," *Chemical Society Reviews*, vol. 47, no. 11, pp. 4073–4111, 2018.

[114] X.-H. Yang and J. Liu, "Liquid metal enabled combinatorial heat transfer science: toward unconventional extreme cooling," *Frontiers in Energy*, vol. 12, no. 2, pp. 259–275, 2018.

[115] H. Tang, Y. Tang, Z. Wan, *et al.*, "Review of applications and developments of ultra-thin micro heat pipes for electronic cooling," *Applied energy*, vol. 223, pp. 383–400, 2018.

[116] R. Wrobel and R. J. McGlen, "Heat pipes in thermal management of electrical machines – a review," *Thermal Science and Engineering Progress*, vol. 26, p. 101053, 2021.

[117] P. Klemens and R. Williams, "Thermal conductivity of metals and alloys," *International Metals Reviews*, vol. 31, no. 1, pp. 197–215, 1986.

[118] H. Aoki, N. Shioya, M. Ikeda, and Y. Kimura, "Development of ultra thin plate-type heat pipe with less than 1 mm thickness," in *2010 26th Annual IEEE Semiconductor Thermal Measurement and Management Symposium (SEMI-THERM)*, 2010, New York, NY: IEEE, pp. 217–222.

[119] S. Xu, R. J. Lewis, L.-A. Liew, Y.-C. Lee, and R. Yang, "Development of ultra-thin thermal ground planes by using stainless-steel mesh as wicking structure," *Journal of Microelectromechanical Systems*, vol. 25, no. 5, pp. 842–844, 2016.

[120] D. Lee and C. Byon, "Fabrication and characterization of pure-metal-based submillimeter-thick flexible flat heat pipe with innovative wick structures," *International Journal of Heat and Mass Transfer*, vol. 122, pp. 306–314, 2018.

[121] Z. Chen, Y. Li, W. Zhou, L. Deng, and Y. Yan, "Design, fabrication and thermal performance of a novel ultra-thin vapour chamber for cooling electronic devices," *Energy Conversion and Management*, vol. 187, pp. 221–231, 2019.

[122] G. Zhou, J. Li, and L. Lv, "An ultra-thin miniature loop heat pipe cooler for mobile electronics," *Applied Thermal Engineering*, vol. 109, pp. 514–523, 2016.

[123] S. Hong, S. Wang, and Z. Zhang, "Multiple orientations research on heat transfer performances of ultra-thin loop heat pipes with different evaporator structures," *International Journal of Heat and Mass Transfer*, vol. 98, pp. 415–425, 2016.

[124] P. Szymanski, R. Law, R. J. McGlen, and D. A. Reay, "Recent advances in loop heat pipes with flat evaporator," *Entropy*, vol. 23, no. 11, p. 1374, 2021.

[125] F. Wu, A. M. EL-Refaie, and A. Al-Qarni, "Additively manufactured hollow conductors integrated with heat pipes: design tradeoffs and hardware demonstration," *IEEE Transactions on Industry Applications*, vol. 57, no. 4, pp. 3632–3642, 2021.

[126] R. Wrobel, B. Scholes, A. Hussein, R. Law, A. Mustaffar, and D. Reay, "A metal additively manufactured (MAM) heat exchanger for electric motor thermal control on a high-altitude solar aircraft – experimental characterisation," *Thermal Science and Engineering Progress*, vol. 19, p. 100629, 2020.

[127] R. Wrobel, B. Scholes, A. Mustaffer, S. Ullah, D. Reay and B. Mecrow, "Design and experimental characterisation of an additively manufactured heat exchanger for the electric propulsion unit of a high-altitude solar aircraft," in *2019 IEEE Energy Conversion Congress and Exposition (ECCE)*, 2019, New York, NY: IEEE, pp. 753–760.

[128] F. Wu and A. M. EL-Refaie, "Additively manufactured hollow conductors with integrated cooling for high specific power electrical machines," in *2020 International Conference on Electrical Machines (ICEM)*, 2020, New York, NY: IEEE, vol. 1, pp. 1497–1503.

[129] R. Wrobel and R. McGlen, "Opportunities and challenges of employing heat-pipes in thermal management of electrical machines," in *2020 International Conference on Electrical Machines (ICEM)*, 2020, New York, NY: IEEE, vol. 1, pp. 961–967.

[130] S. Ayat, B. Dagusé, and R. Khazaka, "Design considerations of windings formed with hollow conductors cooled with phase change material," in *2019 IEEE Energy Conversion Congress and Exposition (ECCE)*, 2019, New York, NY: IEEE, pp. 5652–5658.

[131] C. Wohlers, P. Juris, S. Kabelac, and B. Ponick, "Design and direct liquid cooling of tooth-coil windings," *Electrical Engineering*, vol. 100, no. 4, pp. 2299–2308, 2018.

[132] S. Haller, J. Persson, P. Cheng, and K. Bertilsson, "Multi-phase winding with in-conductor direct cooling capability for a 48V traction drive design," in *2020 International Conference on Electrical Machines (ICEM)*, 2020, New York, NY: IEEE, vol. 1, pp. 2118–2124.

[133] H. M. Kim, K. W. Lee, J. H. Park, and G. S. Park, "Design of cryogenic induction motor submerged in liquefied natural gas," *IEEE Transactions on Magnetics*, vol. 54, no. 3, pp. 1–4, 2017.

[134] A. Chengliu, H. Yuanfeng, and W. Haifeng, "Main losses study of cryogenic induction motor for submerged liquid natural gas pump," in *2015 18th International Conference on Electrical Machines and Systems (ICEMS)*, 2015, New York, NY: IEEE, pp. 133–136.

[135] M. Biasion, F. J. Fernandes, P. J. da Costa Branco, S. Vaschetto, A. Cavagnino, and A. Tenconi, "A comparison of cryogenic-cooled and superconducting electrical machines," in *2021 IEEE Energy Conversion Congress and Exposition (ECCE)*, 2021, New York, NY: IEEE, pp. 4045–4052.

[136] G. V. Brown, A. F. Kascak, B. Ebihara, *et al.*, *NASA Glenn Research Center Program in High Power Density Motors for Aeropropulsion, Cleveland: National Aeronautics and Space Administration Cleveland*, 2005.

[137] A. Perez, R. Van Der Woude, and R. Dekker, "Rotor cooling concept for the ASuMED superconductive motor," in *IOP Conference Series: Materials Science and Engineering*, vol. 502, no. 1, p. 012139, 2019.

[138] A. Minichiello, A. Glezer, J. G. Hartley, and W. Z. Black, "Thermal management of sealed electronic enclosures using synthetic jet technology," in *Advances in Electronic Packaging, 1997: Proceedings of the Pacific Rim/ASME International Intersociety Electronic & Photonic Packaging Conference*, vol. 19, no. 2, p. 1809, 1997.

[139] R. Wrobel, "Thermal management of electrical machines for propulsion – challenges and future trends," *Archives of Electrical Engineering*, pp. 175–187, 2022.

[140] U. Automotive Council, The Roadmap Report – Towards 2040: A Guide to Automotive Propulsion Technologies, ed: Jun, 2018.

[141] R. Del Rosario, "A future with hybrid electric propulsion systems: a NASA perspective," in *Turbine Engine Technology Symposium*, 2014, no. GRC-E-DAA-TN17600.

[142] J. F. Tullius, R. Vajtai, and Y. Bayazitoglu, "A review of cooling in microchannels," *Heat Transfer Engineering*, vol. 32, nos. 7 and 8, pp. 527–541, 2011.

[143] P. Smakulski and S. Pietrowicz, "A review of the capabilities of high heat flux removal by porous materials, microchannels and spray cooling techniques," *Applied Thermal Engineering*, vol. 104, pp. 636–646, 2016.

[144] D. Zhao and G. Tan, "A review of thermoelectric cooling: materials, modeling and applications," *Applied Thermal Engineering*, vol. 66, nos. 1 and 2, pp. 15–24, 2014.

[145] Y. Cai, Y. Wang, D. Liu, and F.-Y. Zhao, "Thermoelectric cooling technology applied in the field of electronic devices: updated review on the parametric investigations and model developments," *Applied Thermal Engineering*, vol. 148, pp. 238–255, 2019.

[146] G. Liang and I. Mudawar, "Review of spray cooling – Part 1: single-phase and nucleate boiling regimes, and critical heat flux," *International Journal of Heat and Mass Transfer*, vol. 115, pp. 1174–1205, 2017.

[147] J.-X. Wang, W. Guo, K. Xiong, and S.-N. Wang, "Review of aerospace-oriented spray cooling technology," *Progress in Aerospace Sciences*, vol. 116, p. 100635, 2020.

[148] G. Liang and I. Mudawar, "Review of spray cooling – Part 2: high temperature boiling regimes and quenching applications," *International Journal of Heat and Mass Transfer*, vol. 115, pp. 1206–1222, 2017.

[149] A. Arshad, M. Jabbal, and Y. Yan, "Synthetic jet actuators for heat transfer enhancement – a critical review," *International Journal of Heat and Mass Transfer*, vol. 146, p. 118815, 2020.

[150] G. Krishan, K. C. Aw, and R. N. Sharma, "Synthetic jet impingement heat transfer enhancement – a review," *Applied Thermal Engineering*, vol. 149, pp. 1305–1323, 2019.

[151] K. Xiong, Y. Li, Y.-Z. Li, J.-X. Wang, and Y. Mao, "Numerical investigation on the thermal performance of nanofluid-based cooling system for synchronous generators," *Entropy*, vol. 21, no. 4, p. 420, 2019.

[152] M. C. Kulan, S. Şahin, and N. J. Baker, "An overview of modern thermo-conductive materials for heat extraction in electrical machines," *IEEE Access*, vol. 8, pp. 212114–212129, 2020.

[153] F. Wu and A. M. EL-Refaie, "Toward additively manufactured electrical machines: opportunities and challenges," *IEEE Transactions on Industry Applications*, vol. 56, no. 2, pp. 1306–1320, 2019.

Chapter 6
Passive devices

Richard Gibson[1] *and Min Zhang*[2]

The role of passive components in a motor drive system is often overlooked at the early stages of the design and in many academic research papers. The size, mass, cost and often temperature rating and lifetime expectation of the drive system may be dominated by the passive components rather than the semiconductor components and control topology. Passive components play two important roles in an integrated motor drive where the exact selection of the components depends on the power system design and the application:

(a) *Energy storage*
(b) *Filter components*

To understand the requirements for energy storage, we can work backwards from the motor and inverter towards DC bus voltage. Several design topologies and control methods have been presented in Chapters 3 and 4 which require a steady DC voltage which is then chopped by the active switching of the semiconductor devices to provide the required output currents and voltages to the motor windings. The quality and stability of the power supply to a motor drive must be designed and managed within the control parameters of the inverter control algorithms, in particular, the maximum and minimum voltage and the frequency content of any superimposed ripple voltages. The first step is to understand the energy supply and system characteristics. The following power systems configurations are considered in this chapter:

(a) *Battery systems*
(b) *DC power network*
(c) *AC mains supply*

Traditional motor design assumes the motor will be supplied with sinusoidal voltages and currents which are adjusted in magnitude and phase to control to motor speed and torque. To obtain efficient variable frequency, power semiconductor transistors are switched in a pattern which applies a pulsed voltage to the

[1]Nidec Control Techniques, UK
[2]Mach One Design EMC Consultancy, UK

motor winding. Due to the inherent inductance in the motor winding, the currents cannot change instantly but will rise and fall according to the L/R time constant of the winding. The fundamental switching frequency often referred to as the PWM frequency is chosen to minimise this current ripple and optimise the heat losses in the inverter and motor. For motors with power less than 500 kW, the typical range of switching frequencies for silicon-based transistors is 2–16 kHz. As the motors get larger, the switching frequency may be reduced to minimise the associated switching losses in the semiconductors. The inertia of the rotor and the driven load provide damping so these lower switching frequencies are rarely observed in the mechanical system, however many problems such as electromagnetic interference (EMI) may be observed throughout the electrical system. Passive components can be selected to contain these high-frequency noises within the integrated drive and avoid interference to neighbouring equipment.

The close positioning of the drive, motor and cooling systems offers an opportunity to integrate passive components into this drive system to optimise efficiency and space saving. The challenge arises when the properties of the passive components are compromised by their surrounding components and environment. The latest technology in passive components is presented in this chapter with a discussion on how these can be integrated within an integrated drive and motor system including sharing inductor magnetic circuits with that of the machine, through to minimisation and customisation of capacitive filters.

It is possible to reduce the current ripple in the DC link by altering the motor control algorithm. The need to reduce the ripple current depends on the operating mission profile, required lifetime, and motor control stability. DC bus ripple currents depend on the machine topology where switched reluctance machines (SRMs) particularly with single pulse operation and low phase numbers for high efficiency SRM [1] required the most ripple as they must charge the magnetic circuit, provide torque producing current, then discharge the magnetic circuit before repeating for the subsequent poles. These are important aspects for the integrated design and the passive component discussions in this chapter can be adapted to suit these topologies.

6.1 Energy storage needs

The energy supplied to the motor must match the needs of the load and application. This included constant current for constant torque, constant voltage for constant speed and bidirectional power flow to match the acceleration or deceleration of the load. The control algorithms can compensate for small fluctuations in each of these conditions by sensing parameters within the drive altering the motor operating characteristics, however larger deviations require passive components to store energy local to the motor. Three distinct stored energy requirements can be considered based on their time duration:

(a) PWM switching frequency (10 ms^{-1} ms)
(b) Switching transients of the inverter (5–150 ns)
(c) Peak power demands (a few ms to a few seconds)

6.1.1 PWM switching frequency

As the motor control algorithm transitions between the various output vectors, the power demanded from the DC bus is discontinuous. The demand can vary by inverter topology but for simplicity a 2-level inverter is considered here. The diagram shown in Figure 6.1 illustrates the phase current and capacitor current for a PWM switching period where approximately constant current is flowing into the motor U phase returning equally on V and W phases.

In period A, the output voltage vector is zero and no current is drawn from the DC bus. The load current is continuous determined by the previous currents and the inductance of the motor windings. In period B, the DC bus voltage is applied and current increases at a rate determined by the inductance of the loop which includes U phase winding, V and W windings in parallel, the series resistance of the power semiconductors, and the ESR and ESL of the DC bus capacitor. In period C, the output vector changes the voltage on the V phase from 0 V to the DC bus voltage. Instantaneously, the current in the V phase is constant but must now flow through the diode, which gives a slight increase in voltage due to the forward voltage and resistance, into the positive DC bus. As the V phase current has remained the same as period B (which was approximately half of U phase current), the current into U can be supplied, from both the returning V phase current and the DC link. This has the effect of halving the DC link current. In period D, the three output voltages are held at the positive DC bus voltages therefore no power is drawn from the DC bus and the motor currents circulate via the diodes and their interconnecting nodes. The DC bus current for this period is shown as I_{DC} where it is discontinuous.

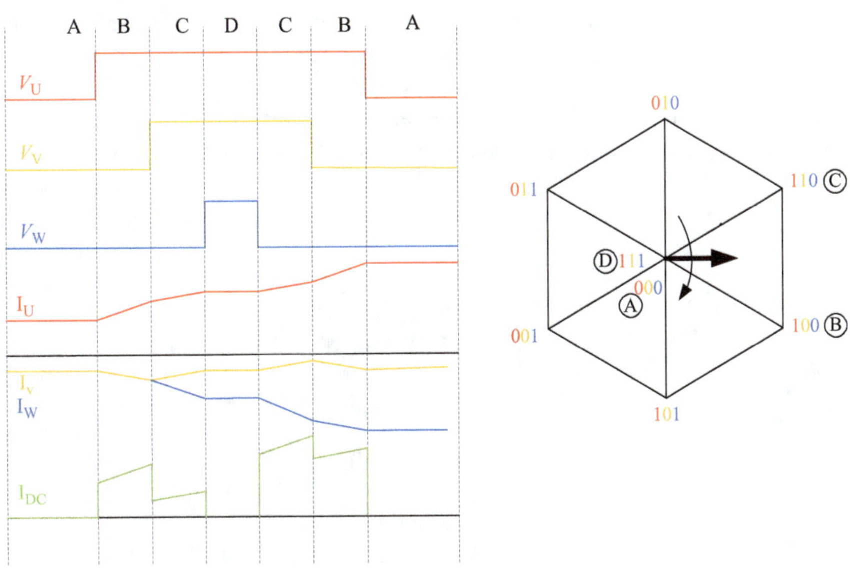

Figure 6.1 PWM period and DC capacitor currents

There are two important features to observe in this waveform. The first is that the fundamental frequency of the current drawn from the capacitors can range from the switching frequency up to twice the switching frequency depending on the load conditions and this will contain many harmonics. These high current pulses are delivered with high power as they are supplied via the full DC bus voltage regardless of the drive output rms voltage. The designed capacitance must deliver this energy and maintain the DC link voltage within its control limits. The recharge time of the capacitors depends on the supply design where for uncontrolled rectification the capacitors must deliver many pulses of motor current before being recharged. If the capacitance is too small, then the dc bus voltage would drop too far, leading to stability issue with the motor control algorithm.

6.1.2 Semiconductor switching transient

The capacitor currents illustrated in Figure 6.1 show an instantaneous rise and fall between the desired levels. In reality, the physical construction of the drive and DC link results in a current loop with several hundreds of picohenrys of inductance rising to several nanohenries for higher voltage systems due to the insulation spacing requirements and the geometry of the DC link capacitors. The current transition between on and off is controlled by the switching time of the transistor and is typically 5–15 ns for wide band gap (WBG) devices or up to 150 ns for Si devices. This rapid di/dt induces voltage drops across the parasitic inductance of this loop which extends the voltage stress on the semiconductors. It is common for a nominal 800 V DC link voltage to contain short duration pulses up to 1,200 V. The most severe condition occurs during the control of a Type 1 short circuit of the semiconductor where the current will rise to several orders of magnitudes higher than normal levels and, unless special controls are in place, will be turned off in the same short duration resulting in very large *di*/*dt*.

To minimize this effective loop area, decoupling capacitors should be fitted as close as possible to the switching devices and across the DC link as illustrated in Figure 6.2. These decoupling capacitors should have low equivalent series

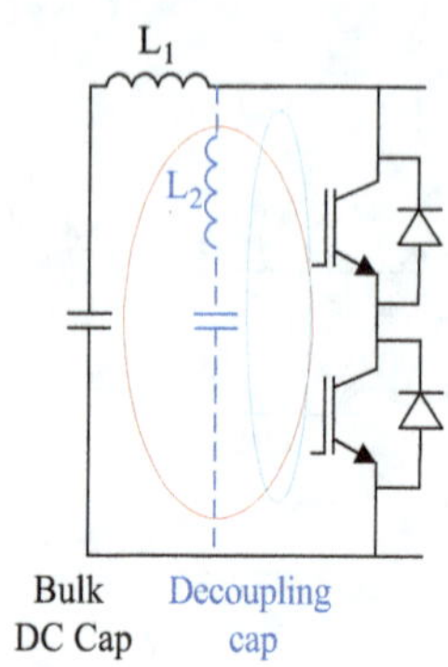

Figure 6.2 Loop inductance

resistance (ESR) and equivalent series inductance (ESL) and should be physically small. The small size of the components can allow these to be positioned close to the required phase leg, reducing the effective parasitic loop inductance ($L_2 << L_1$). In normal operation, voltage across this capacitor will be equal to the DC bus voltage. The capacitance should be large enough to store the energy in the parasitic loop inductance even with DC bias voltage. Increasing the capacitance value of the decoupling capacitors will reduce the voltage stress on the semiconductor devices and often give reduced EMI. However, increasing the capacitance also means increased physical size of the decoupling capacitors, which results in higher loop inductance. It is up to the designer to optimise these parameters for each specific design. Having smaller ESL also means the capacitors could resonate, however, resonance with the bulk DC capacitance is unlikely because the higher ESR of the bulk DC link capacitors can damp the resonance very effectively, however this should be validated in the design.

The very short duration and localization of this voltage overshoot when the device switches off will not have any influence on the motor control but is a source of high-frequency content responsible for EMI in the conducted and radiated frequency range. When the devices switch on, the reduced parasitic inductance may marginally increase switching losses but further reduce EMI.

6.1.3 Peak power demands

The application using an integrated motor drive determines the required mission profile. Motors are often sized to provide for a short duration current overload which is determined by the thermal capacity of the motor materials before any critical temperatures are reached. This is often during the rapid accelerating or decelerating of the load to the required speed but can also happen if additional load stress is placed on the motor at a constant speed for example additional mass suddenly placed onto a conveyor system. It is very important to differentiate between these two conditions with regards to the passive devices and the power supply needs.

In the first instance of an accelerating load, the motor requires a large torque. The torque is proportional to the motor currents and the speed is proportional to the applied voltage. In this scenario, the power drawn from the supply can be less than the rated power as most of the acceleration is from a low speed towards rated speed. Acceleration performance is limited by the torque and hence current rating of the semiconductor devices and is not determined by the passive components. It is only when operating close to rated speed with high overload current that the drive will see an overload in power. It is up to the designer of the motor control to determine if this additional power should be allowed under these operating conditions.

Similarly with the conveyor system, operating at the required speed with a sudden step change in load will draw additional power if the speed is to be maintained. The increase in power will have the immediate effect of increasing the voltage ripple on the DC link capacitance. The impact on the passives will depend on the power system design where low-frequency discontinuous supplies such as rectified AC mains supplies will have large periods where no power is supplied, resulting in the passive devices delivering all the power requirements. Therefore, it is the maximum power condition rather than a motor maximum torque condition which is important for the passive devices.

6.2 Passive component technology

The semiconductor transistors, control methods and topologies for a motor drive are a continued source of academic research to reduce component cost and increase the efficiencies. Surrounding these devices are a range of passive components used to store and deliver energy and to divert and filter unwanted high-frequency signals. They consist of the basic building blocks that engineers will study when first approaching electronic engineering: capacitors, inductors, and resistors. The selection and use of these components is fundamental to the operation of the motor drive and there are a wide variety of properties to consider to fully optimise the design. In addition, both capacitance and inductance are present as unwanted parasitic elements which depend on the materials used and the three-dimensional geometry of the layout.

This section will present and discuss the passive components in their various types and form factors and identifies pros and cons for their use in an integrated drive.

6.2.1 Capacitors

In its most basic form, a capacitor consists of two conductors separated by an insulator or dielectric material. A vast array of different capacitor technologies has been created to suit different applications. This section will show the differences in the available capacitors and compare the important parameters to observe during design of an integrated drive.

Capacitance is given by $C = 8.854 \times 10^{-7}\, \varepsilon A/d$ (μF)
where:

- ε is the dielectric constant
- A is the surface area (cm^2)
- d is the thickness of dielectric (mm)

To maximise the capacitance, the surface area and the dielectric constant must be large and the thickness of the dielectric small [2]. When the electric field is applied across the 2 parallel plates, the material in between experiences the electric field. Electrons do not flow in this material as it is an insulator and tightly bonded in the atomic structure, instead, the electrons in atoms and compounds shift their position at little aligning with the electric field. The ability of the material to align is determined by its dielectric constant or sometimes called *k*-value. Unfortunately, the dielectric constant is not a constant in practical capacitors and will vary depending on temperature, applied voltage bias and frequency. To minimise the size of the integrated drive, the designer will need to maximise the surface area of the parallel plates, minimise the thickness of the dielectric (but not too thin or the break down voltage will reduce) and select a stable dielectric material for the operating environment. There are four major types of capacitors used in power electronics and their key properties are given below.

6.2.1.1 Electrolytic capacitor

The electrolytic capacitor consists of a thin metal aluminium foil anode which is etched on both sides to substantially increase the surface area by a factor of 30–100 which increase the capacitance density. An aluminium oxide layer is formed through electrolytic oxidation on this surface which provides the dielectric. The etching process and oxide growth are energy intensive which dominate the cost of the capacitor. The thickness of the oxide must be controlled to give the required voltage rating of the capacitor, this is achieved by applying a DC forming voltage where a higher forming voltage will form a thicker oxide. The cathode does not contain an intentional oxide layer causing the capacitor to be polarised. To avoid incorrect placement during drive manufacturing, it is common to have three legs to provide an asymmetry preventing incorrect assembly.

The paper electrolyte between the aluminium anode and cathode foils cannot provide close contact to the etched pattern of the oxide and therefore cannot provide a high capacitance on its own. A liquid electrolyte is needed which is conductive, highly permeable, chemically stable, suitable for the required operating temperature and is used to fill the void between the oxide and the cathode (see Figure 6.3). Due to the ionic structure of the electrolyte, it will evaporate and separate over time with an applied electric field resulting in failure of the capacitor. This results in a build-up of hydrogen gas which must be vented off to avoid excess pressure which could result in an explosion. This gives one of the major problems associated with electrolytic capacitors – limited lifetime. The component manufacturers

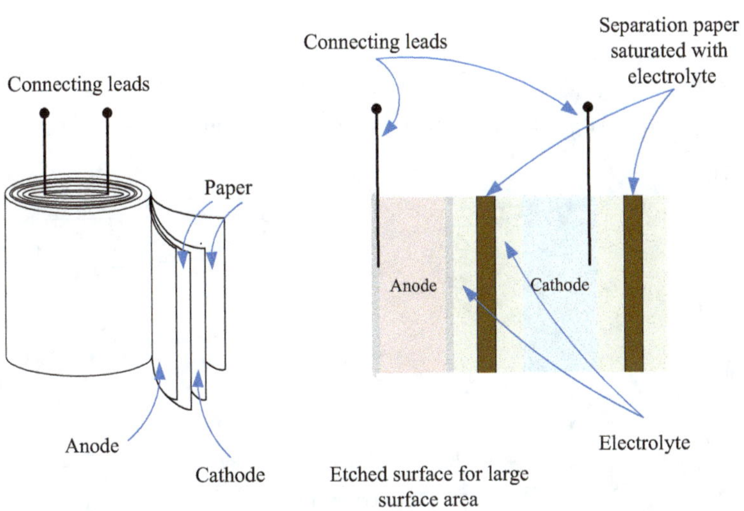

Figure 6.3 Construction of an electrolytic capacitor

offer slightly different equations to calculate the lifetime of the capacitor and many offer online web tools and circuit simulators to estimate the expected lifetime. In each case, the lifetime will reduce exponentially with increasing voltage and internal temperature. With rising temperature, the dielectric strength of the oxide is lowered [3] and chemical reactions between the electrolyte and the oxide take place, therefore hot spot temperatures must stay below the datasheet spot temperature regardless of lifetime requirement.

Power losses and hence heating, occur in the capacitor due to the resistance of the electrodes and the leakage current through the dielectric. This is represented as an equivalent series resistance (ESR) in most datasheets. Electrolytic capacitors suffer from a high ESR which raises the internal temperature due to the dissipated power from the ripple current, bringing the electrolyte towards its maximum rating and limiting the capacitor current ripple capability. Due to the relatively low current ripple rating, additional capacitors are needed in parallel to meet the lifetime requirements for a converter. When comparing component cost, the cost per amp of ripple current rating, is a valid metric [4]. Aluminium electrolytic capacitors are the most common capacitor used within a separated drive and motor where the ambient operating temperature of the capacitors is carefully controlled.

ESR reduces with temperature for modern caps therefore it is important to both simulate the mission profile to determine the ripple current and factor in the expected capacitor operating temperature to optimise the component selection. The change in ESR is shown for a typical 400 V, 440 μF capacitor normalized to 100 Hz ripple current. Figure 6.4 shows that the ESR only reduces at the lower frequencies which is important for losses associated with rectified mains, however the ESR increases with temperature at higher frequencies (10 kHz shown) associated with switching frequencies.

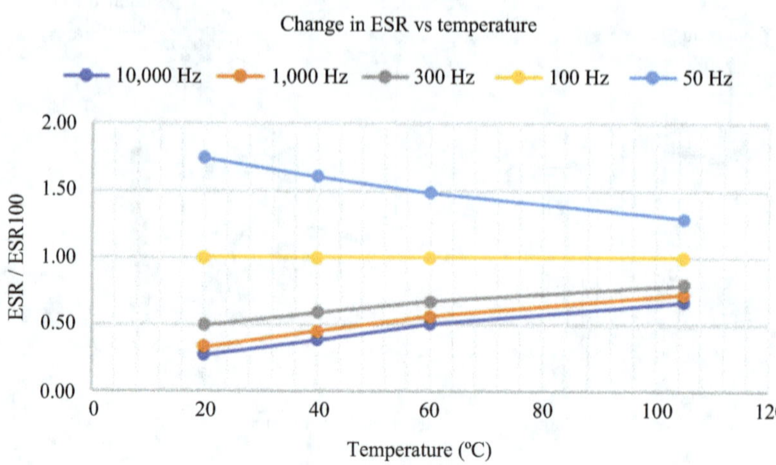

Figure 6.4 Change in ESR versus temperature

The electrolytic capacitors are formed from long strips of the etched metal and paper separator which are then rolled into cylinders and placed in a protective can (see Figure 6.3). The placement and connection of the electrodes has a significant influence on the equivalent series inductance (ESL). As electrons enter the capacitor, they may have a considerable distance to travel to along the length of the foil, to utilise the available capacitor area. At ripple frequencies greater than 10 kHz, this large distance and geometry of the electrode connections leads the ESL to dominate the impedance causing the capacitor to appear inductive.

The voltage rating is determined by the dielectric strength and its thickness. The voltage rating of commercial parts is limited to approximately 450 V so these must be connected in series to achieve the voltage rating for the designed DC voltage. This has the unwanted effect of dividing the rated capacitance and multiplying the ESR and ESL.

Due to its relatively large ESR, electrolytics are sometimes used to damp the resonance in a system where the best practice is to place them in front of ceramics. The aluminium housing also provides some field shielding preventing high frequency signals from coupling to other circuits. Where most industrial drives use electrolytic capacitors due to their very high density, their limited lifetime at moderate temperature operation makes them unsuitable for integrated machines.

The other metal anode commonly used in electrolytic capacitors is tantalum. It is not currently possible to achieve the same increase in surface area compared to aluminium therefore the available range of capacitance is lower. When used with a manganese dioxide, these capacitors can operate up to 200°C. When tantalum caps are taken outside their design limits, they can fail catastrophically resulting in fire prohibiting its use in many applications. While the operation within their limits is very stable, they have a limited operating voltage of 100 V which makes them unsuitable for many purposes in an integrated drive. Tantalum is a very rare metal with an estimated 50 years of material available at the current extraction rates. There has been an increased effect to recycle the capacitor to recover the material. The summary of key properties of electrolytic capacitors are shown in Table 6.1.

Table 6.1 Summary of key properties of electrolytic capacitors

Dielectric	**Electrolyte**	**Dielectric constant**	**Dielectric strength (kV/mm)**	**Temperature rating (°C)**	**Capacitance range**
Aluminium oxide	Liquid Ethylene glycol Boric acid dimethylformamide and dimethylacetamide	7–10	16.7	85–105	330 pF–2.2 F
Tantalum oxide	Solid Manganese dioxide	24	4–8	175–200	0.011 pF–100 mF

6.2.1.2 Film capacitors

There are two capacitor constructions used for film capacitors: Film/Foil and metalized film. This gives rise to different properties which make the capacitors suitable for different applications. For the Film/Foil capacitors, the metal foil is stacked either side of the plastic film layer (dielectric) to form the capacitor with an additional layer of film under the stack. This assembly is then rolled to fit the required packaging and with careful rolling, the second layer of film creates a double-sided capacitor increasing the capacitance density. The individual layers of material need to be mechanically robust to withstand the assembly processes which requires a minimum material thickness. A single point failure in the dielectric layer will destroy the capacitor and to avoid this, the dielectric layer must remain relatively thick limiting the capacitance that can be achieved. The think metal layers provide a low resistance therefore can support high current carrying capacity and good surge ratings.

Metalized film capacitors are constructed by vacuum deposition of metal onto plastic film (dielectric) to create thin electrodes on each side with a very good surface contact onto the dielectric. To avoid the failure mechanism found with the film/foil capacitor, patterns are created in the deposited metals to create a large array of individual cells of capacitors linked together with very fine connections. The patterns used slightly reduce the total capacitance for the given available area of the film. However, if the dielectric film has defects or become damaged in use, the fine connections to the cell act as a fuse, clearing the fault from the remaining capacitor. This is often referred to as a self-healing property. By utilizing this property, a small defect rate can be tolerated without significant loss of capacitance. This allows the dielectric layer to be significantly reduced in thickness, increasing the achievable capacitance. The details of the patterns are often patented by the owning companies as this determines key properties of the capacitor. The cell pattern results in a small reduction of the capacitance but allows for high production yields (lower costs) and avoids the catastrophic failure of the device in early use.

As the capacitor is formed entirely on the film, there are many packaging options available to manufacturers. The metalized film is often available to buy giving the option of custom packages to match the geometry and cooling system of the application.

There are several choices of film material available which determine the operating voltage and temperature limits of these capacitors. As the properties of the film can depend on their manufacturing methods, they are often sold under various brand names however the chemical names are used here. The summary of key properties of film capacitors are listed in Table 6.2.

Polypropylene (PP) film is used in both film/foil and metalized film capacitors. During manufacture, the polymers are stretched in two different directions to align the molecular chain and influence the mechanical and electrical properties. This is referred to as biaxially oriented (BOPP). The material has a high dielectric constant and the highest electric strength allowing very thin films capacitors to be created

Table 6.2 Summary of key properties of film capacitors

Dielectric	Dielectric constant (@1 MHz)[1]	Dielectric strength (kV/mm)	Melting temperature (°C)	Typical temperature rating (°C)	Typical capacitance range	Typical capacitor voltage rating
Polypropylene (PP) Biaxial-oriented polypropylene (BOPP)	2.1–2.3	2,300–2,500	165	105	10 pF–4,500 μF	16–3 kV
Polyethylene terephthalate (PET)	3.2–3.5	17	260	125	1 uF–10 uF	16 V–400 V
Polyphenylene sulphide (PPS)	3.0–3.3	11–24	280	125	100 pF–12 μF	16 V–1 kV
Polyethylene naphthalate (PEN)	2.95	400	263	150	470 pF–5.6 uF	16 V–1 kV

with high capacitance and high voltage withstand capabilities. The material melts at a low temperature of 160°C with typical capacitor rating of 105°C. The capacitors are often wrapped in a flame retardant tape and sealed with epoxy resin.

Polyethylene terephthalate (PET) is from the polyester family. It has a higher melting temperature (260°C) and higher dielectric constant than PP however its electric strength limits the voltages that the capacitors can be used. While PET can operate at higher temperatures than polypropylene, it's ESR increases significantly with temperature and frequency which causes additional self-heating. The high coefficient of thermal expansion limits operation at higher temperatures due to mechanical breakdown of the capacitor.

Polyphenylene sulphide PPS capacitors are only available as metalized film caps and have largely replaced polycarbonate which was discontinued by the supplier in the year 2000. They are widely used for their relatively high temperature stability. The capacitors can withstand temperatures of up to 270°C for short periods making them suitable for SMT soldering processes. The materials cost significantly more than PP however they are increasingly offered a long with polyethylene naphthalate (PEN) and acrylic for SMT capacitors.

Polyethylene naphthalate (PEN) is the most promising film material for integrated drives. While the melting point is similar to PET, the material maintains shape and size up to higher temperature (190°C) allowing reliable operating temperatures of 150°C. The electric strength is lower that PP limiting its use for high voltage capacitors. The cost of the material remains high but for an integrated drive, this can be offset by reducing the system cooling requirements.

The position of capacitors in a circuit can be critical for safety. When some capacitors wear out and breakdown they will cause the integrated drive to malfunction, other capacitors, particularly those used as EMC filter components may lead to dangerous conditions in the event of a failure. There are two capacitor classes which are specified in standard IEC 60384-14 which must meet a defined reliability criterion, specifically X class and Y class capacitors. In most cases, these are film capacitors with defined failure modes.

X class capacitors are intended for line-to-line capacitors where they are designed for a high peak voltage capability. The dielectric is an insulation and can act as a functional or basic barrier, however careful analysis is required according to the appropriate application standard to ensure the material thickness is sufficient. These capacitors can be used where a failure of this capacitor will not lead to an electric shock however it will usually result in a fuse or circuit breaker opening.

Y class capacitors have an electrical and mechanical construction which significantly increases their reliability. These are formed by connecting two capacitors in series within the same package. Unfortunately, this has the effect of increasing cost, increasing the volume and halving the effective capacitance for some filter solutions. These capacitors are often connected between line and ground connections where a failure would result in an electric shock if the ground connection was lost. Some application standards require additional ground connections or large conductor sizes to reduce the possibility of opening this connection.

The capacitors are divided into subclasses depending on their peak pulse voltage rating and rated voltage as shown in Table 6.3. When selecting the correct component, it is important to consider that the peak voltages experienced by these capacitors are more likely to occur from the supply rather that the integrated drive operation. The capacitors can offer some protection to the drive in the event of surges however the capacitance value is often very low compared to the DC link capacitance reducing this benefit.

Film capacitors are particularly sensitive to humidity. Moisture in contact with the metal surfaces can cause the metal to corrode reducing the capacitance. In addition to this, the film materials can absorb moisture increasing the capacitor leakage current and loss in the capacitor. The packaging which contains the capacitor often slow the transfer of atmospheric humidity which may be sufficient if the humidity is only present for a short duration. The film thickness during the capacitor manufacturing process thus becomes very important. Reputable capacitor brands use a thicker film, resulting in a larger capacitor than other brands. The

Table 6.3 X and Y capacitor limits (IEC60384-14)

Sub-class	**Insulation**	**Peak pulse voltage**	**Rated voltage**
X1	Basic	2.5 kV $< V_p \leq$ 4.0 kV	
X2	Basic	$V_p \leq$ 2.5 kV	
X3	Basic	$V_p \leq$ 1.2 kV	
Y1	Double or reinforced	8.0 kV	$V_R \leq$ 250 V
Y2	Basic or supplementary	5.0 kV	150 V $< V_R \leq$ 250 V
Y3	Basic or supplementary		150 V $< V_R \leq$ 250 V
Y4	Basic or supplementary	2.5 kV	$V_R \leq$ 150 V

thinner film makes the capacitor less robust to overload (voltage, current, or temperature) and less likely to self-heal as well. For installations with permanent high humidity, additional protection methods need to be used. This is particularly important for SMD capacitors which do not have protective cases [5].

6.2.1.3 Ceramic capacitors

The ceramic dielectric materials are hard materials which are shaped then sintered at very high temperature. The elemental materials in the mixture give the capacitor its insulating and dielectric properties and as such there are a vast variety of materials which are commonly used. The voltage rating depends on the material thickness and construction method [6]. The capacitance is limited by the available surface area and therefore not suited to high power, bulk energy storage in a DC link. The summary of key properties of film capacitors are listed in Table 6.2.

Ceramic capacitors can be grouped into three different categories based on their electrical properties:

(a) Ionic capacitors have a constant permittivity and have a very stable operation across the rated voltage and temperatures. The range of supported voltages are low and the capacitance is limited. These are often used in timing circuits, analogue and resonate applications. IEC60384-1 assigns these as Class 1 capacitors, and they are often identified by a 3 letter system which defines their stability criteria based on the now replaced EIA standards.

(b) Ferroelectric capacitors demonstrate a reduction in permittivity (dielectric constant), hence capacitance with the applied voltage and temperature. The permittivity is higher than the ionic capacitors and therefore higher capacitance values per volume can be achieved. These are low-cost component and are widely used for decoupling capacitors. For many years, these components have had extremely reliable operation with expected lifetimes of >100 years, however, as efforts have been made to reduce the package size, the oxide vacancy migration may reduce expected lifetime to <5 years therefore close attention is required to the manufacturer's datasheet.

(c) Anti-ferroelectric capacitors are relatively new components developed from piezo device technology. The ceramic material structure contains opposing dipoles in the lattice structure which give a voltage dependent permittivity. Several commercially available components are tuned to display peak permittivity at 400–800 V which correspond to the DC bus voltage for an integrated motor. The performance of these devices is extremely non-linear with both temperature and voltage. Care must be taken that they are evaluated across the full range of mission profiles to avoid unwanted consequences. The high permittivity and breakdown voltage enables these components to be offered in compact surface mount packages which can provide a very low effective ESL. These components are suited for snubber components when located close to the semiconductor switches limiting voltage overshoot and ringing.

The high permittivity of the ceramic allows for small package capacitors which are constructed in alternating layers of ceramic and electrode material. The

Table 6.4 Summary of key properties of ceramic capacitors

Typical dielectric	Dielectric constant	Dielectric strength (kV/mm)	Temperature rating (°C)	Typical capacitance range	Voltage range (V)
Titanium dioxide	83 (ionic)	4	85	0.1 pF–47 mF	–
Barium titanate	500–20,000 (ferro electric)	2–10	85	0.01 pF–220 mF	–
PLZT lead lanthanum zirconate titanate	200–1,300 (anti-ferro electric)	270	125 (150 peak)	0.25–10 mF	500–900

electrodes are terminated at opposite ends of the component and then capped for PCB soldering. To support higher voltages, the components are elongated to increase the distance between the solder pads. The ceramic material and interconnections are brittle and while robust in a fixed position, care is required to protect the components during manufacture from mechanical damage. Flexing of PCB applies significant stress to these components and must be avoided to allow reliable operation. Alternative soft electrode materials or longer package leads are available which allow a small tolerance in movement however it is best practice to prevent movement from occurring. The stray electric fields are not contained within the capacitors and if two are used in series, they should be positioned at 90° to each other to avoid unwanted coupling however this can be difficult to achieve in practice.

6.2.1.4 Ultracapacitors/supercapacitors

Ultracapacitors or supercapacitors offer the highest capacitive energy density. Components are available with hundreds of Farads, and it is worth understanding their structure to determine their suitability for integrated motor drives. Unlike previous capacitors the supercapacitors do not have a solid dielectric layer, instead an ionic depletion layer is formed in the electrolyte at the surface of both electrodes (as shown in Figure 6.5). This layer is very thin, often between 0.3 and 0.8 nm and supports the electric field forming the capacitor. The electrodes are made from porous carbon which has a very high surface area onto which the electrolyte can make contact. The combination of these properties is responsible for the very large capacitances that can be achieved. However, the depletion layer cannot withstand a high voltage before the electrolyte begins to break down. The maximum voltage that can reliably be supported across this structure is 3.3 V, too low for most motor drive applications. The voltage rating of a component can be increased using the series connection of many individual capacitors. Even though this configuration reduces the capacitance, the effective capacitance can still be very high compared to other technologies.

As the individual capacitors fail if their voltage is exceeded, additional measures must be used when connected in series to ensure the voltages are balanced

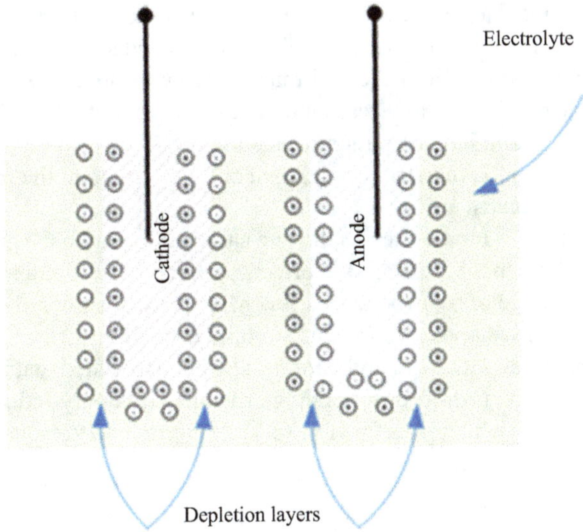

Figure 6.5 Depletion layer formation in a supercapacitor

along the string. This can be achieved using passive components however this will result in a very high leakage current. The resulting power loss reduces the efficiency of these components and increases their heat dissipation.

While these capacitors have very high power density, the low voltage rating, low temperature rating and significant leakage current makes them an unsuitable technology for use within an integrated motor drive. However, due to the rapid charge and discharge capabilities, they are often used within the power system. The capacitors can store energy recovered during braking cycles at a rate where battery technology cannot, they can also be used for backup power systems where the relatively compact size provides dense energy storage.

6.2.2 *Inductors*

Inductors store energy in a magnetic field around a conductor. The presence of this magnetic field presents a back e.m.f. which opposes the change in current which induced it. The characteristics of the magnetic field depend upon the material which surround the inductor and many materials are available which are useful for integrated motor drives.

Where the integrated motor drives are supplied from a.c. mains sources, inductors establish a magnetic field preventing the current from increasing and reducing rapidly. This has the effect of elongating the current pulse, causing a phase shift to the supply voltage, and reducing the peak and therefore R.M.S. current. The phase shift introduced, enables the switching rectifier topologies to control currents and therefore the power flow into the d.c. bus. At these design frequencies, either mains or rectifier switching frequency, it is essential that the losses within the inductor are low. There are significant costs/trade-offs to achieve a compact size with low losses. A simple coil suspended in air is an effective inductor however, the length of the coil must be large to provide sufficient inductance. Copper wire used for low resistance is expensive and while the wire has a low resistance, the long lengths required increase the losses. To reduce the size of this coil, core material is inserted around the coil to concentrate the magnetic field and increase the inductance.

A basic first-order model such as the one shown in Figure 6.6 cannot represent more complex phenomena such as skin effect, proximity effect and saturation. But in the frequency range of a few kHz to a few MHz, which is often the spectrum of interest for motor drive applications, this simplified model is often sufficient. The important point here is that an inductor is always associated with its parasitic capacitance due to the turn-to-turn winding structure. It is only inductive until its self-resonance point (which is defined in (6.1)). After the self-resonance point, an inductor behaves more like a capacitor.

$$f_{res} = 1/(2\pi\sqrt{LC}) \tag{6.1}$$

where L is the inductance value and C is the capacitance value.

In Figure 6.6, L1 is the inductance value of the inductor, C1 represents the turn-to-turn winding capacitance, R1 is the inductor winding resistance, which is associated with copper loss (i^2R loss), R2 is what we call leakage resistance.

As the frequency increases, not only does the parasitic capacitance become dominant, there are also skin effect and proximity effect which are associated with eddy current. Most of the time, from the filter design perspective, both skin effect and proximity effect increase the lossy component in an inductor and reduce the leakage inductance, so that the filtering impact is affected.

The biggest factor that could significantly reduce the effectiveness of an inductor is the saturation of the magnetic core due to excessive current in the winding that is higher than the rated current. When an inductor is saturated, it loses much of its inductance value and its filtering capability is significantly compromised. Another factor that often leads to an ineffective inductor is near field coupling due to its surrounding electromagnetic environment.

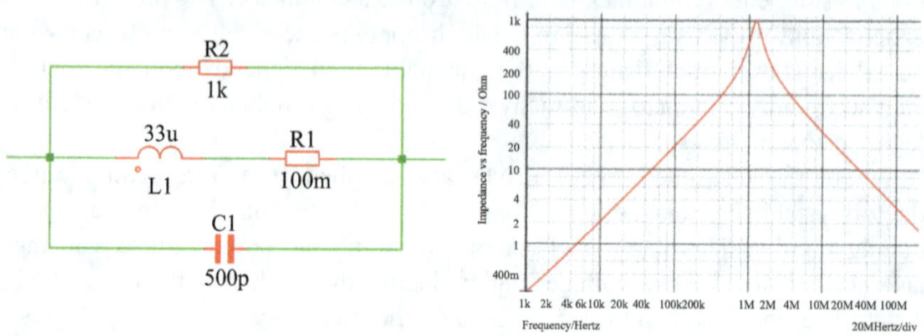

Figure 6.6 Basic model of an inductor and its impedance curve

The shape of the core can be toroidal, E-shaped or many other custom shapes to meet mechanical design needs. The winding can be a single strand conductor, multi-strand (rope-type) winding, or even a Litz wire where the choice depends on the frequency of operation and specific operating characteristics.

For low frequencies up to a few kHz electrical steels which contain a low carbon content and silicon (up to 4%) are used to provide a low loss core. There are many design properties which must be considered to select the correct grade and grain orientation of this material where the suppliers of the material will offer advice. Since integrated motor drives must move to higher switching frequencies to reduce the size and weight, electrical steels are rarely the correct choice.

For motor drive application applications, where EMI issues often start in the few kHz range, the manganese–zinc and nanocrystalline cores are first choice for inductors/common mode chokes, as shown in Figure 6.7.

The relationship between the inductance and the number of turns is defined in (6.2). As can be seen, the inductance is proportional to the n^2:

$$L = n^2 A k \mu_0 \tag{6.2}$$

where n is the number of turns of a winding, A is the cross-sectional area, k relates to the geometry of the coil of an inductor and μ_0 is the permeability of free space.

However, the inductance of an inductor does not have the proportional squared relationship with the number of turns. As the number of turns increases, so does the turn-to-turn capacitance of the winding. This will shift the resonant frequency to a lower frequency point, meaning the capacitance part of an inductor starts to dominate. This is the main reason that an inductor may fails to attenuate the noise.

Skin effect, eddy currents and proximity effect are all related to frequency. As frequency increases, the RF current tends to travel on the very outer thin layer of a conductor, hence the name 'skin effect'. Engineers should be aware that these

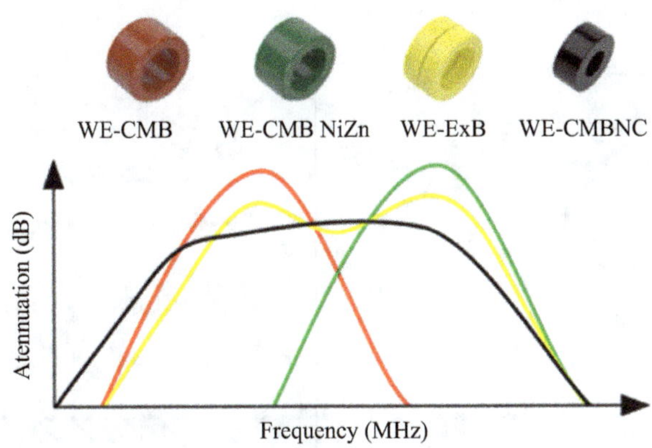

Figure 6.7 Attenuation vs frequency for common core materials. Courtesy of Würth Elektronik (https://www.we-online.com/).

effects not only significantly increase the loss of an inductor, but they also have an impact on the EMC performance.

For instance, eddy currents in the magnetic core increase core loss, which results in a temperature rise. The magnetic core tends to saturate with less current when the temperature increases. Eddy current and the proximity effect between adjacent layers of the winding can also lead to a reduction of leakage inductance.

Since inductors store a magnetic field in space, rather than in the winding or the core, any components nearby could then potentially couple with the field in the same space. A shielded or semi-shielded inductor is designed to contain the magnetic field, thus reducing the potential coupling. In other words, a shielded inductor has much less leakage inductance compared with a non-shielded inductor.

6.2.3 Cable

The traditional cable which connects the drive to the motor represents a significant proportion of the cost of installation of a discrete motor drive system. The selection, of this cable including the length and construction have a significant impact on the operation of both the motor and the drive. It is worth examining properties of the cable to illustrate technical advantages of the integrated motor drive and illustrate where additional stresses can be placed on the motor windings without this cable.

The typical cable used in industry consists of three copper phase conductors, and a protective earth (or ground) conductor each with an individual layer of insulation. These are then surrounded by a further layer of insulation, a shield and a final layer of insulation (see Figure 6.8). The cost is dominated by the copper content where the cross-sectional area is often determined by National Standards with the intention of fire prevention (such as UL758, IEC60364-5-52, IEC61800-5-1) to limit the external temperatures of the cable due to their self-heating from resistance

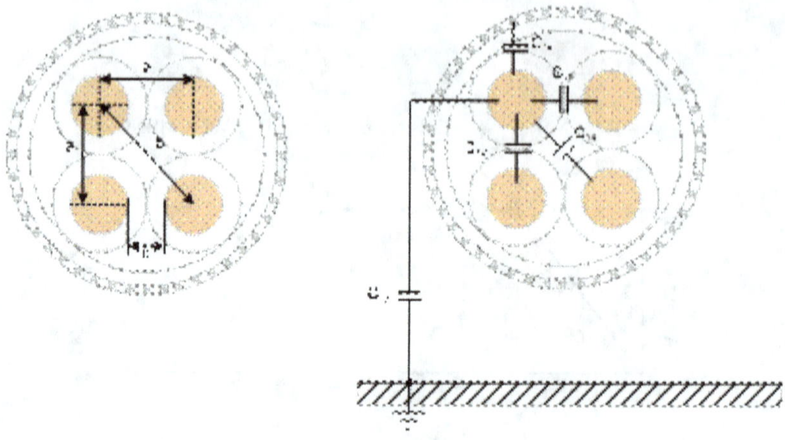

Figure 6.8 Parasitic elements in a typical motor cable

losses. The double insulation, required to keep personnel safe who may come in contact with the cable, prevents the heat from dissipating effectively.

For many industrial installations, the cable is at risk of damage due to fatigue where the motor may be mounted on moving equipment or from damage such a forklift trucks colliding with them. In these circumstances where the cable may have a short circuit, it is undesirable for the cable fault to cause damage to the drive. The drive must detect this short circuit and turn off the power transistors often within 5 μs to prevent catastrophic failure. This requirement imposes a cost penalty and increases conduction losses in the power semiconductor components which required a guaranteed short circuit withstand time.

To comply with electromagnetic compatibility (EMC) emission standards, a shielded cable shield is essential between the drive and motor. The PWM voltage pattern provided from the drive to control the motor current through its inductance contains a broadband spectrum of unwanted frequencies. The cable screen if properly terminated contains most of this noise and prevents it from radiating to the environment. However, this cable introduced an additional unwanted passive component, capacitance due to the proximity of the shield to the phase conductors is a capacitor. This capacitance can be several nano-farads per meter and must be charged and discharged with each voltage transition. This charging current and can be observed in the transient switching of voltages and currents, reducing the rise time but adding up to 40% additional switching loss in the power semiconductor.

The motor cables must be considered as a transmission line due to the high frequency content within the switching edges. With silicon devices, where rise times are in the order 100s of nanoseconds, several meters of cables are required to highlight the problem of impedance mismatch. This can result in three times the DC bus voltage being applied to the motor terminals for short periods of time and damaging the insulation as illustrated in Figure 6.9 which shows the switching

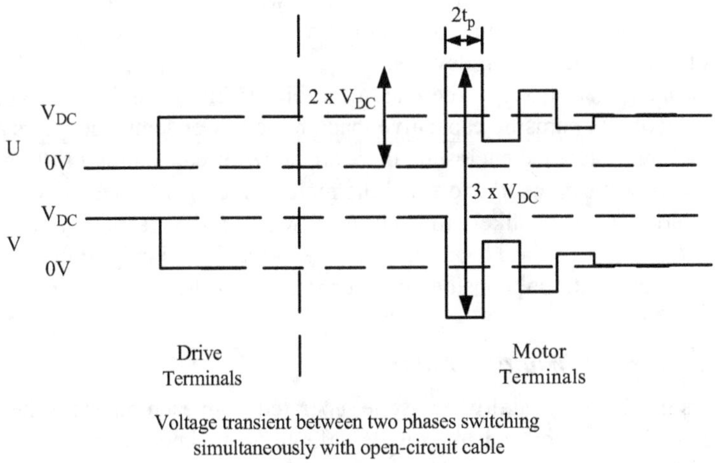

Figure 6.9 Illustration of voltage overshoot on motor terminals

transition of two phases. As the cable length become very long (hundreds of metres), the dispersion in the cable can limit the amplitude of the voltage overshoot.

A time period is inserted between upper and lower device switching in VSI to primarily to avoid current shoot through within the phase leg and the associated rapid temperature increase and high-frequency emissions. This time period is referred to as the deadtime can be up to 5 μs depending on the current rating of the devices and their associated turn off tail and gate drive tolerances. A secondary consideration for this dead time is the phase to phase delay to limit the voltage overshoot observed at the motor terminals. By extending the dead time period with considerations of the flight time for the motor cable length, these overshoots will not coincide limiting the voltage overshoot to a maximum of twice the DC bus voltage.

Long deadtimes are not desirable for the motor control performance. They can be compensated for to a small extend within the control algorithm however this will distort the voltage signals applied to the motor and will have small reduction in efficiency with the introduction of harmonic currents.

Drive output filters are sometimes used with a motor cable for two purposes. First a ferrite ring may be placed around all the output phases where the desired motor current should sum to zero. The high-frequency (often > 1 MHz) components of a signal can be contained in the drive and help with EMC compliance as they no longer radiate from the cable. These ferrites are not necessary within the integrated drive and motor as the connection is so short as to be ineffective at radiating and the connections are enclosed within the metal case of the motor which provides a shield.

These output ferrites are not sufficient to reduce the losses associated with the cable charging. To reduce these, a large inductor is required at the output of the drive. The cost, size and heat dissipation of these inductors are often prohibitively expensive however they can offer advantages with very long cable lengths (> 100 m) where the cable charging pulses of current would otherwise cause nuisance tripping of the current sensing devices.

For an integrated drive, as there are no cables to charge and discharge, a significant reduction in parasitic capacitive load on the power semiconductors and the system is within a shielded enclosure, there are no requirements for a filter network between the power semiconductors and the motor windings. Voltage overshoots at the motor terminals are unlikely to occur even with very fast switching transitions in the order of a few ns. The deadtime can be short and used only to prevent shoot through without additional consideration for staggering the overshooting voltages.

6.2.4 Electrical insulation materials

Conductors in an integrated drive must be separated from each other for the correct operation of the product and to provide safety for anybody who may come in contact with any of the circuits. The simplest insulation is air where the shortest path between two conductors is referred to as a clearance distance and the shortest distance along a surface between conductors is called a creepage distance. The required distance depends on the voltages between the two conductors and whether the circuits for those conductors are accessible. The important consideration for an integrated drive is the quality of the air and any pollution it contains. Different degrees of pollution are defined in IEC60664-1 which depend on the environment that the motor circuits will be exposed to. The required spacings are defined in the relevant product standards and will vary significantly by country and application. In addition to creepage and clearance spacing, solid insulation can be used between conductors, but additional considerations must be taken into consideration. Often a minimum material thickness is specified, multiple layers of material may be needed as additional precautions against damage and the performance of the solid insulation is frequency dependent.

There are several advantages within an integrated drive and motor to minimise the required distances and therefore construct a compact solution:

The first to consider is the removal of power terminals at both the drive and the motor for the connection of cables. In as separated system, these terminals must allow additional spacings due to the possibility of stray cable strands during installation. Second, separate drive and motor solutions require double insulation on the motor cable which adds cost and size. These connections are contained within the integrated motor drive and reduce this complexity.

Potting materials may be installed during the final construction stage of the integrated motor drive. This is common practice for motor windings as the potting material can provide mechanical support for the end-windings, secure temperature or position sensors and improve thermal conduction of heat into the stator. Additional care is required when potting electronic components. While potting can reduce the pollution degree applied to the spacings and could provide some benefits for cooling, It can also apply stress to components as the material cures and contracts. The stress will lead to eventual failure of components where repair work is not practical. A stacking of soft flexible layers directly onto the components followed by stiffer layers can offer improvements however processing time to achieve this is significantly extended increasing the cost of the final product.

Partial discharge is a phenomenon often associated with high-voltage power network with several thousands of volts separated by solid insulation. It results from the partial breakdown of the insulation in a localized area where the electric field strength exceeds the material capability. This may be a small void or defect in the material. Repetitive breakdown in a local spot expands the affected area eventually resulting in a total breakdown and failure of the insulation system. Inhomogeneous electric fields occur in insulation materials due to high $\mathrm{d}V/\mathrm{d}t$ even at modest voltage of several hundred volts. The adoption of wide-bandgap semiconductor take advantage of the rapid $\mathrm{d}V/\mathrm{d}t$ to minimise the switching losses in the devices and allow high-frequency operation however the impact on the insulation must be understood for a reliable integrated motor drive operation. In particular, the motor winding insulation will experience the full $\mathrm{d}V/\mathrm{d}t$ from the switching device on the first few turns in the motor. After these few turns, the distributed parasitic elements shield the remaining windings. Additional insulation should be placed over the windings in this scenario to ensure reliable motor operation over the designed lifetime [7].

6.3 Power systems

The energy to rotate a machine and its load must be supplied and converted to a DC voltage where it is chopped by the inverter to control the motor. The source of this energy will determine the energy storage requirements of the passive components which make up the largest volume electronic components within the drive. In addition, the motor and load must also be considered as an important energy store, and this places additional demands on the power system configuration. A motor driven by an external action or a decelerating load with a large inertia will return energy to the DC bus. The drive topology must be capable of distributing or dissipating this energy to maintain the system within its design parameters to avoid overstressing the components. Three typical power systems are described here with their impact on the passive component selection.

6.3.1 Battery system

Battery technology has received significant focus with the global adoption of electric vehicles. The cost, energy density, charge and discharging times have improved, making this a viable power source. However, even with these technological breakthroughs, the battery is too large and heavy to be included within the integrated motor drive in its present form. In addition, battery performance is sensitive to its environmental conditions particularly the high temperatures found in an integrated motor would reduce its useful lifetime. The battery must be located at some distance from the integrated motor. The battery must be capable of delivering power to the inverter at a rate demanded to control of the load, however this can result in an overdesigned battery for the most dynamic applications.

The battery behaviour under dynamic load depends on the battery chemistry, electrode design and geometry. In addition to providing energy to the motor load, additional frequency components may be superimposed on the DC voltage due to the inverter topology and filter design. These higher frequency components in the order of 1–100 kHz must be taken into consideration to maximise the system efficiency. The battery state of health (SOH) and state of charge (SOC) are often monitored using high-frequency injection [8] to the terminals of the batteries and it is important that these are not disturbed by the load machine. With some battery chemistry, the total loss in the battery may reduce with increasing frequency of the load current however the higher frequency components can stimulate internal resonances and lead to hot spots resulting in early failure. In addition, the oscillating power due to the load conditions results in extra losses due to the high series resistance [1], however some battery chemistries become more capacitive as the frequency increases [9]. Where high-frequency components are used for battery monitoring risk the battery health, additional passive filtering elements are required. Depending on the system point of common coupling, system design and

EMC requirements, these filter components may be placed within the integrated motor or installed as part of the battery pack.

During the deceleration of a motor, energy is returned from the motor to the drive. The rate of the energy transfer depends on the motor deceleration rate, inertia of the load or mechanically driven load. This energy must be removed from the drive by either returning it to the battery or by dissipating it as heat energy. Heat dissipation is undesirable within the integrated motor system as this increases the requirements and hence cost of the cooling system. For applications where the energy recovery is slow, it is possible to directly return the energy to the battery by allowing slight increase in the DC voltage which will control the power flow direction. For rapid energy recovery, the peak DC voltage rise on the DC bus must be limited to protect the connected components. Mechanical brakes on the wheels can dissipate excess energy as heat. The brakes may be required to be installed as part of the system safety and can therefore avoid additional costs of passive components and limit the peak power recovered.

6.3.2 DC power grid

The increase in awareness of climate change has encouraged the development of multiple renewable sources of electricity generation. These are plentiful sources of energy however they are not often available at the same instant that the energy needs to be consumed. The DC power grid can offer efficient solutions to distribute, manage and store this energy and at the same time provide sufficient power to motor loads. A good example is an elevator application in a domestic building. Solar panels are often available on the roof top, the building may have a high impedance power supply in some older urban areas so peak demand to accelerate the elevator would cause the supply voltages to sag. A trickle charger for battery storage can be used with an energy management system. A DC fed integrated motor drive can fit elegantly within this application and the functioning of the inverter is identical to a battery fed system.

A significant additional consideration are the high-frequency currents from the inverter and motor windings. If these currents are allowed to flow through the long cables in the elevator shaft and throughout the building, they will cause interference with nearby equipment. These high-frequency voltages and currents must be contained within the integrated motor. The frame of the motor will be connected to a protective earth for safety in the event of a fault conditions. The DC supply will also have a reference point to ground. Again, two paths of the high-frequency current must be considered: noise generated within the integrated motor; and any disturbances relative to ground on the DC supply. A knowledge of the system is required to prevent disturbances on the DC power grid from saturating or damaging any filter networks at the input of the motor. See Section 6.5.5.1 for methods to minimise and divert these currents.

6.3.3 AC mains supply

The power for most industrial applications is provided from a mains power network. This is usually a three-phase supply operating at a frequency of 50 Hz or

60 Hz depending on the national standards. This ac voltage must be converted to DC and most industrial drives rectify this voltage within the drive in a cabinet. The system designer for the integrated drive must make a choice on how to organise this converter stage when combined with an integrated motor due to the volume and weight of the passive components. The following options must be considered:

(a) Rectifier with DC bus capacitors
(b) Rectifier with DC inductor and DC bus capacitors
(c) Rectifier with AC inductor and DC bus capacitors
(d) Controlled rectifier with inductor and DC capacitors
(e) Bi-directional power flow

The lowest cost rectifier solution is six diodes arranged in a full bridge configuration. These diodes do not require any sensors, control electronics or external power supplies to operate and are therefore the lowest cost solution. With a balanced three phase mains supply, the DC voltage will vary between a minimum of the supply r.m.s. voltage multiplied by $\sqrt{3/2}$ and a maximum of the supply r.m.s. voltage multiplied $\sqrt{2}$. The fundamental frequency on the DC bus will be 6 times the supply frequency. The DC bus voltage relative to earth an important consideration for filter design. For a typical star (wye) supply, the star point is connected to earth, therefore the potential difference between earth and either +DC or −DC is half the DC voltage. This includes half of the ripple voltage which, depending on the design of an EMC filter can return the higher frequency components via earth to the supply. There are defined limits for this leakage current in national standards and problems are often see with residual current devices (RCDs) if used with this type of filter design.

The minimum voltage will reduce significantly towards zero for an unbalanced supply greater than 3% negative phase sequence. To maintain the DC bus voltage within a reasonable band to enable continuous power to be delivered to the motor and the control algorithm to function, capacitance is added across the DC bus. This capacitance must be chosen based on the peak power delivered to the motor. As the capacitance is increased, the DC bus ripple voltage will reduce. This is good news for the control algorithm; however, the current drawn from the supply is delivered in shorter bursts with a very high amplitude. The high input current places additional stress on the rectifier, supply cables, fuses and in combination with the supply impedance, introduces harmonic distortion into the supply network.

An inductor is often added in series with the DC link to smooth current and reduce the peaks. For low powers, the lowest cost solution is to place the inductor on the DC bus after the rectifier where 2 windings are required. As the power increases, it can become more cost effective to place an AC inductor on the three phase supply before the diode rectifier. When placing a DC inductor, care must be taken to avoid resonance with the capacitors occurring at the same frequency as low order harmonics of the rectified supply voltage to prevent excessive ripple current and wear out of the capacitors.

There are various options for controlled rectifiers, the simplest using thyristors to limit inrush current during charging of the DC link capacitors then behaving in a similar manner to a diode bridge. With more sophisticated designs such as the Vienna rectifier [10], the diode rectifier is replaced with transistors with switching frequencies of several kilohertz, the inductors and capacitors can be significantly reduced in size, the input currents and harmonics reduced and the efficiency increased. These solutions have had little adoption to date in industrial applications due to increased cost of the power electronic components however the demand for improved efficiencies, cost reduction in semiconductor components and increase in cost of metals is moving towards a change in technology.

While novel controlled rectifier can offer very high efficiency operation, the power can only flow in one direction towards the inverter. To get full four quadrant operation, back-to-back inverters are used. One inverter must synchronise with the supply voltages and control loops are designed to maintain the DC bus voltage at a defined setpoint. If energy is sent to the motor, the voltage will attempt to fall where the controlled rectifier will provide the energy from the mains supply, conversely, if the energy is recovered from the motor, it will boost the DC bus voltage which will then be sent back to the mains supply.

In both the controlled rectifier topologies, the size of the passive components is determined by the switching frequency of the semiconductor devices. Higher switching is preferred however this results in higher power loss in the devices. Wide bandgap devices provide a suitable method to reduce the switching loss however they can be vulnerable to surge voltages and current which can commonly occur on mains supply lines.

6.4 Design trends of integrated drive systems

The introductory chapter in this book discussed the benefits of integrated drives which include smaller size, weight reduction and increased efficiency. The selection and design of the passive component technology is essential in guiding the topologies of choice for these machines.

There are several examples of industrial motors and drives which are fed directly from the main supply however these products are often a separately designed drive mounted directly onto a motor chassis. Each has their own separate cooling system and often manufactured in separate locations. As discussed earlier, the large size of the rectifier and the associated passive components prohibits real integration. The focus in the industrial markets has been on the functionality of the motor performance and the lowest cost to achieve this. European directives and international standards are close to demanding higher efficiency motor drives. These efficiencies are initially targeted at the motors only with drives required to publish efficiencies rather than meet targets. When the combined efficiencies of both the drive and motor are demanded, a significant move towards integrated drives and motors will take place at this point, the rectifier topology and associated passive components will be under review. These may have a higher capital cost, however these could offer a payback with the increased efficiencies.

Electric vehicle designs have integrated several parts for the inverter and drive system, for example shared cooling fluid, which is passed through the electronics first at a lower temperature, then reused to cool the motor. Others have demonstrated the mounting of industry standard discrete packages for power semiconductors onto the motor frame, however the passive components are maintained at a distance from the motor to take advantage of lower temperatures. The trend is towards higher operating voltages and higher frequency switching which reduces the current and the weight of the associated cables.

In both of these applications, the increase in switching frequency, the reduction in stray inductance by increasing integration reduces the need for purpose designed passive components. While these trends may initially increase the cost of the design, the weight reduction and efficiency improvement will allow the end user to recover these costs in energy savings.

6.5 Introduction of EMI and EMC

Like all electrical systems, an integrated motor drive module needs to be designed in a way that it can withstand electromagnetic interference (EMI) in its working environment – Immunity. In the meantime, it must not affect the adjacent products/equipment in operation – Emissions. Within the integrated motor drive itself, good design will ensure that power and signal integrity are achieved.

Most common immunity issues are electric fast transients, high-frequency noise conducted on the connected cables, electrostatic discharge (ESD) and immunity against radiated emissions. Most common emissions are conducted emission, radiated emission and low-frequency magnetic field.

In order to make sure that each product can work without interfering with or being interfered by other products/equipment nearby, the Electromagnetic Compatibility (EMC) standard committee have defined many EMC standards for products to comply with. Depending on the application, integrated motor drive system will need to comply with the relevant EMC standard. For instance, an integrated motor drive module in an electric vehicle will need to meet CISPR 25 emission requirements, while in an industrial environment, a variable speed drive system will need to meet the emission limits defined in EN 61000-6-4. The standards are often created to balance what is reasonably possible to achieve with the likely operating environment.

The subject of EMC is often related to power and signals in the spectrum of radio frequency (RF), although very low-frequency magnetic field could also be an issue as they can disrupt magnetic field sensors (such as a Hall effect sensor) or cause harm to human beings.

Generally speaking, the demarcation between low and high frequency is defined as follows: when the parasitic components cannot be ignored and become important enough to affect circuit operation, the line from low frequency to high frequency is crossed [11]. This can happen from a few kHz to several GHz depending on the product. For example, a grid-tied motor drive needs to ensure that the total harmonics distortion (THD) of the unit meets regulation standards. When it comes to mains related harmonics, we use the term 'low frequency'. A modern motor drive will have a switching frequency in the order of tens of kHz, the EMI impact of the switching can extend to a few hundred MHz, if not GHz. We call the tens of kHz the 'low-frequency range' and any frequency above 1 MHz is called 'high frequency'.

6.5.1 Differential mode and common mode

EMI noise can be separated into differential mode and common mode. For an integrated motor drive system, differential mode noise dominates the low frequency range (a few kHz to 1 MHz), while common mode noise contributes to the interference above the MHz range. Differential mode noise is mostly related to the DC link design in an integrated drive system, a low impedance DC link is desirable to keep the emissions to a minimum, as shown in Figure 6.10. At frequency f, above 1 MHz, the impedance of the parasitic capacitance in the system starts to reduce as $Xc=1/2\pi fC$, where C is the parasitic capacitance between the noise source and common ground, leading to a lower impedance path for the noise to close the loop. Therefore, common mode noise starts dominating the noise spectrum. A typical common mode current loop is demonstrated in red line in Figure 6.10. Note that the common ground could be earth, cables nearby, circuit next to the drive, or simply a metal frame. In fact, any conductor will have coupling effect to the drive circuit which can be characterized as parasitic capacitance and inductance [12].

Filter design choices must be optimized for both differential and common mode. A typical front-end and output filter configuration for integrated motor drive is shown in Figure 6.11. The common-mode choke (CMC) and Y-class capacitors are common mode filter while the π(C–L–C)-filter and L–C filters are typically designed for differential mode.

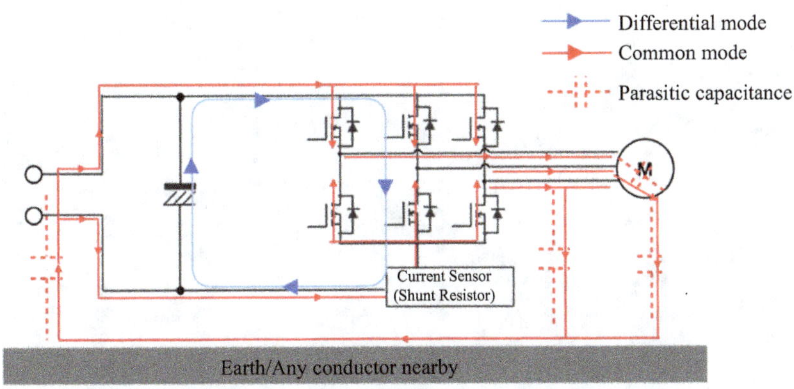

Figure 6.10 Demonstration of differential mode current path and common mode current path

6.5.2 The RF content in a typical motor drive system

The frequency contents of the of a motor drive system due to the switching events can be observed in both voltage and current waveforms. Here, a typical motor phase current is shown in Figure 6.12 where a single phase in a three-phase motor can be represented by (6.3):

$$I_{ph} = I_{fundamental} + I_{switching} + I_{RF} \tag{6.3}$$

where $I_{\text{fundamental}}$ is the fundamental current content associated with the electrical frequency of the motor drive; $I_{\text{switching}}$ is the PWM switching current content; I_{RF} is the radio frequency content which are associated with high dV/dt and dI/dt events, such as fast switching of the power electronics devices or partial discharge observed in the form of bearing current or cable discharge.

Phase currents measured during a motor operation present useful information.

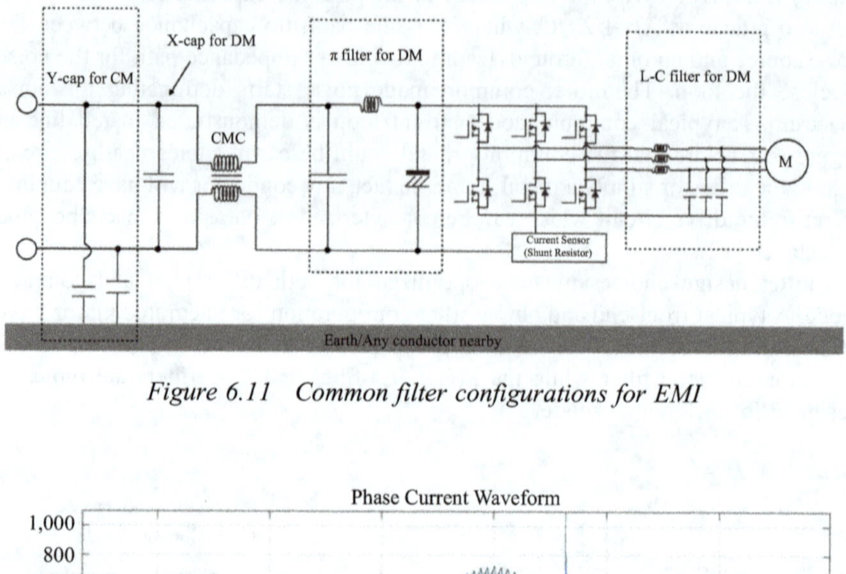

Figure 6.11 Common filter configurations for EMI

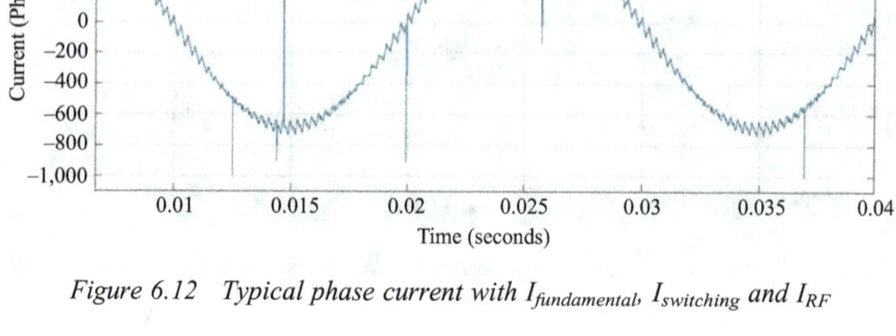

Figure 6.12 Typical phase current with $I_{fundamental}$, $I_{switching}$ and I_{RF}

In order to measure the motor phase current accurately, special wide bandwidth current probes are often required.

Harmonic content of the fundamental current waveform contribute to the low-frequency harmonics (<9 kHz) in a typical motor drive system. The motor drive switching frequency generates the low to medium frequency noise, often sits between a few kHz to tens of MHz. The radio frequency content can be further divided into two types of noise: Broadband noise (10–300 MHz) where the energy is spread across all frequencies reducing in magnitude as the frequency increases. This arises from the fast rise time and fall time of the switching devices and high frequency noise (>300 MHz) as a result of events such as reverse recovery charge from the body diode of a power electronics device [13], or bearing current due to an occasional electrical breakdown [14].

6.5.3 Noise source

6.5.3.1 Switching frequency and the transient of a modern switching device

The emerging wide-band-gap (WBG) material power semiconductors such as silicon carbide (SiC) and gallium nitride (GaN) devices are game-changing devices. They can exceed the limit of silicon (Si) and are used to pursue ground-breaking high frequencies, high efficiencies, and higher power density power conversion. Recent years have seen SiC or GaN MOSFETs slowly replace Si devices such as IGBTs and MOSFETs in integrated drive designs. For instance, SiC MOSFETs from STMicroelectronics™ are found in Tesla Model 3's electric drive unit [15].

The faster switching characteristics of WBG allow improved system efficiency due to reduced power electronics device loss (both conduction and switching losses). It can also bring great benefits to the electric motors. For current control of a motor drive, with the availability of high-speed switching devices, low inductance motors become feasible. The winding configuration of low inductance motor has a smaller number of turns, so it is easier to design, and also has benefits for cost, weight, and manufacturing ease points of view.

The problems that are associated with high switching frequency WBG devices are mainly EMC and thermal. On the thermal side, the WBG devices have very low switching and conduction loss and generally can withstand higher temperature compared with Si devices. However, packaging materials and associated installation and mounting have limited the component maximum operating temperatures. The thermal issue can also be solved via current sharing by multiple devices in parallel distributing the loss and temperature rise across a larger area. The more challenging design obstacle is the EMC aspect, because switching devices act as a noise source.

Both the switching frequency and the switching speed contribute to the overall noise spectrum. The switching frequency of an integrated drive module (kHz to 100s kHz) is often much lower than a switched mode power supply (100s kHz to MHz). The switching speeds of typical WBG devices can easily exceed 20 V/ns [16].

An FFT analysis of a typical PWM controlled phase voltage waveform of a motor drive is often called spectrum analysis and it is frequently used in

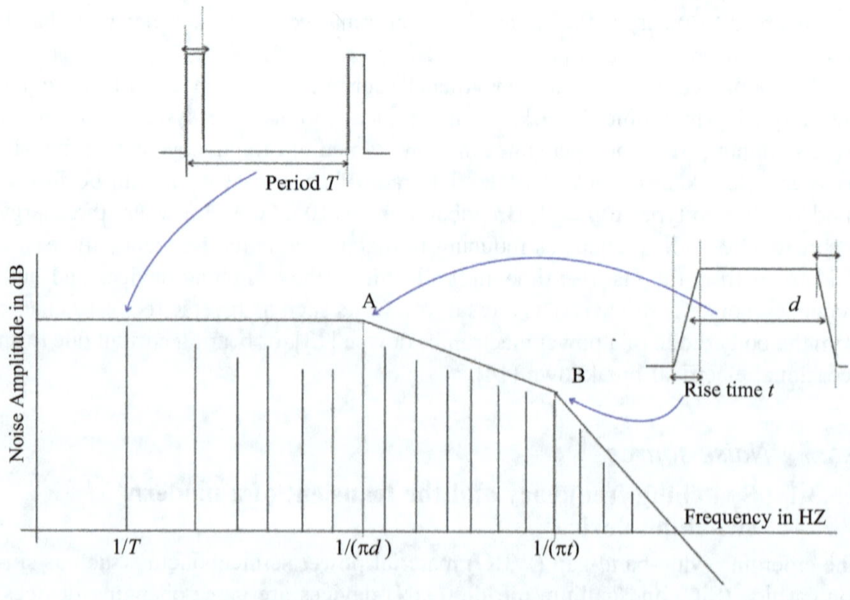

Figure 6.13 FFT analysis of a typical switching voltage pulse

high-frequency measurement and EMI tests. As shown in Figure 6.13, there are two distinct corner points (marked A and B) in the spectrum plot of a typical voltage pulse waveform. The cut-off frequency at point A is defined as $1/\pi d$, where d is the pulse width of one switching event, whereas the cut-off frequency at corner B is defined as $1/\pi t$, where t is the rise time of the switching event (assuming rise time equals the fall time). The slope between the two corner points has an attenuation of 20 dB/dec (a 20 dB/dec attenuation means harmonics decay linearly with the frequency) whereas the slope beyond frequency at corner B has a 40 dB/dec attenuation.

As it was mentioned before, both switching frequency and switching speed contribute to the overall spectrum of a switching event. Two examples are demonstrated there to show the effect of switching frequency and switching speed. This amplitude of the noise is directly related to the DC bus voltage being switched and not to the r.m.s motor voltage therefore hard switched 2 level converters which apply the full voltage will have a higher noise signature for the same transient time.

6.5.3.1.1 Switching frequency

If switching frequency is doubled, the number of transitions within a given amount of time is doubled, this means the EMI energy is doubled. Considering power is $10\log_{10}(2)$, this is means 3 dB more energy shown on the spectrum analysis. The measuring bandwidth is fixed by the EMC standards such as CISPR 16 [17] which has a bandwidth of 120 kHz at –6 dB between 30 MHz and 1 GHz. Doubling the frequency also means only half of the bins available to locate this energy, however this is dependent on the frequency band and could result in 6 dB more amplitude per doubling the frequency. This situation rarely occurs with a few kilohertz increase available to Silicon devices however for WBG whose switching frequencies are orders of magnitudes higher, uncontrolled emissions could be increasingly high.

Figure 6.14 shows the FFT analysis of a 600 V switching event with two switching frequencies. The two switching frequencies, 30 kHz and 60 kHz, are shown to demonstrate the noise profile. In both cases, the rise and fall time of the switching event are kept the same at 60 ns.

6.5.3.1.2 Rise time

As it can be seen from Figure 6.15, faster switching means shorter rise time, which leads to a shift of corner B towards higher frequency range. This effectively

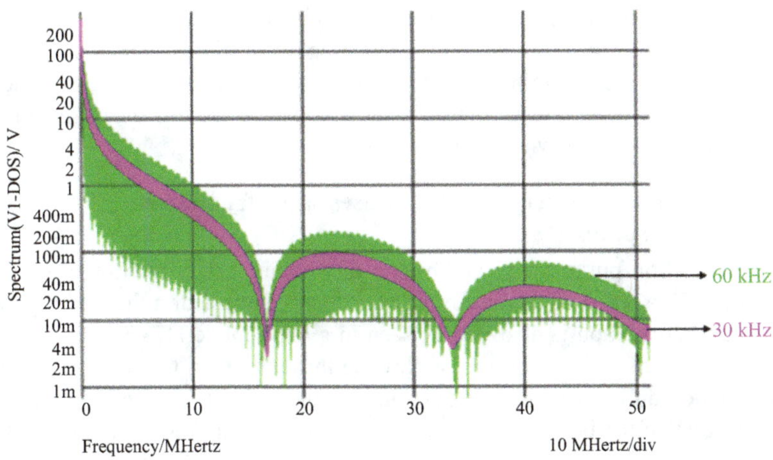

Figure 6.14 FFT analysis of 600 V Si device switching event

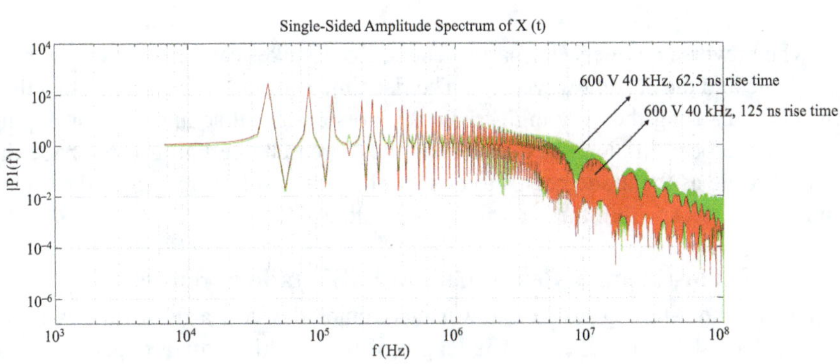

Figure 6.15 FFT comparison of rise time

extends the frequency range of the 20 dB/dec attenuation, meaning the noise level of a faster switch is much higher than that of a slower switch.

6.5.3.2 Relationship between the time domain and frequency domain measurement

There is a correlation between the time domain waveform and the frequency domain spectrum. A fast speed switching waveform can never be a clean trapezoidal waveform. Overshoot and ringing are common because there is always parasitic capacitance and inductance in the system design and non-linear switching of the semiconductor will give rise to variation in the oscillations across the DC voltage, current and temperature range. For instance, a via between two adjacent layers on a PCB will introduce at least 1 nH inductance. Depending on the power electronics package types, at least 1 nF parasitic capacitance will exist for a MOSFET. Package type with long lead such as TO-247 will have much larger (often with 10 nH) inductance.

A 20 MHz ringing is observed on a switching event (time domain) correlates to a frequency resonance point measured in the frequency domain. If the overshoot is relatively high, the resonance region in the frequency domain measurement will show a 'peak' as it is shown in Figure 6.16.

6.5.3.3 Reverse recovery charge causes high-frequency noise spikes

For demonstration purposes, a half bridge of a motor drive inverter is shown in Figure 6.17. This could be the half bridge for a single phase or a three-phase motor drive. The phenomenon is also widely seen in a buck-mode DC–DC converter. In the free-wheeling stage, the current is flowing through Q2 in the reverse direction, Q2 is switched on because the conduction loss of a MOSFET is often lower than that of the loss of the body diode. In Figure 6.17(b), before the next commutation, Q2 is switched off, as a result the current is diverted to the body diode of the MOSFET during the dead time. In Figure 6.17(c), the upper device Q1 is switched on however the body diode of Q2 takes time to deplete its stored charge to allow it to block the high voltage. This stored charge is rapidly removed as a diode negative or reverse recover current due to the applied DC bus voltage and a very low impedance in this loop.

When reverse recovery discharge occurs, sharp spikes can be observed on both the voltage and the current waveform. The rise time is of the spike is often less than 1 ns, hence causing the very high-frequency noise. Careful semiconductor doping profiles have been implemented to reduce the stored charge but also to shape the voltage profile as the diode begins to block voltage in an effort to reduce the higher frequency content within these currents and voltages [18].

6.5.3.4 Shoot-through, deadtime and IGBT tail current effect

To prevent shoot-through of the power electronics devices, a dead time is introduced in the switching scheme. The length of the deadtime often depends on the device turn on/off delay, rise/fall time and system efficiency requirement and gate drive component tolerance.

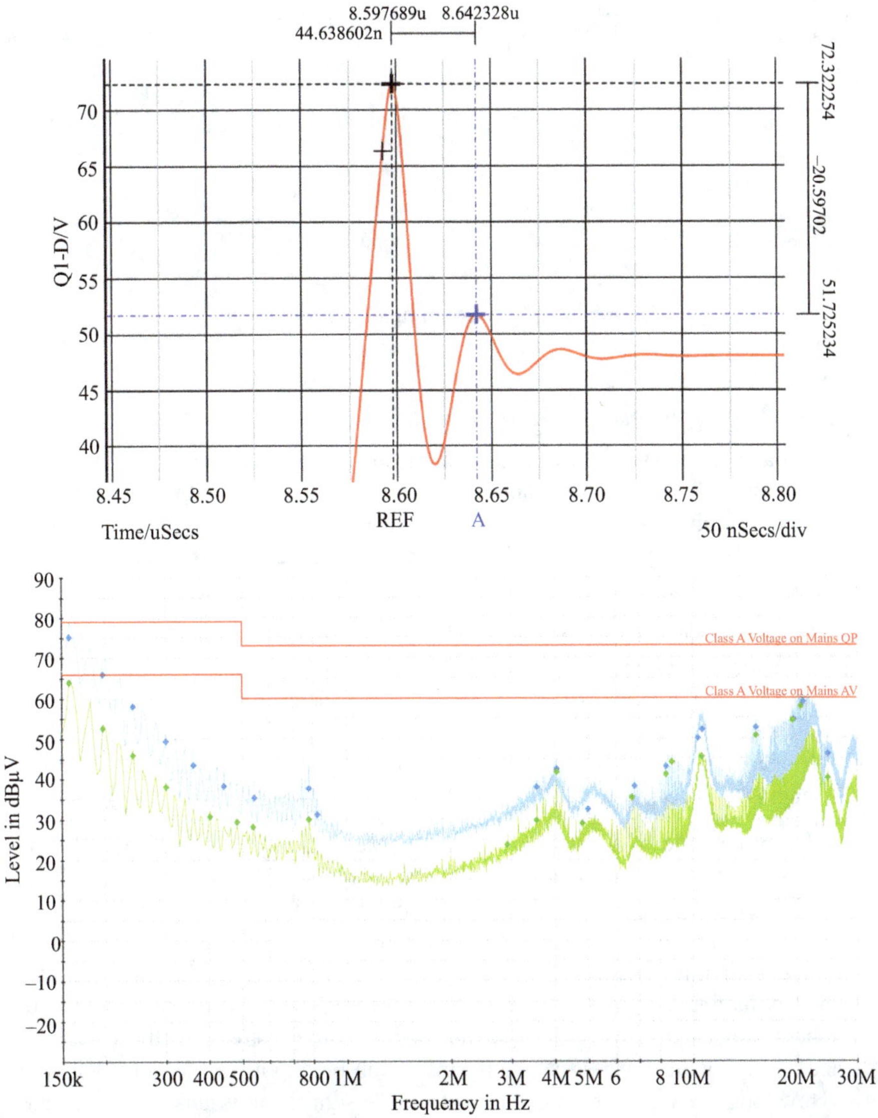

Figure 6.16 Time domain to frequency domain effects

It should be noted that too short a deadtime will result in a shoot-through of a half bridge. A shoot-through event will cause a very high short-duration current to be drawn from the DC link. For IGBT based voltage source inverter, this is most likely to happen with the devices (IGBTs) with a long tail current. Tail time is also related to temperature, and this may not be easily detected during inverter characterization but becomes obvious during conducted emissions measurements with very high levels of emissions.

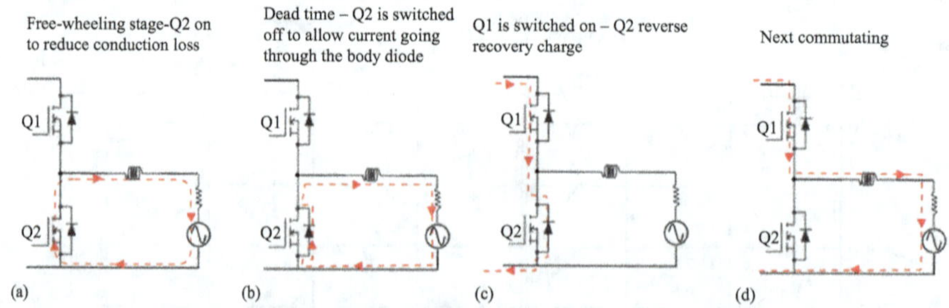

Figure 6.17 Single-phase current loop

6.5.4 Propagation path and coupling

An integrated drive system consists of an electric motor and the drive circuit. The drive circuit includes a power stage, a control unit, together with communication lines. Such highly integrated systems are built with compact design, therefore the distance between the motor windings and the drive circuit is very small compared to discrete systems. This gives many advantages:

(a) It reduces cable loss (i^2R).
(b) It reduces the ringing and oscillation on the switching waveform, hence reduces the voltage stress on both power electronics devices and motor windings.
(c) It achieves better EMC performance due to reduced overshoot and damped ringing.

The importance of the cable distance on the system design will be discussed here. As a rule of thumb, a path of reasonable sized conductor contributes approximately 10 nH/cm and is a good indicative value to estimate the loop inductance of a system. As mentioned previously, to avoid overshoot and ringing for a fast-switching event, one should always aim to reduce the parasitic parameters. By having a long cable between the power stage and the electric motor, the overshoot and ringing can be large, increasing voltage stress on the electronics devices. In the worst case, it can cause insulation breakdown of a motor winding.

How long does the cable need to be to cause significant issues in an integrated drive system? To answer this question, we need to borrow some RF engineering knowledge and see the problem from a transmission line point of view. The *Rise-Time Rule* states that if the length of the transmission line is less than one-third the distance a wave travels in one rise time, then it is not necessary to take the transmission line design into consideration [12].

The propagation velocity depends on permeability and permittivity of the dielectric materials between the conductors. The typical dielectric constant is about 4.5 (very close to FR4 material), for simplicity, we assume the wave travels at 1.5×10^8 m/s. For a rise time of 5 ns, this distance is 0.75 m, a third of this distance is 0.25 m. Therefore, for an integrated machine the voltage reflection problem can be ignored, and the insulation can be rated for the maximum DC voltage. This is a significant improvement over discrete system.

Cables between the power stage and the motor can be treated as a lumped L–C network. Cables can be shielded types, twisted pair, ribbon cable or bus bar depending on the power level and application. But regardless, cables can radiate, and the resonance frequency depends on the length of the cable.

There are two coupling mechanisms in general, capacitive (electric field) coupling and inductive (magnetic field) coupling. Coupling depends on the parasitic capacitance and inductance in the integrated drive system as shown in Figure 6.18.

On the driver side, the main coupling path is through the power electronics devices. The package of the device introduces its intrinsic inductance and capacitance. The layout of the device also contributes to more parasitic parameters. This is particularly important where the semiconductor devices are often separated around the motor diameter to improve cooling. The inductance in the DC bus between devices, which is very important during periods where the phase currents are decaying will be higher. Each loop during the switching period must be considered separately and minimized.

Another common coupling path in an integrated drive system is the space between the device and the heat sink. Sometimes, the heat sink radiates if it is not grounded properly. As discussed previously, the parasitic parameters of a switching device may lead to overshoot and ringing measured in a switching event.

On the machine side, both amplitude and waveform shape of generated voltages and magnetomotive force (MMFs) of an electric motor are determined by the winding arrangement. There are many different types of winding configurations, among which, single-tooth winding has become more and more popular with switched reluctance machines and permanent magnet machines due to its simplicity and ease of manufacturing.

This type of winding also achieves very short end windings, which limits the leakage flux coming out of the motor which could couple to the integrated drive located in close proximity. Therefore, the single-tooth winding is a preferred option from an EMC point of view. A short end winding means less parasitic capacitance between the winding and the motor chassis ('ground'), hence less common mode noise.

The magnetic flux generated by motor operation is mostly contained within the motor frame, but the leakage flux from end windings is not contained. Exposed to free space, the flux can easily radiate. This needs particularly important consideration when the switching and transient frequencies increase by using WBG devices. The smaller openings that may be provided for cooling become more effective at allowing higher frequency noise to radiate.

Motor position sensors are generally located close to the stator end windings or rotor ends. There are two types of sensors, namely the optoelectronic type and the hall effect type (more details in Chapter 3). Optoelectronic position sensors offer high accuracy and more immune to electromagnetic interference, but they can have a high cost. The cabling to the drive control processors must be protected from

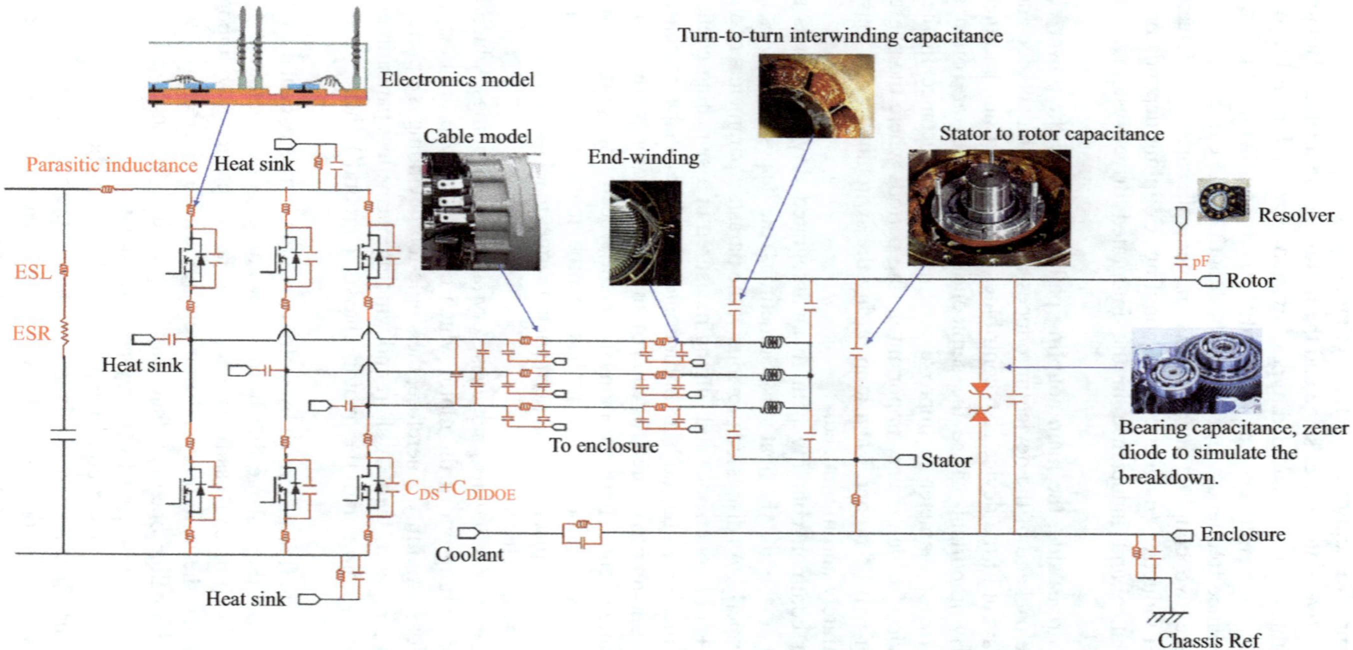

Figure 6.18 Inductive and capacitive coupling

noise pickup. The opto-encoders are very susceptible to dusty/dirty environments, hence not a good option for automotive application without additional protection. The hall effect sensors, on the other hand, do not mind such environments, but they are susceptible to stray or low-frequency magnetic fields. Therefore, having a short winding configuration can also reduce the noise coupling effect of leakage flux, which helps achieve better accuracy of position sensors. Sometimes, shielding is needed to protect the hall effect sensors. Static and low-frequency magnetic field needs special shielding material, i.e. mu-metal, a very high permeability material [19].

Bearings have a moving metal ball or roller and a fixed metal shell. Grease sits between the two parts, but it is very thin and therefore has a high capacitance and can carry high displacement current. Because it is thin, under certain fault condition, there can be an occasional electrical breakdown and even direct touching of the two metal parts. Therefore, the bearing current is partly capacitive, which gives a pulse of current every switching transition and partly random high current spike [20].

6.5.5 Design for EMC and mitigation scheme

If a system is designed with EMC considerations in the beginning, there will only be a slim chance of needing noise mitigation schemes later.

At low-frequency range (from a few kHz to 1 MHz), the noise level is predominantly differential mode rather than common mode. Low-frequency noise is most likely differential because capacitance to common ground is always smaller than capacitance across the DC link. At low frequency, the common mode current

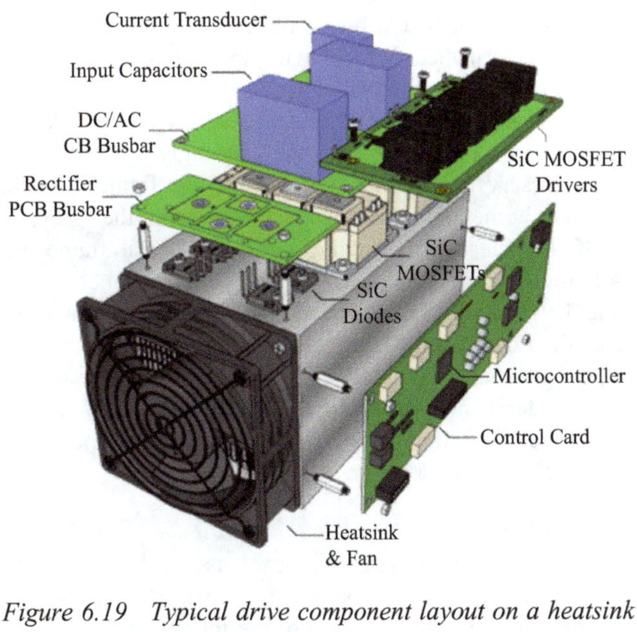

Figure 6.19 Typical drive component layout on a heatsink

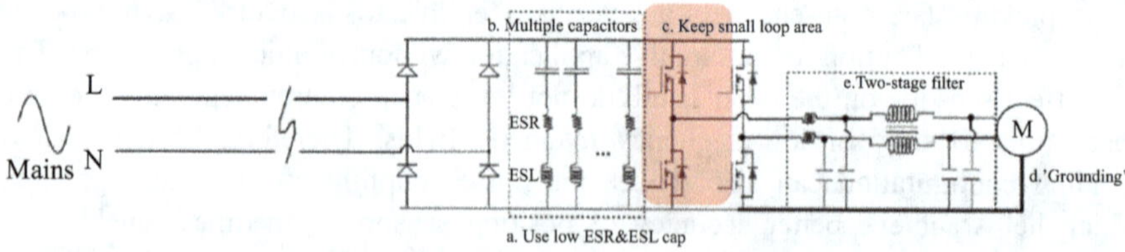

Figure 6.20 Cost-effective EMC design techniques in a motor drive system

in the small capacitance to earth is probably an order of magnitude (or more) lower than the differential mode current.

Low-frequency conducted emissions are always related to the DC link design (impedance of the DC link). One should always aim to select low impedance capacitors for the DC link, the connection to the switching device should also avoid increased parasitic inductance. This is generally achieved by using laminated bus bars and minimizing the loop area for currents with high-frequency content. The connections within the packages of the semiconductors also need to have a low inductance connection for good signal integrity and to reduce stray fields.

As shown in Figure 6.20, commercial drives often use multiple capacitors rather than a single large capacitor. There are many good reasons of doing so. First is the cost. The cost of capacitors increases exponentially with capacitance value for the same package size. Using one single capacitor is perhaps more expensive than using a few smaller capacitors. Second is the size. The limiting factor for product design is often the height of a component. For capacitors with large capacitance, the height is often too big to be fit in the product design. The third reason is the EMC performance. When having multiple capacitors in parallel, the ESR and ESL are in effect parallel, which results in lower impedance. In addition, it is a good practice to keep the hot loop area as small as possible. A larger loop is a large inductor. A larger loop also radiates more strongly.

It is often necessary to 'ground' the motor if the frame is metal for safety reasons. In a typical home appliance motor drive system, the 'ground' is the 0 V reference (or DC link capacitor negative end). The metal frame of a motor is connected via a low impedance connection to the 0 V reference, most of the conducted noise can be returned to the 'ground'. This method serves as an alternative cost-effective method compared with shielding the motor frame.

Multi-stage filters should be applied rather than single-stage filter. Multi-stage filter achieves a wider bandwidth filtering effect while keeping a smaller size compared with single-stage filter. From the cost point of view, a multi-stage filter might be more expensive than a single stage filter.

6.5.5.1 Mitigation scheme

While it is desirable to design a system that can pass all the EMC test for the first time and comply with the EMC standard, in real life this seldom happens. Mitigation schemes therefore often consist of the following:

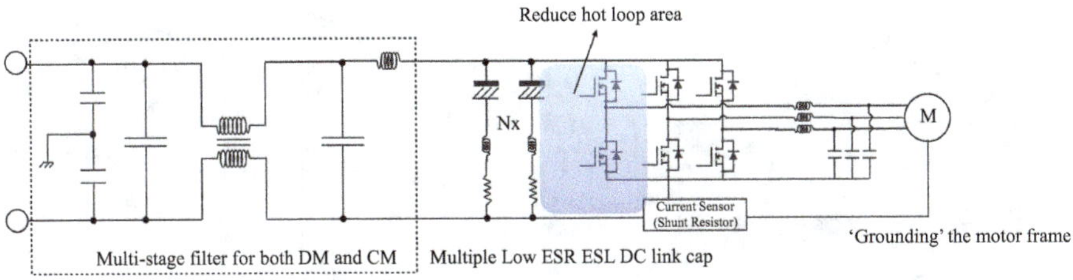

Figure 6.21 Multi-stage filter design in a motor drive system

(a) On the noise source side, identify the noise source and contain it using improved shielding and grounding techniques.
(b) On the coupling and propagation path, identify the coupling path, reduce the coupling either magnetically (inductively) or electrically (capacitively). This often means changing the layout of a PCB, adding more layers, and improving the grounding scheme.
(c) On the victim side, isolate the victim by either provide shielding protection or keeping the victim away from the noise source.

For motor drive EMC design, there are still many ways to achieve a cost effective solution, even with all the design constraints. Here lists a few:

(a) For PCBs that do not have a solid, continuous ground plane, attention shall be paid to DC-link capacitors design, as they serve the low impedance path for differential mode noise. The capacitor negative is effectively the 0 V reference (or ground plane). Therefore, selecting low ESR and ESL capacitors help achieve a low impedance 0 V reference for single phase applications.
(b) Sometimes, as a last resort, engineers will find a proper filter design is the only option to achieve an EMC pass. In this case, multi-stage filter should be applied rather than single-stage filter. Multi-stage filter achieves a wider bandwidth filtering effect while keeping a lower cost and smaller size compared with single-stage filter. A two-stage filter is shown in Figure 6.21.

6.6 Filter simulation

Simulation can give design engineers a good confidence level when designing a customized filter solution. However, a simulation model needs to be accurate enough so that the result does not over-predict or under-predict the filter performance (which is usually characterized as insertion loss). Unfortunately, when it comes to RF signals, due to the prevalence of parasitic parameters and the non-linearity associated with frequency, simulation modelling often takes a long time to develop and may not predict a result that agrees with the experiment.

A SPICE-based simulation tool provides a quick and cost-effective way of filter performance analysis. In this section, a filter is modelled to help the readers grasp the skills needed for filter simulation.

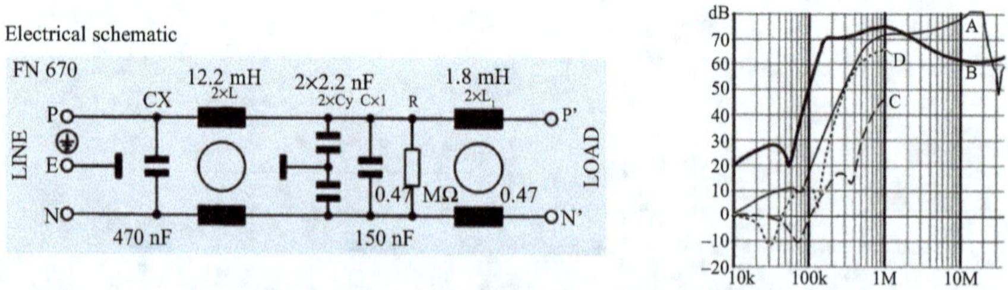

Per CISPR 17; A= 50Ω/50Ω sym, B = 50Ω/50Ω asym, C = 0.1Ω/100Ω sym, D = 100Ω/0.1Ω sym

Figure 6.22 Simulated filter insertion loss

An off-the-shelf filter is used to validate the simulation model. The schematics and the insertion loss of the filter can be found in Figure 6.22. CISPR 17 defines the measurement set-ups for filters. Ref. [17] argues that the 0.1 Ω/100 Ω and the 100 Ω/0.1 Ω set-ups give better filter performance prediction since the impedance of neither the source nor the load can be 50 Ω. As a result, manufacturers sometimes provide insertion loss of four different set-ups, as shown in Figure 6.22.

A simulation model is built using Simetrix (a SPICE simulation tool) as shown in Figure 6.23. Symmetrical (differential mode) and asymmetrical (common mode) performance can be analysed by using the AC analysis in the simulation tool. Rsource and Rload have four combinations of impedances: 50 Ω/50 Ω symmetrical (mode A), 50 Ω/50 Ω asymmetrical (mode B), 0.1 Ω/100 Ω (mode C) symmetrical and 100 Ω/0.1 Ω symmetrical (mode D).

The AC analysis sweeps between 10 kHz and 30 MHz for mode A and B. For modes C and D, the AC analysis is swept between 10 kHz and 1 MHz, same as the datasheet. This is because when the frequency goes above 1 MHz, asymmetrical mode (common mode) starts dominating.

The common-mode choke (CMC) is modelled by defining a coupling factor between the two inductors in the simulation. To start with, a reasonable value of coupling is set as 0.99. When adjusting the coupling factor, the leakage inductance of the CMC changes. As one will soon find out, the leakage inductance of the CMC resonates with the capacitor Cx (C1 in the simulation file), it determines the first order response of the filter performance.

The parasitic parameters of each passive component are estimated and tuned in the following steps:

1. Initial parameters are estimated. At high frequencies, a capacitor's equivalent series inductor (ESL) starts taking effect (correlated with an inductor's inter-winding capacitance). Depending on the capacitor type and the connection to the PCB, the ESL and equivalent series resistance (ESR) values can be estimated.
2. Damping of the capacitor and inductors are important; damping resistors are necessary to minimize resonance-induced overshoot.
3. Once the insertion loss profile is formed, one can then tune the inductance and capacitance value to match the resonance(s).

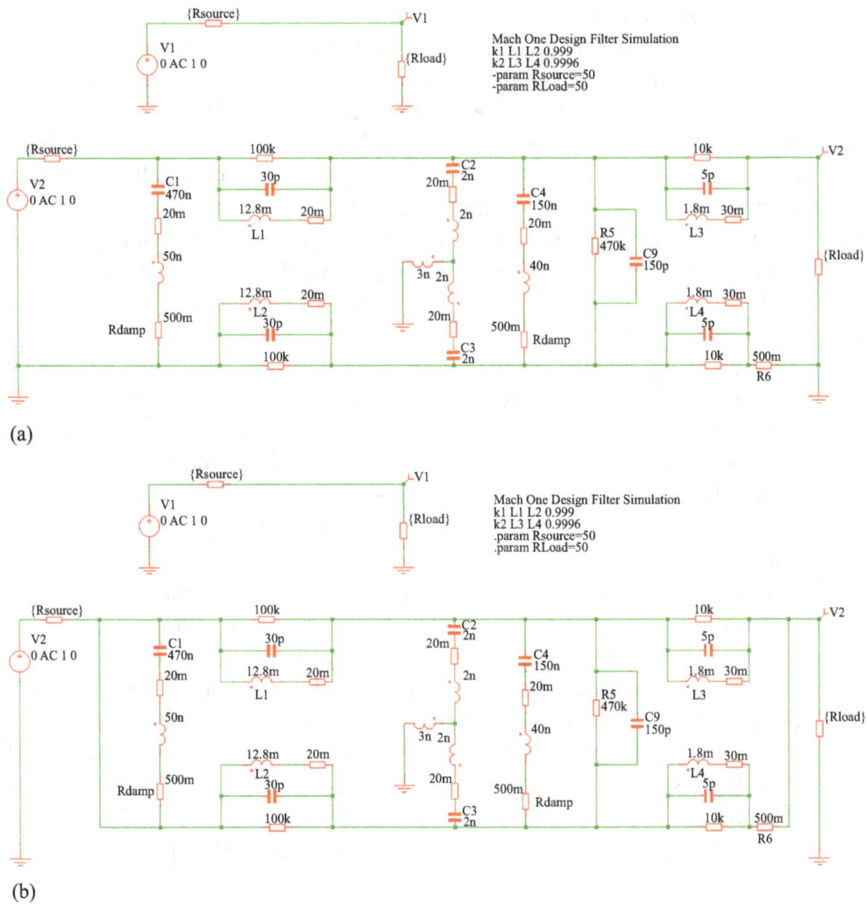

Figure 6.23 Simulating insertion loss (a) symmetrical mode and (b) asymmetrical mode

It should be noted that the ESL of a capacitor is often overlooked by design engineers while the connections of the passive components are often ignored. Connections often introduce far more inductance than the internal ESL of a capacitor [21]. In the simulation model shown in Figure 6.23, the ESL includes the inductance of the connections.

A 470 nF X type film capacitor often has about 20 mΩ ESR with a resonant frequency that sits between 1 and 10 MHz. An estimated ESL of 20 nH generally goes well with this type of capacitor; another 30 nH inductance introduced by the connection is also considered. (So, the lumped inductance value is shown as 50 nH.) Often, capacitors need to be damped to avoid oscillation (caused by LC tank resonance with the inductive component in the filter). In this case, a 500 mΩ damping resistor is put in series with the capacitor (shown as Rdamps in the simulation).

The coupling factor of both CMCs was tuned to be 0.999 to give a very strong common mode coupling. This leads to a leakage inductance of around 20–30 μH for the first stage CMC. Note that the leakage inductance depends on how the CMC is wound (sectional or bifilar), number of turns, and so on [22].

In this case, the leakage inductance is quite important for the differential mode filtering performance, particularly in the frequency range of a few MHz. Together with the capacitor C1, it also determines the poles of the filter transfer function. Another important parasitic parameter of the chokes is the interwinding capacitance, which depends on the number of turns and the distance between each winding. The values for CMCs are tuned to be 30 pF and 5 pF to match the resonance point.

It should be noted that the interwinding capacitance values seem to be extremely low in this case, as we were expecting an order of 100 pF for a 12.8 mH CMC. This is due to the over simplicity of the CMC model we used. In reality, the common mode and the leakage inductance reduce with frequency and current. Therefore, the same resonant frequency in the insertion loss (measured) means the interwinding capacitance is much larger than that of the simple simulation model.

Figure 6.24 shows the simulation results compared with the results in the datasheet. While the profile of the 50 Ω/50 Ω filter setup simulation matches the datasheet measurement well, there is a 10 dB over prediction of the simulation model between the frequency range of 1 and 10 MHz. It is not clear what caused the difference, but we assume this is due to the frequency dependency of the leakage inductance of the CMC. Both high-frequency eddy current and the proximity effect between adjacent layers could lead to reduction of leakage inductance. In fact, this drop could be significant, as it was found in [23], where the leakage inductance of a transformer dropped from 200 nH to 140 nH between 100 kHz and 1 MHz.

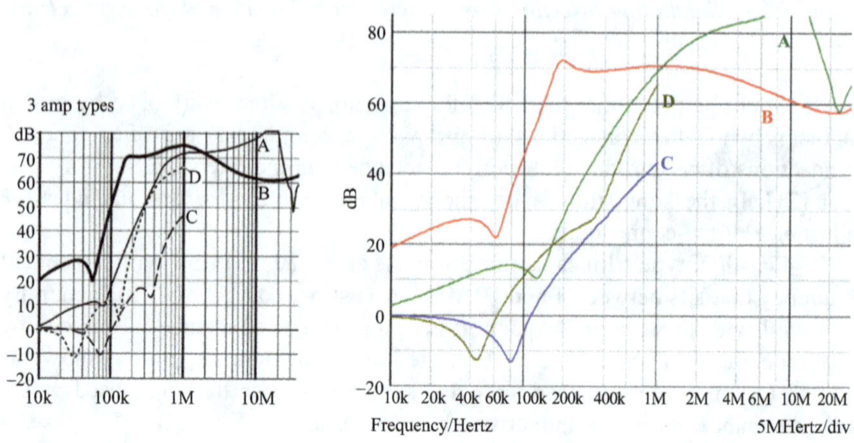

Figure 6.24 Simulation results compared with the results in the datasheet

6.7 Conclusion

The use of passive devices within an integrated drive has been discussed in this chapter. The largest components used are for energy storage are largely determined by the supply from which the motor is fed and are used to provide a stable DC link voltage required for motor control. The materials used in these components become more expensive and can be difficult to achieve robust operation and long lifetime operation due to the high temperatures experienced within the integrated drive.

While the drive and motor are integrated, careful power system design can improve the size, weight, and cost of the integrated machine. Special attention is required to minimize the parasitic capacitances and loop inductances within the fast switching nodes in the circuit. This will reduce the number and size of additional filter components which must be used to ensure reliable operation and to prevent interference with other equipment. The use of WBG semiconductor devices enables the efficient and compact motor drive, but the higher frequencies involved place additional importance on designing for EMC and testing to confirm compliance. More integrated filter configuration for motor drive application should be considered. Further material research is required to continue to push the limits of the passive devices and develop components which can operate at higher temperatures with ever more compact size.

References

[1] A. Klein-Hessling, K. Felsing, and R. W. De Doncker, "Active DC-power filters for switched reluctance drives based on dual-active bridge converters," in *2019 IEEE International Electric Machines & Drives Conference (IEMDC)*, 2019, New York, NY: IEEE, pp. 1694–1699.

[2] N. Corporation. General Description of Aluminum Electrolytic Capacitors. Available: https://www.nichicon.co.jp/english/products/pdf/aluminum.pdf

[3] J. W. Kolar and S. D. Round, "Analytical calculation of the RMS current stress on the DC-link capacitor of voltage-PWM converter systems," *IEE Proceedings-Electric Power Applications*, vol. 153, no. 4, pp. 535–543, 2006.

[4] H. Wu, D. Winterborne, M. Ma, V. Pickert, and J. Widmer, "DC link capacitors for traction SRM drives in high temperature automotive environments: a review of current issues and solutions," in IET Hybrid and Electric Vehicles Conference 2013 (HEVC 2013), 2013.

[5] H. Wang, P. D. Reigosa, and F. Blaabjerg, "A humidity-dependent lifetime derating factor for DC film capacitors," in *2015 IEEE Energy Conversion Congress and Exposition (ECCE)*, 2015, New York, NY: IEEE, pp. 3064–3068.

[6] K. Bray, R. Wu, S. Fries-Carr, and J. Weimer, "Aluminum oxynitride dielectrics for multilayer capacitors with higher energy density and wide temperature properties," *Thin Solid Films*, vol. 518, no. 1, pp. 366–371, 2009.

[7] M. S. Diab, W. Zhou, C. Emersic, X. Yuan, and I. Cotton, "Impact of PWM waveforms on partial discharge in SiC-based motor drives," in *IECON 2021–47th Annual Conference of the IEEE Industrial Electronics Society*, 2021, New York, NY: IEEE, pp. 1–7.

[8] P. García, Á. Navarro-Rodríguez, S. Saeed, and J. García, "Soc estimation in Li-ion batteries exploiting high-frequency model properties," in *2018 IEEE Energy Conversion Congress and Exposition (ECCE)*, 2018, New York, NY: IEEE, pp. 1103–1110.

[9] A. Jossen, "Fundamentals of battery dynamics," *Journal of Power Sources*, vol. 154, no. 2, pp. 530–538, 2006.

[10] J. W. Kolar, U. Drofenik, and F. C. Zach, "VIENNA rectifier II-a novel single-stage high-frequency isolated three-phase PWM rectifier system," *IEEE Transactions on Industrial Electronics*, vol. 46, no. 4, pp. 674–691, 1999.

[11] D. C. Smith, *High Frequency Measurements and Noise in Electronic Circuits*, New York, NY: Springer Science & Business Media, 1992.

[12] R. Morrison, *Fast Circuit Boards: Energy Management*, New York, NY: John Wiley & Sons, 2018.

[13] "EMI specifics of synchronous DC-DC buck converters," in *2005 International Symposium on EMC*, 8–12 Aug. 2005.

[14] M. Zhang, *Interference Technology*, ed: Interference Technology, 2020.

[15] G. Luke, Available: https://www.idtechex.com/en/research-article/teslas-innovative-power-electronics-the-silicon-carbide-inverter/23080

[16] S. Satpathy, P. P. Das, and S. Bhattacharya, "Study of switching transients based on dv/dt and di/dt for a GaN-based two-level pole," in *2021 IEEE 12th Energy Conversion Congress & Exposition-Asia (ECCE-Asia)*, 2021, New York, NY: IEEE, pp. 19–25.

[17] CISPR 17 Measurements, Available: https://docs.rs-online.com/7c66/0900766b815237ea.pdf

[18] A. Chernikov, V. Gubarev, A. Stavtsev, A. Surma, and I. Vetrov, "One more way to increase the recovery softness of fast high-voltage diodes," in *Proceedings of the 2011 14th European Conference on Power Electronics and Applications*, 2011, New York, NY: IEEE, pp. 1–10.

[19] M. Zhang, "Low-frequency magnetic fields in electric vehicles," *Incompliance Magazine*, pp. 9–15, April 2022.

[20] M. Zhang, "Demystifying EMC in an electric vehicle's drive unit," *Interference Technology,* 2021, available: https://interferencetechnology.com/demystifying-emc-in-an-electric-vehicles-drive-unit/.

[21] EMC Question of the Week 2017–2020. Available: https://learnemc.com/qotw

[22] T. Brander, *Trilogy of Magnetics*, Germany: Würth Electronic, 2018.

[23] Z. Ouyang, J. Zhang, and W. G. Hurley, "Calculation of leakage inductance for high-frequency transformers," *IEEE Transactions on Power Electronics*, vol. 30, no. 10, pp. 5769–5775, 2014.

Chapter 7
Gearboxes

Timothy Lisle[1], *Robert Frazer*[1], *Chris Aylott*[1] *and Brian Shaw*[1]

7.1 Introduction

Gears and gearboxes are used to convert torque and speed to optimize the overall performance of a transmission system. A gearbox reduces the cost of manufacture and optimizes the operating efficiency by matching the performance characteristics of the prime mover, such as a combustion engine or electric motor, with the vehicle torque and speed requirements.

Modern automotive transmissions require high power density, low weight, high efficiency, low noise, and reduced manufacturing costs. This can only be achieved with the application of precise and effective gearbox design, manufacture, high quality material and heat treatment and effective quality control.

Gear technology is challenging and often poorly understood, and the objective of this chapter is to direct engineers to appropriate gear related resources and to improve their understanding in general. Throughout this chapter, we will use International Standards Organization (ISO) terms and symbols which are in effect the international language of gears. ISO standards and their supporting Technical Reports provide a useful source of reference material for those involved with gears and gearboxes and will be referenced throughout.

7.1.1 Gear types

The most common types of gears used in automotive power transmissions are parallel axis spur and helical gears. Spur gears have straight teeth which are parallel to the axis of rotation while helical gear teeth are inclined at an angle, as illustrated in Figure 7.1. Gears are inherently noisy, with a tendency to generate noise at tooth passing frequency and its harmonics. Spur gears are used for slower speed drives because they tend to be noisier, while optimized helical gears are adopted in automotive transmissions where noise is often undesirable. Although this chapter predominantly focusses on parallel axis spur and helical involute gears, the key

[1]Design Unit, Newcastle University, UK

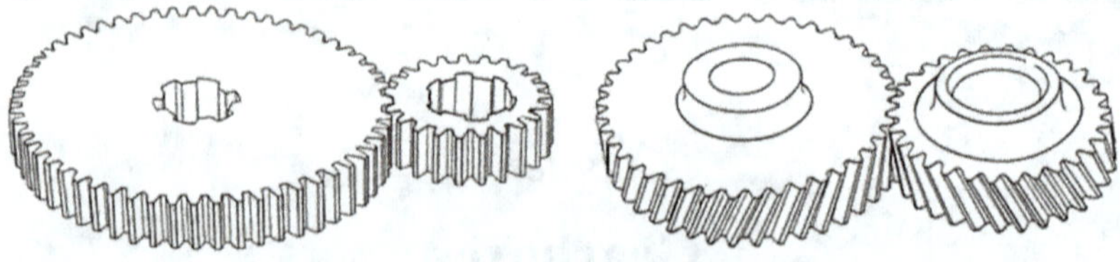

Figure 7.1 Spur (left) and helical (right) gears

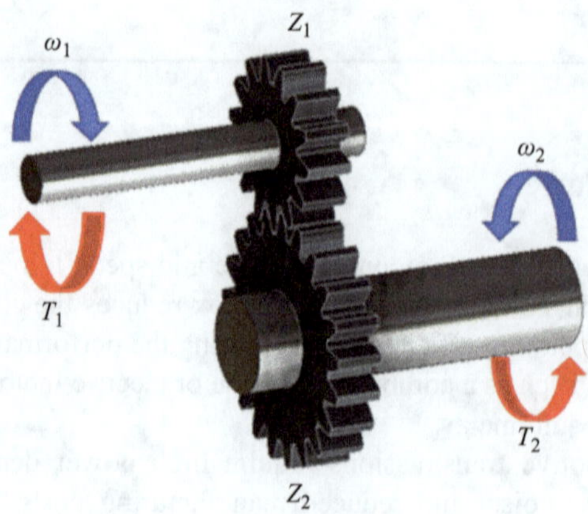

Figure 7.2 Pinion and wheel tooth numbers and rotation speed

principles apply to other types such as straight or spiral bevel gears, as used in many differential gearbox applications.

7.1.2 Gear ratios

Gears change rotation direction, speed and torque by differing the tooth numbers on the pinion – the gear with fewest teeth – and wheel – the gear with the largest number of teeth – as illustrated in Figure 7.2, whereby the gear tooth ratio u_i is function of the pinion z_1 and wheel z_2 tooth number as presented in (7.1). Conversely, the rotation speed ratio can be established based on pinion ω_1 and wheel ω_2 rotational velocities as presented in (7.2).

$$u_i = \frac{z_2}{z_1} \geq 1.0 \tag{7.1}$$

$$\frac{1}{u_1} = \frac{\omega_2}{\omega_1} \tag{7.2}$$

For a simple 2-stage compound gear drive as illustrated in Figure 7.3, the total gearbox ratio u is a function of the first u_1 and secondary u_2 stage ratios as presented in (7.3)–(7.5).

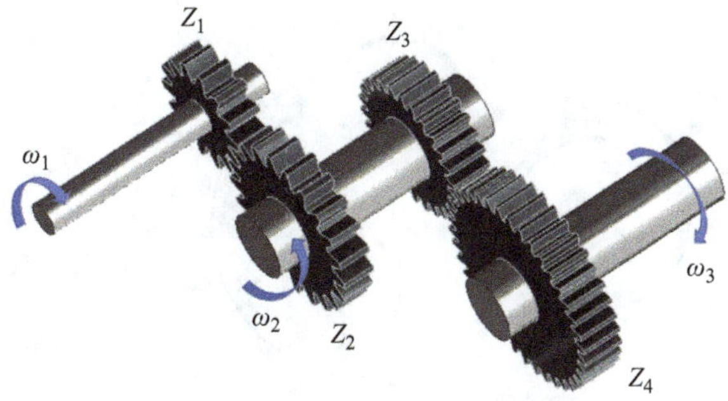

Figure 7.3 A simple 2-stage compound gear arrangement

First-stage gear ratio,

$$u_1 = \frac{z_2}{z_1} \tag{7.3}$$

Second-stage gear ratio,

$$u_2 = \frac{z_4}{z_3} \tag{7.4}$$

Total gear ratio,

$$u = u_1 u_2 = \frac{z_2}{z_1}\frac{z_4}{z_3} \tag{7.5}$$

For the single-stage arrangement illustrated in Figure 7.2, output power is equivalent to $\eta\,\omega_1 T_1 = \omega_2\,T_2$. Most initial designs may assume an efficiency of $\eta = 1$ and thus power is maintained constant throughout the gearbox. In actuality, spur and helical gears achieve an efficiency of approximately 98–99% depending on their design.

7.1.3 Gear theory

The requirement for gears to work without introducing noise or vibration, is such that when the input shaft and gear rotates smoothly and uniformly, the output shaft also rotates smoothly and uniformly, but in a different direction and speed depending on the tooth number ratio, effectively operating like friction discs as illustrated in Figure 7.4. The tooth shape is designed to ensure that when they contact at different diameters, the gears rotate, and operate smoothly without introducing noise and vibration. There are many tooth shapes available to achieve this, but the involute curve as illustrated in Figure 7.5 is the most commonly adopted tooth form used in parallel axis gears.

The involute is a curve that can be created by unwrapping string from a diameter known as the base diameter or generating diameter, as illustrated in Figure 7.5. If two involutes are arranged to contact one another, they will do so

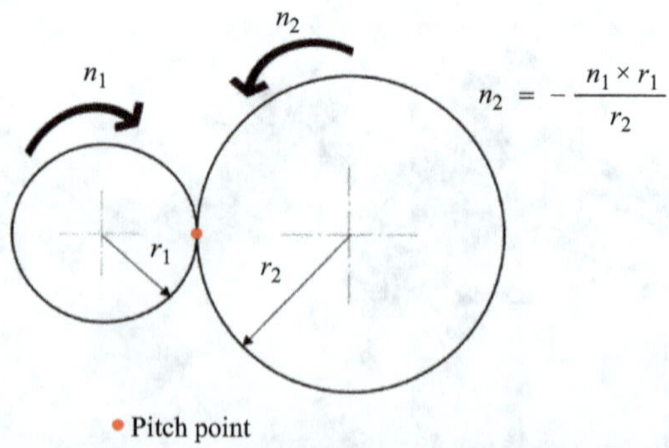

Figure 7.4 The principles of how gears work – the equivalent friction disc

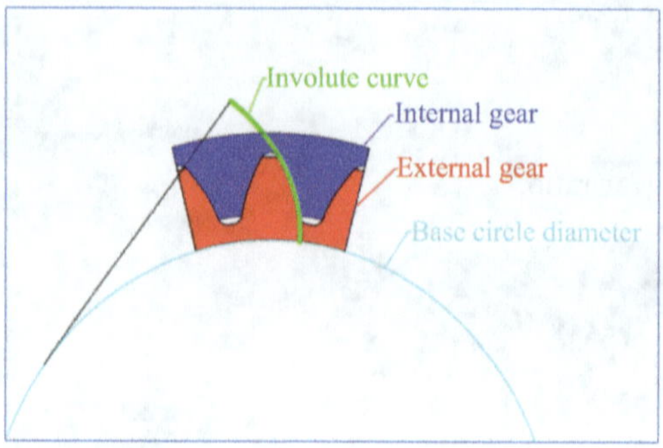

Figure 7.5 The involute curve

along a line of contact which is tangent to the two base diameters, as illustrated in Figure 7.6.

As the drive gear rotates at constant speed, the contact point moves along the line of action as illustrated in Figure 7.7, also at constant speed, and thus the driven gear will rotate uniformly, as desired. However, this seldom occurs due to manufacturing errors in the involute tooth shape and deflections under load. Because all manufacturing processes are imperfect, in addition to the loaded tooth deflections, gear designers account for this by specifying corrections to the involute tooth form, thus minimizing the impact on noise and vibration caused by these errors.

The size of the gear and teeth are defined by the gear module – in millimeters – and the proportions of the gear are defined by the basic rack. Standard tooth proportions are defined in accordance with ISO 53 [1] however automotive designers often use non-standard designs to minimize noise and optimize gear performance.

Accurately manufactured spur gears with standard tooth proportions can still experience noise, due to changes in tooth load sharing. When teeth transfer load from a single tooth pair (Figure 7.7) to double tooth pairs (Figure 7.6) the load intensity is reduced. This reduces the tooth bending and contact deflection, causing

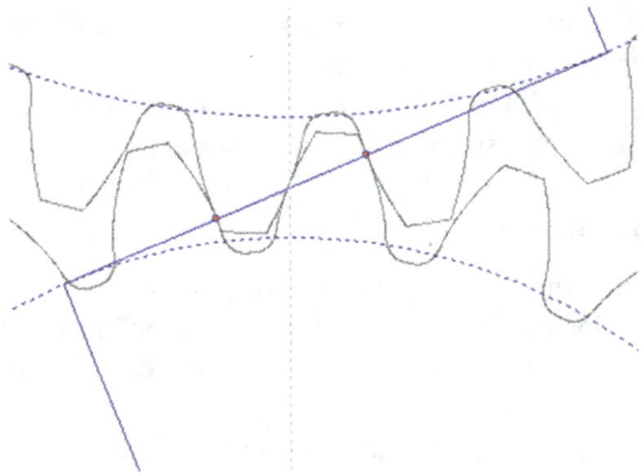

Figure 7.6 The line of contact (solid line) and contact points (red dots) connecting the two base diameters (dotted circles). There are regions where there are two contact points to transfer motion to successive pairs of teeth, as illustrated here

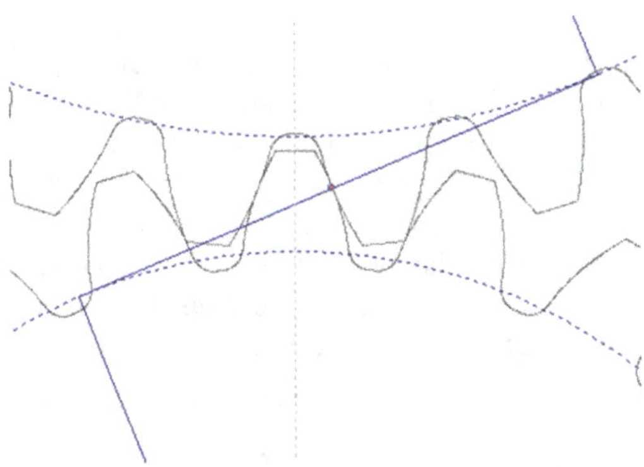

Figure 7.7 A single contact point (red dot) where an individual tooth pair transmit the load in a spur gear

the driven gear to accelerate, subjecting it to dynamic loads and exciting noise. This occurs each time gear teeth enter and exit mesh, causing noise at tooth passing frequency. Helical gears are quieter – although they have the same inherent problem – because the transfer of load between double and single tooth contact occurs at different rotational positions relative to the gear facewidth and therefore the change in tooth deflection is smoother. However, this only occurs if helical gears are made accurately.

7.1.4 Gear geometry

Gear geometry is complex because the traditional methods of manufacturing gears require standard cutting tools such as gear hobs or shaping cutters. These tools generate the finished involute tooth shape when the gear blank is mounted on a dedicated mechanical gear cutting or grinding machine. Modern CNC technology can manufacture gears without using these traditional cutting tools; however, the geometry of the gear is still specified using traditional manufacturing methods.

The methodology commonly adopted to design and define qualitative gear geometry is as follows:

1. Specify the tooth height proportions using a basic rack. This may utilize standard proportions, as specified in ISO 53 [1], or more commonly for automotive gears, it may be based on a custom design to optimize the gear performance.
2. Specify the cutting tool requirements using the basic rack.
3. Select tooth numbers based on the required speed ratio. Normally, the process begins by defining the number of teeth on the pinion, whereby fewer teeth produce a smaller gear, however too few teeth may lead to excessive noise during operation.
4. Select the number of teeth on the wheel to give the required tooth ratio.
5. Select the gear module or scaling factor to calculate the tooth size and gear diameter.
6. Select the gear face width to ensure a sensible gear proportion – too large a face and the gear will misalign due to torsional deflection, too small and there is a risk of increased distortion during cutting or heat treatment.
7. Decide whether spur gears or helical gears are required. Most low noise applications require helical gears, but these generate axial loads and moments which may require suitably larger bearings and more complex arrangements.
8. Specify appropriate tooth thickness and working clearance – or backlash of the gear pair – when mounted between centers.

Each of the above stages requires detailed gear knowledge and calculations to quantify the gear geometry. The most useful reference for gear calculations is ISO 21771 Gears – Cylindrical involute gears and gear pairs – Concepts and geometry [2], the calculations of which are best implemented in a spreadsheet or dedicated software. Most gear designers use commercial packages to perform these calculations, however, the gear designer must still make informed choices and decisions. Gears are often specified on manufacturing drawings using a table of data, as illustrated in Table 7.1, and for consistency, it is considered good practice to define gear geometry using relevant ISO terms and symbols.

In addition to specifying the gear macro geometry as illustrated in Table 7.1, the specification should also include tolerances for allowable profile, helix, and tooth pitch (spacing) deviations. These are often chosen based on performance parameters, but still refer to a tolerance standard such ISO 1328-1 cylindrical gears – ISO system of flank tolerance classification Part 1: Definitions and allowable values of deviations relevant to flanks of gear teeth [3].

The methodology described above relates to specifying only the macro gear geometry, such as tooth number, diameters and face width, as presented in Table 7.1. However, manufacturing tolerances associated not only with the gears,

Table 7.1 Gear geometry specification in accordance with ISO terminology

Gear data		
Macrogeometry		
Tooth number	z	16
Normal module (mm)	m_n	3.9
Facewidth (mm)	b	29
Normal pressure angle (°)	α_n	20.0
Helix angle (°)	β	30.0
Hand	–	LH
Profile shift coefficient (tight mesh)	X	0.29
Reference data		
Reference diameter (mm)	D	72.053
Base diameter (mm)	d_b	66.425
Tip diameter (mm)	d_a	82.115
Root diameter (mm)	d_f	63.395
Basic rack data		
Basic rack addendum	h_{aP}/m_n	1.0
Basic rack dedendum	h_{fP}/m_n	1.4
Basic rack tool tip radius	ρ_{fp}/m_n	0.39
Basic rack module (mm)	m	3.9
Basic rack pressure angle (°)	α_P	20.0
Basic rack protuberance (mm)	–	0
Tooth thickness		
Number of teeth spanned	k	4
Span [max] (mm)	$W_{k\ \max}$	42.280
Span [min] (mm)	$W_{k\ \min}$	42.180
Ball diameter (mm)	D_M	7.0
Measurement over balls [max] (mm)	$M_{dK\ \max}$	83.88
Measurement over balls [min] (mm)	$M_{dK\ \min}$	83.66
Meshing information		
Mating gear	–	0000–001290
Centre distance (mm)	a_w	91.5
Contact ratio	ε_α	1.171

Table 7.2 Flank tolerances for helix, profile and pitch deviations with the additional definition of deviation evaluation ranges

Flank tolerances (ISO 1328-1:2013) (Grade)		
Tip form diameter (mm)	d_{Fa}	82.1115
Profile control diameter (mm)	d_{Cf}	67.288
Helix length of evaluation (μm)	L_β	26.100
Single pitch deviation (μm)	f_P	±6.5 (5)
Cumulative pitch deviation (μm)	F_P	19 (5)
Profile form deviation (μm)	$f_{f\alpha}$	±7.5 (5)
Profile slope deviation (μm)	$f_{H\alpha}$	+6/−6 (5)
Total profile deviation (μm)	F_α	9.5 (5)
Helix form deviation (μm)	$f_{f\beta}$	±6 (5)
Helix slope deviation (μm)	$f_{H\beta}$	+6/-6 (5)
Total helix deviation (μm)	F_β	8.5 (5)

but their respective components such as shafts, bearings, and gearbox housings are such that gears will always be subjected to a degree of misalignment when the pinion contacts the wheel. The problem is often compounded due to loaded deflections, again associated with the gears, shafts, bearings, and housings. To minimize the effect of these errors on noise, vibration, stress and performance in general, small micron size modifications to the gear flank profile and helix are often required. These are commonly known as the gear's micro geometry.

7.2 Gearbox arrangements

A typical gearbox assembly and its associated components are illustrated Figure 7.8. This is a theoretical design that will be used to help the reader gain an understanding of gearbox design and commonly adopted mechanical components. This example is perhaps the simplest gearbox consisting of a single stage spur or helical parallel axis gear pair. Power is transferred into the gearbox via the input shaft (2) and its splined connection, before proceeding to the pinion (4) via a key (21). Similarly, power is transferred from the wheel (5) to the output shaft via another keyway (21) and exits the gearbox via another splined coupling (3). As this is a speed reduction gearbox, the increase in torque often results in a larger diameter output shaft to withstand the resulting stresses.

Keys and keyways (21) are a simple and cost-effective way of transmitting torque via a flat face and are designed to withstand contact and shear stress of the load bearing area and can be analyzed in accordance with AGMA 6101-E08 [4]. Although effective, they are not the most power dense method of torque transfer

Parts List	
ITEM	PART NUMBER
1	Gearbox Base
2	Input Shaft
3	Output Shaft
4	Pinion
5	Wheel
6	Cylindrical Roller Bearing
7	Deep Groove Bal Bearing
8	Taper Roller Bearing
9	Radial Shaft Seal
10	Spacer
11	Lock nut M55
12	Lock nut M75
13	Tab Washer M55
14	Tab Washer M75
15	End Cap
16	External Circlip
17	O-ring
18	Internal Circlip
19	Internal Circlip
20	Gearbox Lid
21	Key
22	Dowel Pin
23	Socker Head Cap Screw
24	Hex Head Bolt
25	Drain Plug
26	Breather Cap

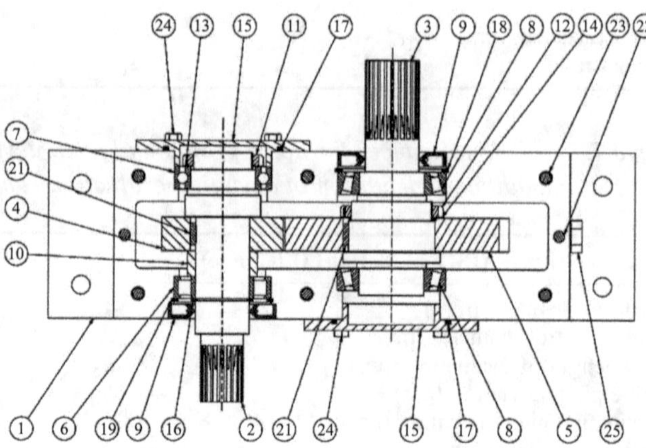

Figure 7.8 Single-stage parallel axis gearbox

and are susceptible to large stress raisers due to the sharp notches generated during the cutting of a keyway into the shaft and hub.

Splines can transmit greater torques than a key because they share load among multiple teeth. It is common for splines to adopt the involute tooth profile with either a flat or fillet root and can be manufactured and measured using the same technology as that adopted for gears, although other profiles such as straight sided and serrations are also available. The design, analysis, and manufacture of splines are more complex than keys, often requires an additional heat treatment process, and there is no internationally recognized ISO standard for spline stress analysis. Commonly, splines can be analyzed in accordance with the Dudley [5] or SAE [6] method, each of which is based on simple contact and shear stress calculations, with additional factors introduced to accommodate tooth load sharing, misalignment, torque overloads, reverse loading and torque and rotational cycles.

The use of keys or splines can be avoided by adopting cylindrical interference fits, whereby the inside diameter of a hub – such as a gear – is smaller than the outside diameter of the shaft to which it contacts. If designed correctly, the resulting pressure generated from an interference fit, in conjunction with the friction at joint interface, is a reliable method for transmitting torque while avoiding the resulting stress concentrations associated with keys and spline teeth (root stress concentrations). The friction and therefore torque that can be transmitted is a function of the mating material properties, their cleanliness, and the method of assembly which can often involve heat or lubrication. Although interference fits are a relatively inexpensive method of torque transfer, the bore and shaft diameters often require precision grinding to achieve the necessary tolerances required for the design to operate satisfactorily. Furthermore, their assembly often requires the use of heat to expand the hub (or cool the shaft) and is more difficult to both assemble and disassemble than a simple keyway or spline connection both of which can be designed with clearance.

Shafts can be constrained radially and axially using various bearing configurations such as cylindrical roller bearings (6) deep groove ball bearings (7) and taper roller bearings (8). Successful gearbox operation requires an understanding of the axial gear load path, such that bearing arrangements are not over or under constrained. In this example, the pinion shaft deep groove ball bearing is fully located such that it can react axial loads in both directions. As a consequence, it is coupled with a non-locating cylindrical roller bearing (6) which is free to float in the axial direction. The wheel shaft adopts a pair of taper roller bearings (8) each of which can only react an axial load in one direction, such that the system is also fully, but not over constrained. An advantage of the floating cylindrical roller bearing is its ability to accommodate large changes in the shaft axial thermal expansion without preloading the bearings.

Helical gears generate an axial force which can act in either direction, depending on their helix hand and direction of rotational input, and therefore must be suitably constrained. Figure 7.9 illustrates how the axial load path is reacted for both the pinion and wheel, depending on the direction of axial gear force.

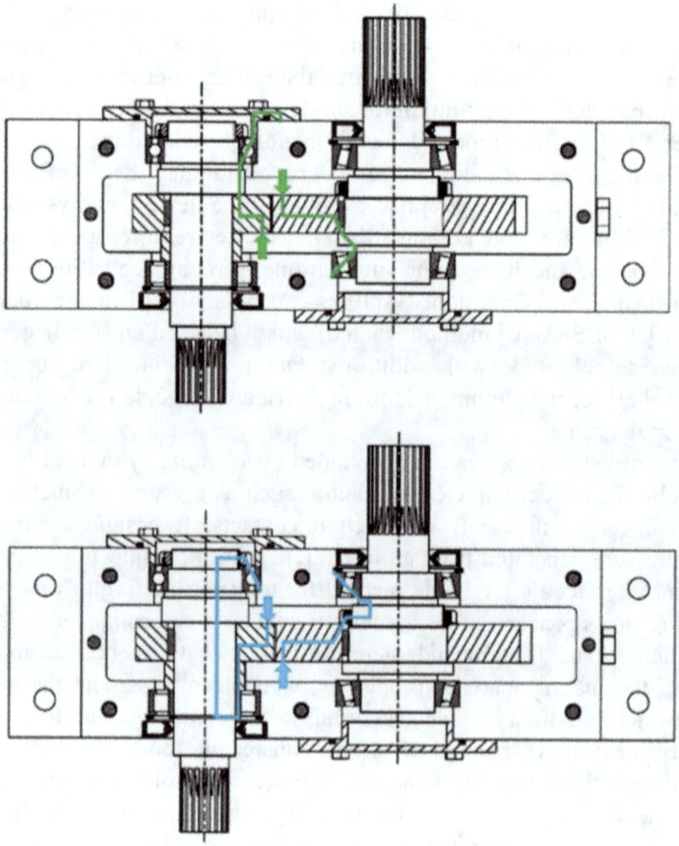

Figure 7.9 Axial load paths

Both gears and bearings are often constrained axially using both internal (18, 19) and external (16) circlips, lock nuts (11, 12) and their associated locking tab washers (13, 14) or simple end caps (15).

Oil is contained within the gearbox via the use of radial contacting shaft seals (9), end caps (15) and O-rings (17) and can be removed from the gearbox via the drain plug (25). Contact seals also prevent contamination from external sources such as dirt and water and are suitable for most industrial gearboxes though there are limitations on the allowable sliding speed at the shaft interface and the maximum temperature that they can withstand. For high-speed applications, suitably designed non-contacting labyrinth seals can be adopted.

For a split housing design such as that illustrated in Figure 7.9, it is important that accurate alignment of the lid and base bores is maintained after manufacture. This can be achieved with the use of precision dowels (22), such that the gearbox lid bolts (23) provide no alignment and only preload. The gearbox base should be fixed using bolts which can withstand the overturning moment of the gearbox, which is a function of the input and output torque.

7.2.1 Multi-stage gearbox layouts

Due to undercutting of gear teeth during manufacture, there are limitations as to how few teeth can be cut onto a pinion, however there is no theoretical upper tooth limit on that of the wheel, and hence the magnitude of the gear ratio that can be achieved in a single stage. However, an infinitely large ratio requires a wheel of infinite reference diameter, effectively becoming a rack and pinion drive as used to convert rotary to linear motion, such as that adopted in a vehicle steering system. Thus size, manufacturing, and cost restrictions seldom see an individual gear pair ratio exceed 5, beyond which multistage gearboxes are adopted whereby additional stages are added, providing a greater overall ratio, as illustrated in Figure 7.10. For example, for a two-stage system, the total ratio i can be split into the first i_1 and second stage i_2 ratio. Olver [7], among others, provide a method of optimizing the ratio split based on minimizing mass. The largest ratio of the split should be

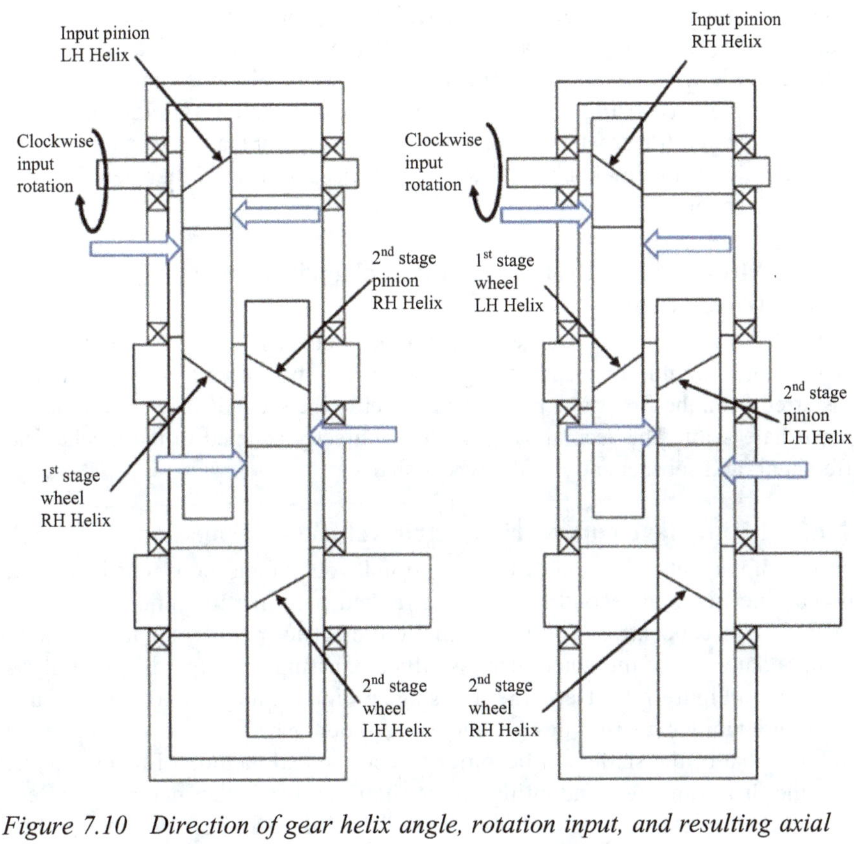

Figure 7.10 Direction of gear helix angle, rotation input, and resulting axial loads

situated on the high-speed shaft, whereby the associated low torque will minimize the tooth module and therefore the size of what is already the largest ratio pair, thus reducing the overall gearbox mass and cost.

When designing multistage systems, the choice of helix angle can be used to optimize the magnitude of the bearing axial force. The direction of the axial force generated at the gear mesh is a function of the helix hand and the direction of rotation input, as illustrated in Figure 7.10, for various hands of helix with a clockwise rotational input, whereby the direction of axial force is illustrated by the blue arrows. Here, the hand of the helix angles on the intermediate shaft are shown equal, such that the axial forces oppose one another, minimizing the axial bearing load.

7.2.2 *Electric vehicle transmission layouts*

Modern electric vehicle technology adopts high speed and power dense motors, which are coupled to the road wheels via a multistage gearbox system, in conjunction with a mechanical differential for torque splitting and accommodating differences in road wheel speed. For example, assuming an input motor speed of 16,000 rpm, then a road wheel speed of 1,000 rpm would require a total gearbox reduction of $i = 16$, and a first and second stage ratio of $i_1 = 5.08$ and $i_2 = 3.15$ would provide a reasonably optimized design. With electric vehicles, this is commonly achieved using (1) standard multi-stage parallel axis layouts, (2) a concentric design for greater compactness, or (3) a combination of parallel axis and epicyclic arrangements, all of which are discussed in further detail in the proceeding sections.

7.2.2.1 Multi-stage, offset parallel axis electric vehicle transmission

A typical 2-stage parallel axis gearbox is illustrated in Figure 7.11 whereby the electric motor would connect to the gearbox via a splined connection. Power is transferred from the first stage pinion and wheel to the second stage pinion via the intermediate shaft. The second stage wheel is directly coupled to the mechanical differential, and subsequently to the wheel shafts.

7.2.2.2 Multi-stage concentric electric vehicle transmission

A more compact electric vehicle transmission layout adopts a concentric design, whereby the first and second stages are arranged such that the input and output power remain concentric. To enable this, the electric motor rotor is hollow to accommodate one of the wheel drive shafts emanating from the differential, as illustrated in Figure 7.12. Here, the first stage pinion is integral to the rotor shaft. This is not unique to this design and is common when the gear diameter is similar to that of the shaft as it becomes an economical method of manufacture. Here, the disadvantage of adopting an overhung pinion is the increased deflection and therefore face load distribution and resulting stresses. However, when compared to the offset parallel axis design (Figure 7.11), the volumetric saving

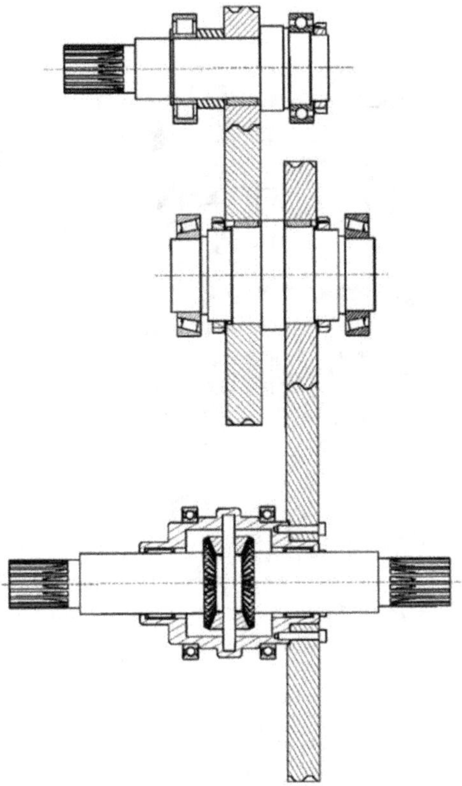

Figure 7.11 Two-stage, offset parallel axis gearbox and differential

of the concentric layout is clear, making this a more compact solution for passenger vehicles.

7.2.2.3 Multi-stage epicyclic design

Compact transmission arrangements often make use of epicyclic gearing which consists of a sun, multiple planets, ring and planet carrier, the latter of which can be single sided, or straddle mounted for greater stiffness. Depending on the required ratio, either the Sun, planet carrier or ring is fixed, with the remaining two members used for power input and output. Epicyclic systems have a high-power density since numerous planets can share the torque, allowing their size to be minimized, while also enabling the input and output shafts to remain concentric. However, there exist significant complications regarding their analysis, specifically surrounding the achievable equality of load sharing amongst the planets, their deflections under load, and design implications for planet bearings and lubrication. As illustrated in Figure 7.13, power enters the gearbox via the sun shaft, and assuming a fixed ring, power transfers out via the planet carrier, and into the second stage which is a simple parallel axis helical gear pair.

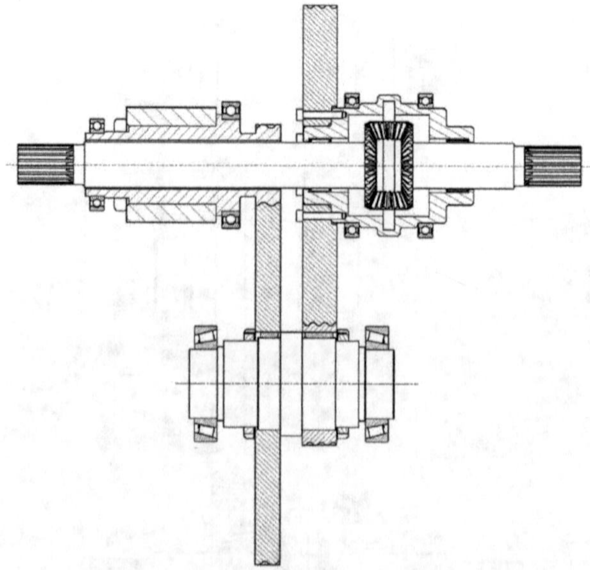

Figure 7.12 Two-stage, concentric gearbox and differential

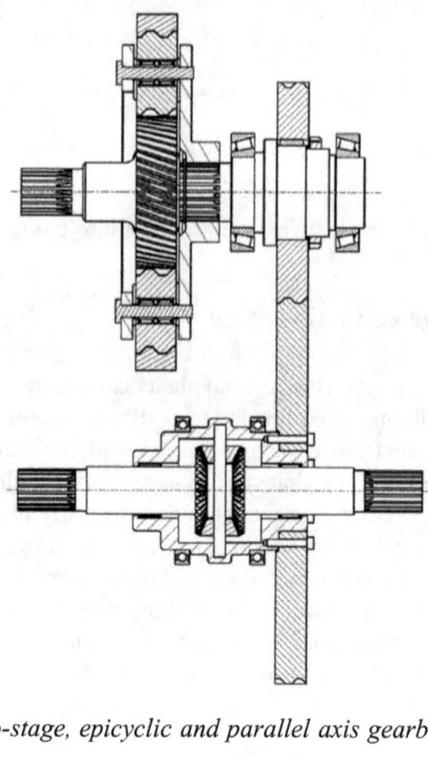

Figure 7.13 Two-stage, epicyclic and parallel axis gearbox and differential

7.3 Gear stress analysis

Although there are many ways whereby a gear may fail, as will be discussed in further detail in Section 8.1, the two most predominant failure modes are associated with contact fatigue – pitting of the flank surface – and bending fatigue, whereby a tooth completely detaches at it base. It is these two modes of failure which often get the most attention from the gear designer, though there are others to consider.

Gears are subject to a fatigue load cycle every time a tooth passes through mesh, resulting in tooth contact (Hertzian) and root bending stresses, which can be difficult to accurately establish due to manufacturing errors, misalignments, dynamic effects and torque fluctuations. The method recommended for gear stress analysis is specified in a collection of internationally recognized ISO 6336 standards, although they are complicated, time-consuming to apply and at times, difficult to understand. Together, they run to hundreds of pages of complicated formula and text and are not 100% accurate – whereby gears that are predicted to be acceptable, fail, and vice versa. The standard is comprised of 10 parts, as follows.

1. ISO 6336-1:2019 basic principles, introduction, and general influence factors [8].
2. ISO 6336-2:2019 calculation of surface durability (pitting) [9].
3. ISO 6336-3:2019 calculation of tooth bending strength [10].
4. ISO TS 6336-4:2019 calculation of tooth flank fracture load capacity [11].
5. ISO 6336-5:2016 strength and quality of materials [12].
6. ISO 6336-6:2019 calculation of service life under variable load (Miners Sum cumulative damage) [13].
7. ISO TS 6336-20:2017 scuffing using flash temperature method [14].
8. ISO TS 6336-21:2017 scuffing using integral temperature method [15].
9. ISO TR 6336-30:2017 calculation examples for parts 1–5 [16].
10. ISO/TR 6336-31:2018(E) calculation examples of micropitting [17].

Due to the complexity of the analytical methods, they can only realistically be implemented in a program or spreadsheet, and it is for this reason that most users purchase a commercial software package. A warning, however: due to their complexity, the standards are difficult to interpret and differences in results established between commercial software packages and users are common.

7.3.1 The principles of ISO 6336

ISO 6336 is based on a comparator principle, whereby it compares the stresses established on the subject gears, to the permissible stresses established historically from tests conducted on real gears, operating in a test rig, running under nominally known conditions. It is a comparator approach since it adopts the same methodology of calculating stress on the subject gears being evaluated, to those of the test gears for which the standard provides a significant material database. Hence, any errors associated with the method of stress analysis are somewhat compensated,

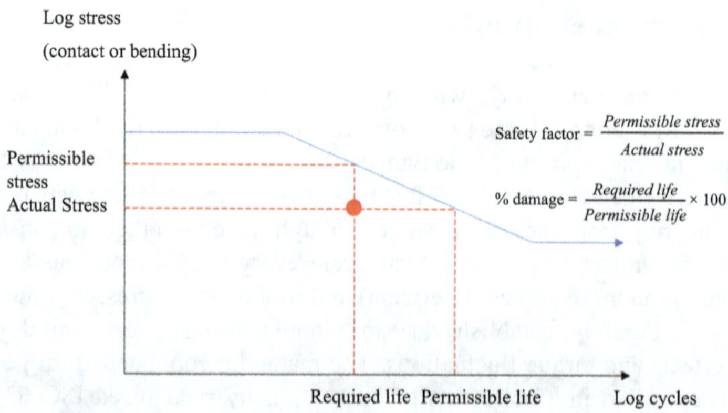

Figure 7.14 Illustration of a S–N fatigue curve for 1% failure

although not entirely, especially when gear geometry deviates significantly from that of the reference test gears. Based on a specific material, quality and heat treatment, permissible fatigue data is generated based on the results from real gears under reference test conditions. This is used to establish material S–N curves which consist of the fatigue endurance limit and slope in the finite life region as illustrated in Figure 7.14. S–N curve data is available for both bending and contact fatigue in ISO 6336-5 [12] based on a 1% failure probability and is thus used to establish safety factors. It should be noted that the allowable stresses are defined as 'stress numbers' as they have the units of stress, but they are representative values only. Furthermore, the standard does not recommend minimum values for safety factors. Design Unit, based at Newcastle University, use default safety factors of 1.0 for contact and 1.4 for bending fatigue; however, some applications define this according to customary practice and experience. Finally, the standard includes different methods, defined as A, B and C, for calculating the actual stresses in the subject gears. It is the responsibly of the user to decide which is most appropriate, with method A the most accurate and method C the least.

7.3.2 ISO 6336 terminology and key factors

Many of the terms and symbols adopted for specifying gear geometry parameters and stress analysis factors have their origins from Germany and are thus are not always logical when converted to English. Familiarization with the ISO system of symbols is essential, as they are used on gear drawings and specifications worldwide and are thus considered the international language of gears. Other gear geometry terms and symbols are presented in accordance with ISO 21771 [2], ISO 1328 [3], ISO 1122 [18], and ISO 701 [19]. A list of the most common key factors is presented below, however, they often have multiple intermediate factors and complex calculations which can only be appreciated by carefully studying the relevant ISO standards:

- The load sharing factor k_γ defines how load is shared amongst planetary gears in epicyclic systems, which although desirable, is not always unity.

- The face load distribution factors for contact $k_{H\beta}$ and bending $k_{F\beta}$ increase the applied load to account for uneven load distribution over the gear face width.
- Manufacturing errors f_{ma} increase the nominal applied load due to random misalignment from manufacturing assembly errors. This effects $k_{H\beta}$ and $k_{F\beta}$.
- Shaft deflections under load f_{sh} affect the face load distribution $k_{H\beta}$ and $k_{F\beta}$.
- Transverse load factors $k_{H\alpha}$ and $k_{F\alpha}$ increase the nominal applied load to account for uneven load sharing between teeth during the meshing process due to errors or tooth deflection.
- The application factor k_A accounts for peak loads that occur external to the gears. Although these peaks may be infrequent, they inevitably increase stress.
- The dynamic factor k_v estimates the increase in load due to self-excited forces transmitted during changes from single to double tooth contact which occurs during the meshing cycle.
- The helix angle factors Z_β and Y_β modify contact and bending stress dependent on the helix angle.
- The rim thickness factor Y_B accounts for the effect of thin rims on root bending stress. For example, when the bore of a gear blank is close to that of the tooth root.
- The grinding notch factor Y_{sg} accounts for the increase in bending stress caused by grinding notches generated during the manufacturing process, which should be avoided.
- The reverse bending load factor is applicable to idler gears and applications with frequent torque reversals, such as electric drives in regeneration mode. Although there is no defined symbol, the increase in stress range is such that the permissible bending stress should be reduced by including a factor of 0.7.

7.3.3 ISO 6336 load-time analysis

Although contact and root bending stresses are established based on single load, the ISO 6336 standards can be applied to gears subject to varying torque and speed conditions over their operating life. Options include applying a simple application factor $K_{A\text{-}B}$ which assumes that the gears are subject to a peak load continually. Although simple, this method may result in over engineered designs which are heavier and less efficient. The preferred method for automotive gears is to apply a Miners sum cumulative damage analysis, $K_{A\text{-}A}$ the details of which are provided in ISO 6336-6 [13]. Here, a duty cycle is defined based on load, speed and time, and converted from time-based loads to a time-load histogram, under the assumption that damage is accumulated for each load, independent of the order in which they are applied.

7.4 Gear optimization

7.4.1 Transmission error and tooth contact analysis

Accurately specifying gear geometry and achieving acceptable safety factors in accordance with an ISO 6336 fatigue stress analysis is not sufficient to assure gears will operate with acceptably low noise and vibration, although it can be minimized with careful design, manufacture and alignment. Though it is difficult to predict noise, it is closely correlated with transmission error (TE) which is the kinematic or angular position error ($\delta\theta_2$) of the driven gear with respect to that of the driving gear, whilst taking into account the gear ratio, as illustrated in Figure 7.15. Thus, by reducing the magnitude of the TE, the dynamic loads and noise of the system are also reduced.

The tooth-to-tooth TE can be established and minimized using a tooth contact analysis (TCA) model which is commonly used for detailed gear design and optimization. TCA models simulate the static tooth mesh conditions of a gear pair at distinct positions of the tooth engagement cycle, by considering the stiffness of the system, either analytically or numerically, as illustrated in Figure 7.16, often including that of the gear teeth and blank geometry. In addition, they can incorporate manufacturing errors, blank misalignments, and micro geometry including tooth flank profile and helix corrections. Based on a known load or torque condition, the TCA can establish the "no load" tooth gap and the resultant load intensity distribution and tooth deflection at distinct tooth engagement positions of mesh.

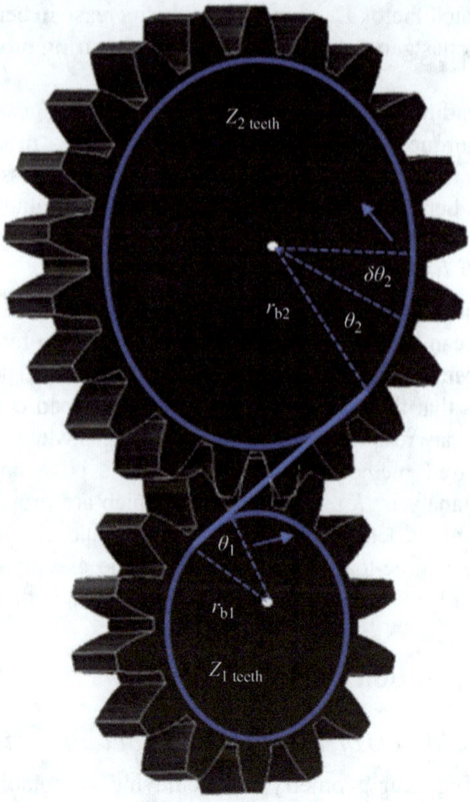

Figure 7.15 Transmission error of the driven gear $\delta\theta_2$

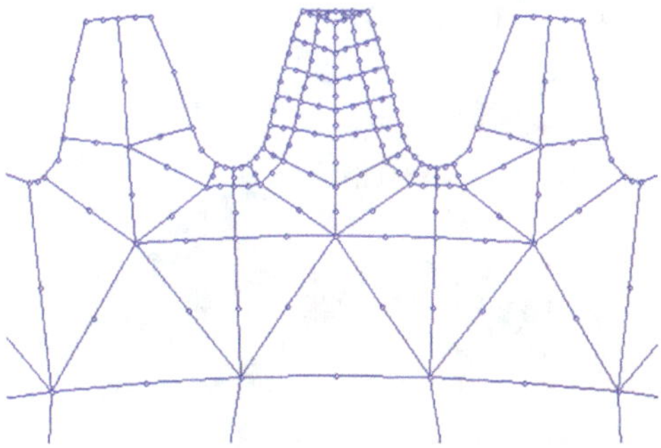

Figure 7.16 TCA finite-element model for establishing tooth stiffness (Dontyne Systems Ltd)

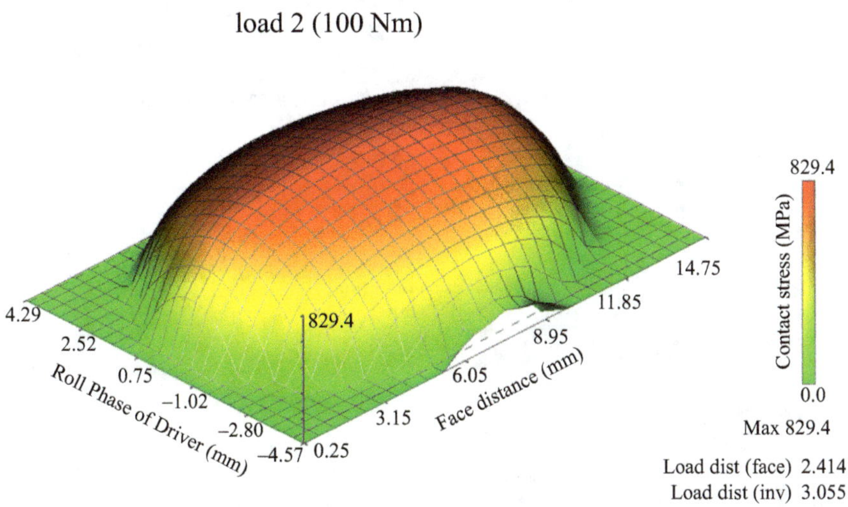

Figure 7.17 TCA surface contact stress (Dontyne Systems Ltd)

This allows the designer to establish performance parameters such as contact and bending stress distribution, tooth temperature friction losses, efficiency, and TE. An example result from a TCA is provided in Figures 7.17–7.19 illustrating the contact stress over the active face width, frictional tooth losses and a plot of TE with an associated FFT analysis, respectively.

7.4.2 Micro and macro gear geometry optimization

TCA models provide an excellent tool to develop and optimize gear macro and micro geometry for maximum load carrying capacity and thus power density, in addition to minimizing TE and therefore noise. The optimization of macro gear geometry can include:

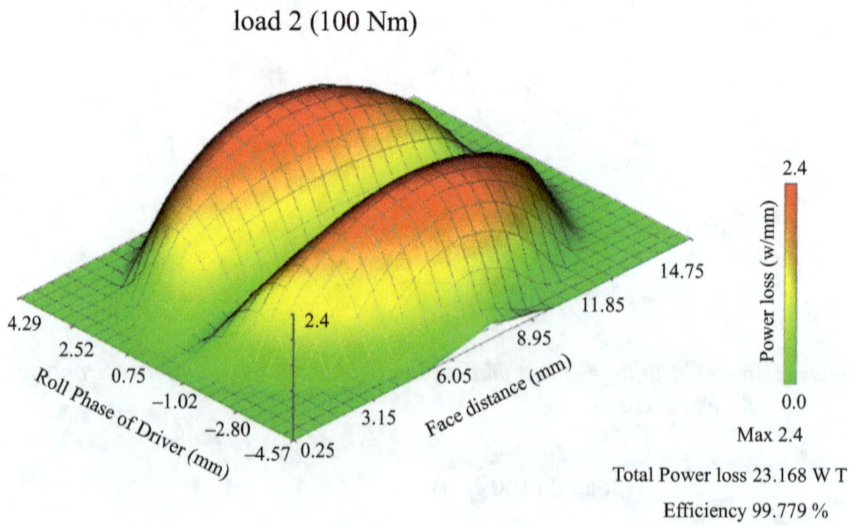

Figure 7.18 TCA frictional power loss over the active tooth surface (Dontyne Systems Ltd)

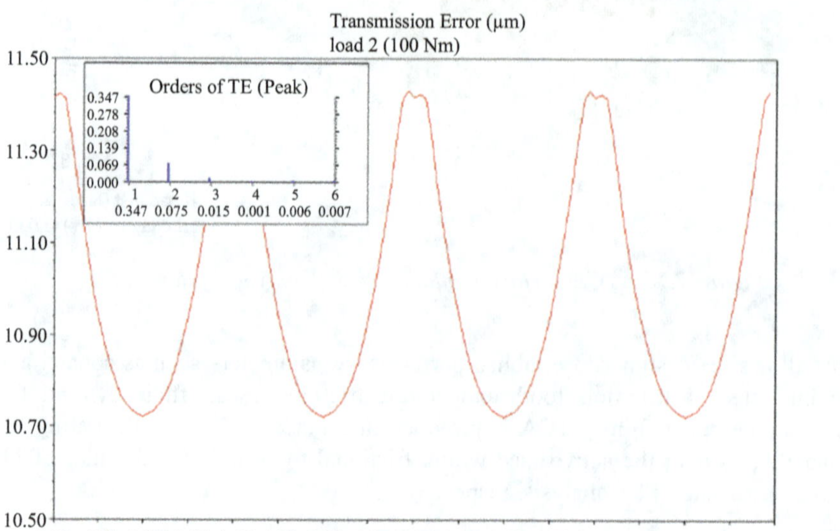

Figure 7.19 TCA TE plot showing a predicted tooth-to-tooth TE of 0.71 μm and FFT first-order component of 0.35μm (Dontyne Systems Ltd)

1. Minimizing the gear blank torsional deflection by controlling the gear diameter to face width ratio.
2. Increasing the tooth depth and changing the pressure angle to assure that there are always 2 pairs of teeth in contact at any position of engagement.
3. Adopt helical gears with an integer overlap ratio, which is that between the face width and axial pitch. This is achieved by changing the helix angle or face width.
4. Adjust the pressure angle, profile shift coefficient, and gear tip diameter to increase gear bending or contact strength, improve efficiency, and minimize the risk of scuffing failure (see Section 7.8).

The optimization of micro gear geometry has the objective of minimizing the effect from random unknown manufacturing errors and systematic, load dependent elastic deflection and can include:

1. The application of relief to the tip of the involute profile to minimize the risk of premature contact caused by elastic deflection or manufacturing errors as illustrated in see Figure 7.20. Typically, this involves the removal of 5–50 μm of material, often assuming a linear or parabolic function. Tip relief also increases efficiency as it reduces the load in regions of high tooth sliding contact.
2. Analogous and often in conjunction with tip relief, material can be removed from the base of the tooth, known as root relief, as illustrated in Figure 7.20.
3. Tip and root relief is sometimes replaced with profile crowning, which achieves the same effect.
4. Over the face width of the gear, micro geometry corrections can include helix crowning, as illustrated in Figure 7.21 to minimize the effect of gear mis-alignment, manufacturing errors and bending and torsional deflection. This improves load distribution and minimizes peak contact and bending stresses.

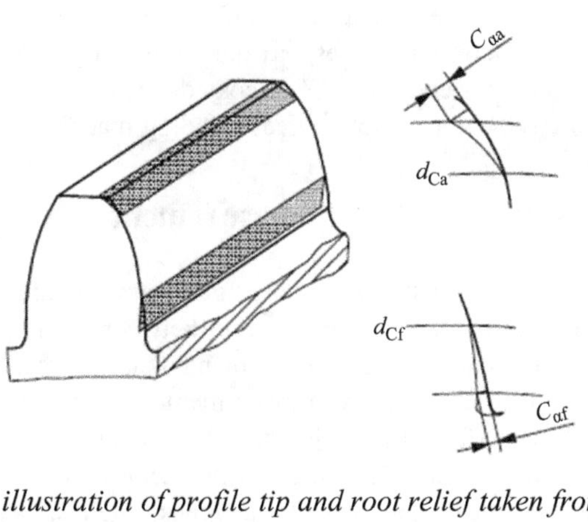

Figure 7.20 An illustration of profile tip and root relief taken from ISO 21771 [2]

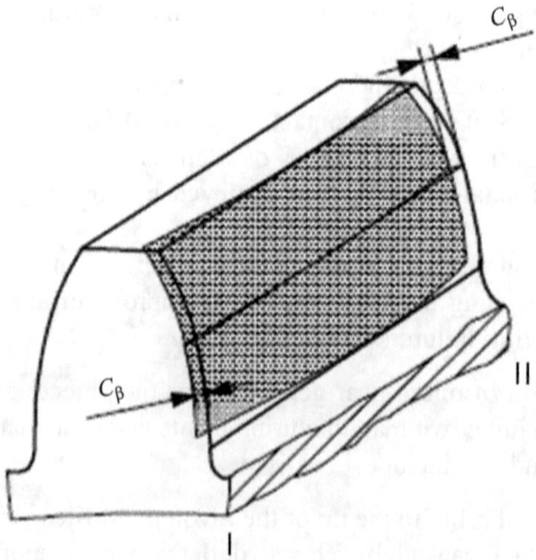

Figure 7.21 An illustration of helix crowning taken from ISO 21771 [2]

7.5 Manufacturing methods

Automotive gearboxes require precise, good quality, hardened gears to meet the demands of high-power density and noise. This requires more than precise manufacture which includes accurate, high quality, functional datum surfaces, with low runout on both the gear blanks, mounting mandrels and tooling. Furthermore, it requires accurate soft cutting, prior to hardening, proceeded by a controlled heat treatment process, accurate finish grinding with low runout and carefully controlled and applied measurement procedures.

There are many methods available for manufacturing gears; however, high precision is usually achieved with form and generation cutting processes, and automotive gears generally require an ISO 1328-1 [3] grade 5 accuracy or better. This limits the manufacturing process available for hardened gears to finish grinding using either vitreous or CBN-coated grinding wheels or honing. Figure 7.22 shows a typical automotive gear grinding machine.

7.6 Gear quality control and measurement

Gears are accurate components with complex geometry and are commonly specified with tolerances on profile and helix form of between 5 μm and 10 μm or less. Conventional measurement practices recommend that the measurement method should have a capability, defined by measurement uncertainty with a 95% confidence interval (U_{95}) of 10% of the tolerance. This requires a measuring machine capability of ±0.5 μm to 1.0 μm, which is not achievable in a manufacturing environment. Modern gear measuring machines such as the Klingelnberg

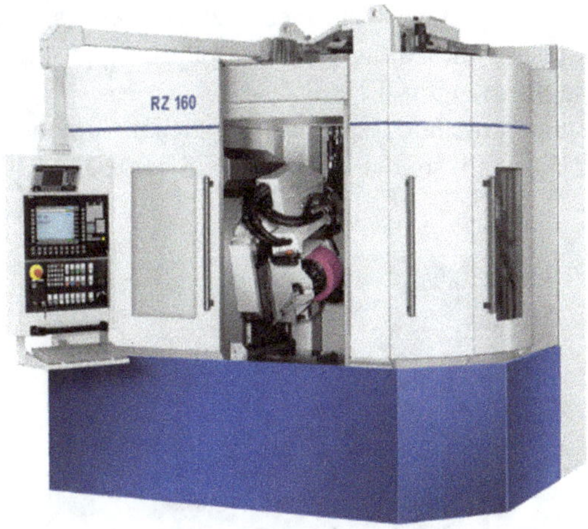

Figure 7.22 Example of a gear grinding machine – Reishauer RZ160 automotive generative gear grinder (from https://www.reishauer.com)

Figure 7.23 A Klingelnberg P65 CNC precision gear measuring machine

illustrated in Figure 7.23 can achieve ±1.0 μm to 2.0 μm which is still considered extremely high performance, but this requires suitable calibration and operator training to follow best measurement practices. Guidance on good measurement practice is provided in ISO TR 10064-1:2019 [20] and is recommended reading for users measuring or interpreting gear measurement results. This also includes information on assessing measuring machine capability using calibrated workpieces such as those illustrated in Figure 7.24.

Other measurement methods such as the radial composite (double flank) measurement methods, as illustrated in Figure 7.25 are used to 100% inspect gears

Figure 7.24 1 mm module master-gear for calibrating gear measuring machines.

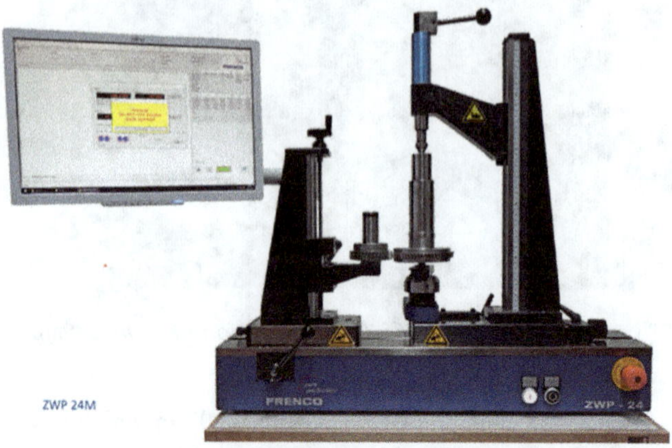

Figure 7.25 Frenco Gmbr – radial composite double flank tester with evaluation software (from https://www.frenco.de/en)

from the manufacturing process. These are commonly used alongside CNC form measuring machines to investigate issues identified by the radial composite measurements.

7.7 Steel selection and heat treatment methods

Gears used in automotive power transmissions require surface hardening to impart the necessary levels of fatigue strength and wear resistance. The two most common methods of surface hardening used in high performance gearing applications are carburising and nitriding.

Carburising, which is often referred to as case hardening, involves the diffusion of carbon into components at elevated temperatures, typically 850–1,000°C. The subsequent quenching and tempering operations generate a hardened surface layer. Carburising is capable of achieving consistent case depths of up to approximately 6 mm, but problems can potentially occur as a result of near-surface oxidation, distortion and/or quench cracking.

Nitriding is similar to carburising, but with nitrogen diffused into the components instead of carbon. The process is carried out at lower temperatures, typically 490–520°C and can generate a higher surface hardness. The nitriding cycle does not include a quenching operation and so distortion is less likely; this makes nitriding an attractive option for surface hardening ring gears used in epicyclic transmissions that would undergo significant distortion if quenched. However, the maximum case depth that can be achieved by nitriding is relatively low, typically less than or equal to approximately 0.8 mm.

In most cases, the key to selecting an appropriate gear steel is identifying the optimum combination of properties that satisfy the requirements of the given application at the lowest cost. The selection process is governed by a number of factors including mechanical properties, metallurgical characteristics, manufacturing route, availability, and cost.

Guidance on the selection and specification of steels that will be surface hardened by carburising or nitriding can be found in ISO 6336-5:2016 [12], ISO 683-3:2016 [21], and ISO 683-5:2017 [22]. ISO 6336-5:2016 [12] also provides guidance on the heat treatment requirements that should be specified for carburised and nitrided gears.

7.8 Gear failure modes

The main failure modes to consider for surface hardened gearing used in automotive power transmissions are bending fatigue, contact fatigue and scuffing. Bending fatigue is a well understood mode of failure that typically occurs as a result of fatigue crack initiation in the root fillet, either from the surface or from non-metallic inclusions present in the steel. Subsequent crack propagation then causes catastrophic failure due to partial or complete tooth detachment, an example of which is shown in Figure 7.26.

Figure 7.26 Root bending fatigue failure on a carburised helical test gear

The risk of bending fatigue failure can be minimized through optimization of the safety factors obtained from ISO 6336-3:2019 [10] and by avoiding stress raisers such as small root fillet radii and grinding notches. Controlled, post heat treatment shot peening is another well-established method of enhancing bending fatigue resistance through the introduction of increased levels of near-surface compressive residual stress. Root grinding can also be used to impart increased bending fatigue resistance through removal of surface weaknesses associated with the surface hardening process and the introduction of increased levels of surface compressive residual stress.

Contact fatigue damage on the flanks of surface hardened gear teeth initially occurs in the form of microscopic, surface initiated pits that are typically less than 10 μm in depth. In some cases, micro-pits formed during early stage running will arrest and remain benign during subsequent operation. However, in many instances, the micro-pitting will gradually increase in severity with further running resulting in sufficient loss of material from the gear tooth profiles to cause increased levels of noise and vibration. Severe micro-pitting can also lead to the occurrence of catastrophic failures in the form of macro-pitting and/or tooth breakages caused by flank initiated bending fatigue. Examples of these failure modes are shown in Figures 7.27–7.29.

The mechanism by which micro-pitting develops is not well understood but is known to be influenced by a number of factors which include gear geometry, surface topography, surface metallurgy, operating conditions and lubricant properties. The risk of micro-pitting and macro-pitting occurring during in-service operation can be minimized through optimisation of the safety factors obtained from ISO/TR 6336-31:2018 [17] and ISO 6336-2:2019 [9] respectively. The application of isotropic superfinishing and/or diamond-like carbon PVD coatings to finish ground gears has also been shown to give increased resistance to micro-pitting and therefore associated failure modes through modification of surface texture and/or friction characteristics.

Figure 7.27 Micro-pitting on a carburised and finish ground spur gear

Figure 7.28 Macro-pitting initiated from micro-pitting on a carburised and finish ground helical gear

Scuffing failures occur when the lubricant film separating mating gear tooth flanks breaks down resulting in metal-to-metal contact. This causes welding and tearing of the mating surfaces to ensue at high speed, an example of which is shown in Figure 7.30. Scuffing occurs instantaneously with no prior warning and, in severe cases, material may be displaced over the tips or into the root fillets causing significant degradation of the tooth geometry.

The risk of scuffing can be assessed and minimised using the calculation methods outlined in ISO/TS 6336-20:2017 [14] and ISO/TS 6336-21:2017 [15]; both analysis methods should be used, and it should be ensured that both methods give acceptable results. The risk of scuffing can also be reduced through the use of isotropic superfinishing and/or diamond-like carbon PVD coatings to give enhanced surface texture and/or friction characteristics. Further guidance on these

Figure 7.29 *Partial tooth breakage initiated from macro-pitting on a carburised and finish ground helical gear*

Figure 7.30 *Scuffing damage on a carburised and finish ground spur gear*

failure modes and the measures that can be taken to minimise the risk of them occurring can be found in ANSI/AGMA 1010-F14 [23].

7.9 Bearings

Typical gearbox bearings can be generally classified into one of two groups: either plain/sliding bearings or rolling element bearings. There exist other types such as magnetic and air bearings though they are uncommon automotive gearbox solutions and will not be discussed further. Plain bearings, in their simplest form, are a cylindrical sleeve with no moving components and rely on low friction at the shaft interface which can be achieved via the use of various materials and lubricants and can be designed to withstand both radial and axial forces. Plain bearings can be designed such that the rotational action of the shaft entrains and squeezes a fluid film in between the shaft and sleeve. This lifts the shaft clear of the sleeve and is known as a hydrodynamic bearing which is often a compact and low wear solution. However, satisfactory design relies on suitable oil properties and cleanliness, and since the shaft will contact the sleeve at zero speed, care is needed at low speeds. Alternatively, fluid can be supplied under pressure using a hydraulic power pack, such that the shaft and sleeve are separated regardless of rotational velocity. Known as a hydrostatic bearing they have good efficiency, running accuracy, stiffness, and load carrying capacity at low speeds, though they require supplementary hydraulic equipment.

Most industrial gearboxes adopt rolling element bearings, consisting of an inner race, outer race, rolling elements, and cage. The rolling elements are either balls or cylinders, albeit the latter can be tapered or barreled for withstanding greater axial loads or shaft misalignments respectively. Size for size, cylindrical roller bearings withstand greater radial loads than ball bearings due a larger Hertzian contact area, though they cannot withstand the same speeds. Almost all rolling element bearings can withstand a degree of axial load, though the taper roller and the angular contact ball bearing are most common for large axial forces in a single direction and are often used in pairs to accommodate bi-axial loads.

The bearing cage separates the elements, preventing sliding, and are typically made from steel, brass or polymers. By removing the cage, more elements can be assembled within a bearing, resulting in a greater load carrying capacity. Known as a full complement bearing, their speed and efficiency are reduced due to sliding losses generated between contacting individual elements. The fit of the inner and outer race is either clearance, interference, or transition, and is driven by the operating characteristics of the machine, the direction of loading and which race is stationary. The most compact gearbox design can be achieved by integrating the inner and outer race into the gearbox housing and shaft, respectively, such that they become the running surfaces. A typical example is the integrated planetary bearing, whereby the planet pin and bore becomes the inner and outer race respectively. Although more compact, it requires careful consideration of the material properties and heat treatment. This is often a compromise, whereby the optimal specification for a gearing steel and heat treatment is not necessarily that of a bearing steel. Furthermore, the failure of an integrated bearing may require all integrated components to be replaced. Careful consideration also needs to be given to the amount of internal clearance designed into system.

Bearings can be sealed for life with either contacting or non-contacting seals, and they are normally supplied pre-filled with grease, though there are limitations on speed and temperature. Open bearings without a seal or grease require a supply of oil but can operate at greater speeds and have a greater ability for heat removal. In a gearbox application, this can be achieved via the gearbox oil which can be either splash lubricated or directly supplied to the bearings.

Bearing life can be analyzed both statically and dynamically in accordance with (7.6) and (7.7), respectively. The first establishes the safety factor relating to the overload or plastic deformation of the bearing under a static load. This is a

Figure 7.31 Contact fatigue damage on the inner race of a taper roller bearing

function of the static load rating of the bearing C_0 and the equivalent static load P_0 which can be a function of both the static radial and axial loads. Equation (7.7) establishes the dynamic life rating L_{10}. This establishes the bearings susceptibility to contact fatigue, an example of which is illustrated in Figure 7.31, based on the number of bearing revolutions, in millions of cycles, at which 90% of the bearings would be expected to survive. It is a function of the bearing's dynamic load rating C_r and equivalent load P_r with the exponent p taken as 3 for ball bearings and 10/3 for roller bearings. The L_{10} life can be adapted to establish the more complex modified rating life L_{nm} in accordance with (7.8) to account for a reliability greater than 90% via the a_1 parameter, and to account for lubrication cleanliness, operating viscosity and the bearing's fatigue limit via the a_{ISO} parameter. The reader is directed to ISO 281:2007 [24] for further guidance on establishing the modified L_{nm} rating life.

$$S_0 = \left(\frac{C_0}{P_0}\right) \tag{7.6}$$

$$L_{10} = \left(\frac{C_r}{P_r}\right)^p \tag{7.7}$$

$$L_{nm} = a_1 \cdot a_{ISO} \cdot L_{10} \tag{7.8}$$

7.10 Lubrication and cooling

Gears and bearings require lubrication to maintain an oil film between moving surfaces thus avoiding metal to metal contact, which would increase friction and the risk of scuffing. Generally, only a few microns of lubricant thickness are required to achieve this condition. It is common that the lubrication specification requirements are different for gears and bearings and therefore a compromise is often required. The lubricant also removes heat generated by gears and bearings due to friction losses, windage losses and lubricant churning, transferring the losses to the gearbox housing and its subsequent surroundings.

The selection of lubricating oil viscosity and extreme pressure (EP) additives to minimize the risk from metal-to-metal contact is a specialist field and often requires close collaboration with lubricant suppliers. Minimizing the quantity of the lubricant and the EP additives reduces weight and minimizes their environmental impact.

Most conventional automotive gearboxes for petrol and diesel engines do not require external pumping, filtering or cooling of the lubricant. They rely on the gears to circulate oil by splash lubrication and the gearbox housing for cooling, often designed with cases and fins for improved heat transfer.

Some electric vehicle transmission gearboxes use the same lubricant to cool the E-motor and gearbox, resulting in the selection of a non-optimized oil. This simplifies the overall design and may reduce cost and weight, but the risk of scuffing type failures may be increased. Controlled testing of candidate oils is often the only feasible solution to validate any key decisions.

References

[1] ISO 53:1998, Cylindrical Gears for General and Heavy Engineering – Standard basic rack tooth profile.

[2] ISO 21771:2007(E), Cylindrical Involute Gears and Gear Pairs – Concepts and Geometry.

[3] ISO 1328-1 Cylindrical Gears – ISO System of Flank Tolerance Classification. Part 1: Definitions and Allowable Values of Deviations Relevant to Flanks of Gear Teeth.

[4] ANSI/AGMA 6101-E08 Design and Selection of Components for Enclosed Gear Drives (Metric Edition).

[5] Dudley, D.W., When splines need stress control, *Product Engineering*, 1957: pp. 56–61.

[6] Cedoz, R.W. and M.R. Chaplin, *SAE Design Guide for Involute Splines,* Warrendale, PA: Society of Automotive Engineers, 1994.

[7] Olver, A.V., The effect of configuration in the design of geared transmission systems, *Journal of Mechanical Design, ASME*, vol. 131, no. 7, p. 074504, 2009.

[8] ISO 6336-1: 2019(E), Calculation of Load Capacity of Spur and Helical Gears – Part 1: Basic Principles, Introduction and General Influence Factors.

[9] ISO 6336-2:2019(E), Calculation of Load Capacity of Spur and Helical Gears – Part 2: Calculation of Surface Durability (Pitting).

[10] ISO 6336-3:2019(E), Calculation of Load Capacity of Spur and Helical Gears – Part 3: Calculation of Tooth Bending Strength.

[11] ISO TS 6336-4:2019 Calculation of Tooth Flank Fracture Load Capacity.

[12] ISO 6336-5:2016(E), Calculation of Load Capacity of Spur and Helical Gears – Part 5: Strength and Quality of Materials.

[13] ISO 6336-6:2019(E), Calculation of Load Capacity of Spur and Helical Gears – Part 6: Calculation of Service Life under Variable Load.
[14] ISO/TS 6336-20:2017(E) – Part 20: Calculation of Scuffing Load Capacity (also Applicable to Bevel and Hypoid Gears) – Flash Temperature Method.
[15] ISO/TS 6336-21:2017(E) – Part 21: Calculation of Scuffing Load Capacity (also Applicable to Bevel and Hypoid Gears) – Integral Temperature Method.
[16] ISO TR 6336-30:2017 Calculation Examples for Parts 1 to 5.
[17] ISO/TR 6336-31:2018(E), Calculation of Load Capacity of Spur and Helical Gears – Part 31: Calculation Examples of Micropitting Load Capacity.
[18] ISO 1122-1:1998 Vocabulary of Gear Terms – Part 1: Definitions Related to Geometry.
[19] ISO 701:1998 International Gear Notation – Symbols for Geometrical Data.
[20] ISO TR 10064-1:2019. Code of Inspection Practice. Part 1: Measurement of Cylindrical Gear Tooth Flanks.
[21] ISO 683-3:2016, Heat-Treatable Steels, Alloy Steels and Free-Cutting Steels, Part 3: Case-Hardening Steels.
[22] ISO 683-5:2017, Heat-Treatable Steels, Alloy Steels and Free-Cutting Steels, Part 5: Nitriding Steels.
[23] ANSI/AGMA 1010-F14, Appearance of Gear Teeth – Terminology of Wear and Failure.
[24] ISO 281:2007 Rolling Bearings – Dynamic Load Ratings and Rating Life.

Chapter 8

Research prototypes and commercial products

Xu Deng[1], Barrie Mecrow[1], Thomas Jahns[2], Liliana De Lillo[3], Lee Empringham[3] and Christopher Hilton[4]

In this chapter, we will review several of cutting-edge integrated motor drive (IMD) research prototypes and a commercial product.

8.1 Research prototype 1: development of a high-speed, permanent magnet, SiC-based drive with integrated input filters

8.1.1 Background

Some IMD research has focused on volume reduction of the passive components, such as the converter chokes, DC-link capacitors and electromagnetic interference (EMI) filters [1–3]. However, standard packaging of these passive components is considered, forming a technological barrier that restricts the total level of integration. This problem becomes worse in high-speed motor applications, as the volume of the low-frequency filters dominates the total volume of the system. The envelope of the filter components in conventional AC motor drives is rarely congruent with the machine, which leads to a poor utilization of the available volume. There is hence a strong motivation to integrate the filters and machine together, sharing magnetic circuits and cooling systems where possible.

A high-speed permanent magnet (PM) integrated motor is proposed in this paper based on the authors' previous studies in [4,5], with integrated LCL input filter inductors in the electrical machine, achieved using a double-slot stator structure. The stator has been designed to reduce the size and loss of the filter inductors without affecting the base machine performance. The AC filters operate at high current density and share the same core back as the armature windings. The integrated motor incorporates a number of important design features, which are not possible with discrete AC filter inductors. Segmental stators simplify manufacture

[1]School of Engineering, Newcastle University, UK
[2]Department of Electrical and Computer Engineering, University of Wisconsin-Madison, USA
[3]School of Electrical and Electronic Engineering, University Park, UK
[4]Chief Technology Officer, Protean Electric Ltd, Surrey, UK

and permit the use of compressed coils, which are pressed directly onto the stator and form solid parts with a high copper fill factor. The prototype benefits from the design features outlined above; consequently it offers superior performance, with a major volume reduction over industrial AC inductors manufactured for the same power rating. The experimental prototype is extensively tested and compared with design predictions. Following the integration of LCL filter magnetic components, a copper bus bar solution is developed from the authors' previous work in [6] and is employed to manage a large number of winding terminals introduced from the integrated LCL filters. An axial machine housing mounted power electronics design, based on SiC devices, is developed. A 34.0 kW IMD prototype is constructed, with a series of static and dynamic tests carried out to validate the FEA, decreasing the total harmonic distortion and maximizing the performance of the PM machine.

8.1.2 Integrated motor design

8.1.2.1 Base machine design

To ensure the integrated filters do not significantly affect the performance of the base machine, the filters must be magnetically decoupled from the base machine. While it is advantageous for the motor and filter to share magnetic paths, they should not be mutually coupled or influence each other through magnetic saturation of the mutual paths. The specifications of the 34.0 kW base machine and the LCL filter inductors are listed in Table 8.1.

An 8 pole, surface mounted neodymium-boron PM rotor is selected for the base machine. A single tooth wound winding structure is employed to minimize the

Table 8.1 Specifications of the base machine, drive and filters

Design parameters of the base machine		
N_{Shaft}	Shaft speed	25,000 r/min
n_{p}	Pole number	8
T	Machine torque	13.0 N·m
n_{t}	Tooth number	12
W_{tb}	Tooth width	10.0 mm
d_{r}	Rotor diameter	62.0 mm
OD_{m}	Machine lamination diameter	109.0 mm
S_{L}	Stack length	70.0 mm
–	Winding type	Concentrated
–	Magnet material	NdFeB (N42UH)
L_{d}	*d*-axis inductance	0.158 mH
L_{q}	*q*-axis inductance	0.163 mH
η	Efficiency	98.0 %
Design parameters of drive and filter inductors		
P_{in}	Drive input power	38.0 kW
L_2	Drive side inductance	160.0 μH
L_1	Grid-side inductance	5.0 –H
i_{g}	Grid-input current	53.0 ARMS
f_{sw}	Switching frequency	40.0 kHz

length of end winding and the overall volume of the machine. The lamination material (JNEX-900) has 6.5% silicon content in a 0.1-mm thick lamination, which ensures the machine achieves a relatively low iron loss. To retain the rotor parts at high-speed operation, a 0.35-mm thick carbon fiber sleeve is employed. Figure 8.1 shows a 3D model of the selected PM machine, while Figure 8.2 shows the magnet and rotor assembly of the base machine.

A range of magnet segmentations was simulated using 3D FEA to examine the effect upon magnet eddy current loss. Results are presented in Figure 8.3. The eddy currents are predominantly generated by asynchronous space harmonics of flux, generated by the winding, with a relatively short pole pitch. When the magnets are axially segmented the segmentation breaks up the eddy current path, adding to the effective resistance and reduces the loss. Figure 8.3 illustrates the time-averaged losses graphically and shows the loss distribution at an instant in time. The loss

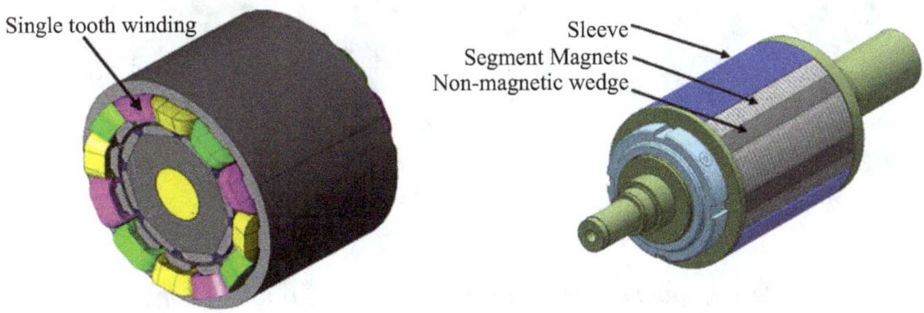

Figure 8.1 3D model of the base machine: (a) stator and (b) rotor

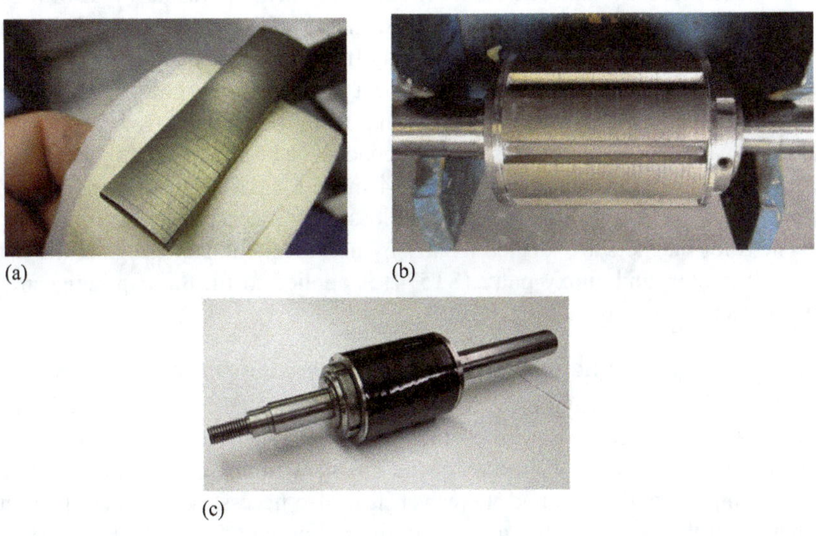

Figure 8.2 Base machine rotor assembly: (a) segmental magnets, (b) assembled rotor magnets, and (c) final rotor assembly

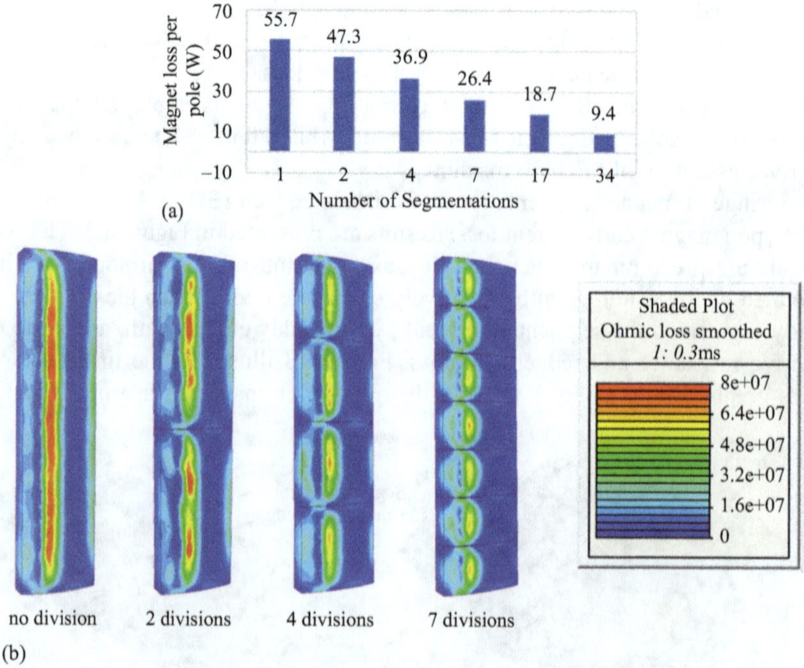

Figure 8.3 Effect of magnet segmentation on losses: (a) losses comparison and (b) eddy currents in each individual magnet

only reduces modestly until the axial length of segment approaches that of a pole pitch of the eddy currents, which coincides with about 30 segments. This is because it is only then that it has a large effect upon effective resistance. For this reason, it was decided to split each magnet into 34 axial segmentations, producing 83% less magnet loss than a single segment. Additional radial segmentation was also investigated but was not adopted because it only reduced the loss by an additional 1%. The magnets were therefore simply split into 34 magnet axial segmentations per pole (Figure 8.2(a)) to reduce rotor eddy current losses to an acceptable level [7,8].

The rotor magnets are located axially by non-magnetic steel rings at the axial ends of the rotor, and epoxy putty (ST574) is applied to fill the gap between the rotor magnetic poles.

8.1.2.2 Electromagnetic design of the integrated filters

A range of different integration methods were investigated for the magnetic filter elements. For power dense machines, the optimum choice is one in which the input filter inductors share part of their magnetic circuit with that of the motor and also have a common cooling method. However, it is also necessary that there is no net magnetic coupling between the motor and filter windings. Figure 8.4(a) illustrates the chosen topology: there are 12 additional outward facing slots to house the filter coils, with a core back shared between the motor and filters. Two coils are wound on each of the twelve outward facing teeth, one for each of the input filter inductors. The number of turns and the width and length of the air-gap between the filter tooth tips are the main design features used to create the desired inductance values.

The flux distribution of the integrated double-slot machine is shown in Figure 8.4(b). The flux density of the shared path in the stator core has been limited to 1.45 T and the outer diameter of the base machine is increased by 23.1%. Table 8.2 lists the double-slot machine dimensions.

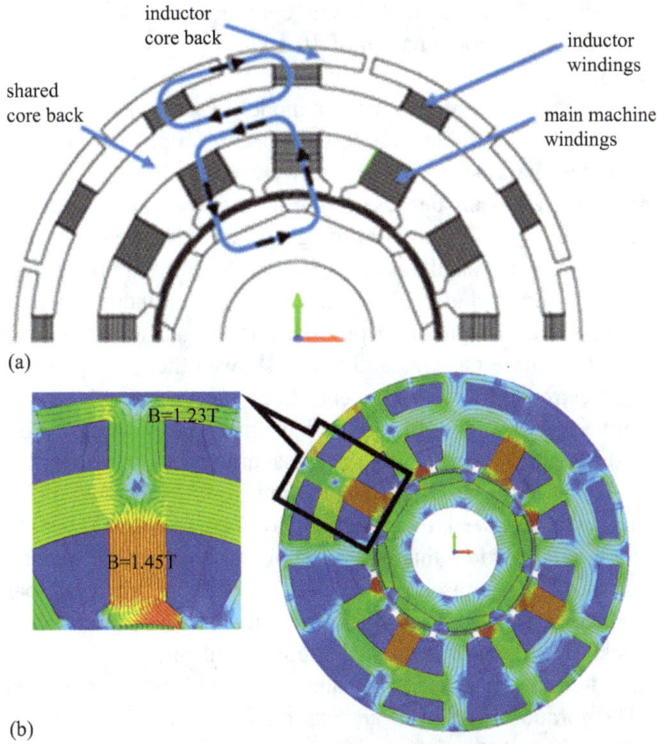

Figure 8.4 Flux distribution of the integrated double-slot machine: (a) flux path and (b) flux contour

Table 8.2 Design of the double-slot machine

Tooth number	12
Tooth width	10.0 mm
Winding type	Concentrated
Main winding slot height	12.3 mm
Filter winding slot height	7.6 mm
Shared stator core back depth	10.0 mm
Filter stator core back depth	3.6 mm
Inductor air gap	2.0 mm

8.1.2.2.1 Winding configuration of the integrated filters

As shown in Figure 8.4(a), the magnetic flux paths of the base machine and integrated filters are independent from each other, therefore there is little coupling between them. In addition, different pole numbers have to be selected for the two sets of inductor windings to eliminate mutual coupling.

A range of pole number combinations were investigated, combining two of 2, 4, 8, 10, 14, and 16 poles. 6 and 12 pole, three phase options were not possible with a 12 coil winding. For each winding, the 12 coils are spaced 30 degrees apart mechanically. By connecting the coils in different combinations, different pole number windings are created. The combination of 10 and 16 pole windings was ultimately chosen for the prototype [9]. The two input filter inductances, L_1 and L_2, were first designed using a simple lumped parameter model, which was then validated by a finite element simulation of the double-slot machine operating with rated drive current. The lumped parameter predictions were typically 20% lower than those predicted using finite elements because of the failure to model fringing fields.

The inductor coils produce rotating magnetic fields at the outer surface of the laminations. These are very small, except when in close proximity with the inductor gaps, where there is some fringing flux. This fringing flux induces eddy currents in the alumina frame and consequently produces additional losses. However, finite element modeling of these currents and losses showed them to be very small.

A second electromagnetic design challenge for this double-slot machine is decoupling between the two filters and the base machine windings. Finite element simulation of the mutual coupling between machine windings and filter windings has been carried out. With rated current flowing in the machine windings and the machine rotating at rated speed, there are very low voltages (0.1 V peak) induced in the filters, which indicates less than 0.1% coupling between the machine and filter windings. This is a direct consequence of having separate magnetic paths for the inductors and main motor windings. It is important to keep magnetic saturation low in the core backs in order to maintain this condition. Figure 8.5 shows FEA predictions of the rated machine torque at rated speed both before and after the filter integration. The torque waveforms are essentially identical in these two cases, which shows that integration of the filters does not affect the operation of the base machine.

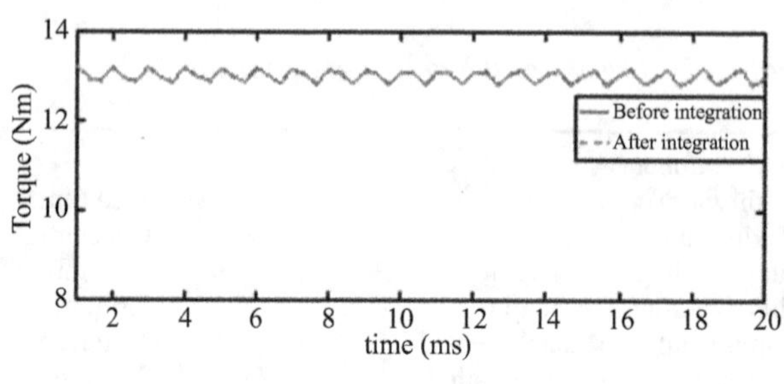

Figure 8.5 Machine rated torque comparison before and after filter integration

8.1.2.2.2 Winding losses and their minimization

The AC winding losses in most low-speed machines are very small due to the low fundamental frequency. However, the additional AC winding loss in this high-speed machine is significant. Different numbers of strands of Litz wire have been simulated to achieve acceptable AC winding losses of the base machine. Table 8.3 shows copper losses of three selected options, where end windings losses are not included. The final choice is 20 strands (30 AWG), where AC loss comprises 10% of the total copper losses. AC winding losses in the Litz wire were calculated using a 2D finite element model of the motor in which every strand of the Litz wire was explicitly modeled, with equal currents in each strand to simulate idealized transposition of the strands.

The inductor windings only have significant currents at the AC supply frequency and so AC losses are insignificant. They were therefore wound with far fewer strands of wire. Table 8.4 summarizes the winding configuration of the double-slot machine. Note that the grid side inductor, L_1, has a very low inductance, which was achieved with a single turn per coil.

8.1.3 Prototype manufacture

This special double-slot stator geometry poses several challenges for manufacture of the machine prototype.

Table 8.3 Base machine winding copper losses 2D FEA comparisons

Type of wire	DC loss (W)	AC loss (W)	Total winding loss (W)
36 solid turns	149.0	166.9	315.9
36 bundles, 4 strands	149.0	32.3	181.4
36 bundles, 20 strands	149.0	16.5	165.5

Table 8.4 Winding configurations of the double-slot machine

	L_1	L_2	Main windings
Number of poles	16	10	8
Coil wound type	Concentrated	Concentrated	Concentrated
Coils internal connection	In series	In series	In parallel
Wire type	Stranded solid wire	Stranded solid wire	Litz wire
Number of turns per phase	1 (turn) × 4 (coils)	7 (turns) × 4 (coils)	36 (turns) × 4 (coils)
Conductor dia.	18 AWG × 6 (strands)	18 AWG × 6 (strands)	30 AWG × 20 (strands)
Phase inductance	5.0 μH	196.0 μH	0.21 mH
Phase resistance	0.0025 Ω	0.022 Ω	0.029 Ω

8.1.3.1 Segmental stator

Due to the double-slot geometry, it is more difficult to place the windings with a single piece lamination. A segmental stator lamination is employed, with each segment composed of a single slot pitch. This significantly increases manufacturability of the windings and provides a higher slot fill factor [10]. The additional outer slots have not affected the symmetry of the stator hence the stator can be constructed from twelve identical segments with interlocking mating geometries, as shown in Figure 8.6.

8.1.3.2 Direct on-tooth compressed coils

A high slot fill factor is required to get a high power density for this machine. Increased slot fill factor gives more MMF for a given loss and increases the effective thermal conductivity in the slots.

High fill factors can be achieved by pre-pressing coils and then inserting the pressed coils over the stator teeth. A direct on-tooth compressed coil technology is used. The coils are wound onto a single tooth segment before pressing. This direct on-tooth compressed coil technology helps the demonstrator get up to 60% fill factor. The manufacturing process is presented in Figure 8.7. After loading the stator segment into the winding jig seat in Figure 8.7(a), the tooth is wound and a thermally activated bonding agent (Ultimag 2002L) is used to impregnate the windings. The coils are then pressed into a regular shape by a pressing punch, as shown in Figure 8.7(c). The fully assembled compressed coils on a single tooth segment are shown in Figure 8.7(d).

8.1.4 Comparison with other discrete AC inductors

Table 8.5 compares the integrated AC filter inductors with equivalent commercial discrete inductors in terms of power, inductance, and volume. Since the windings of the filter inductors share the stator lamination with the base machine, it allows the filters to use the same cooling system as the base machine and have a high current

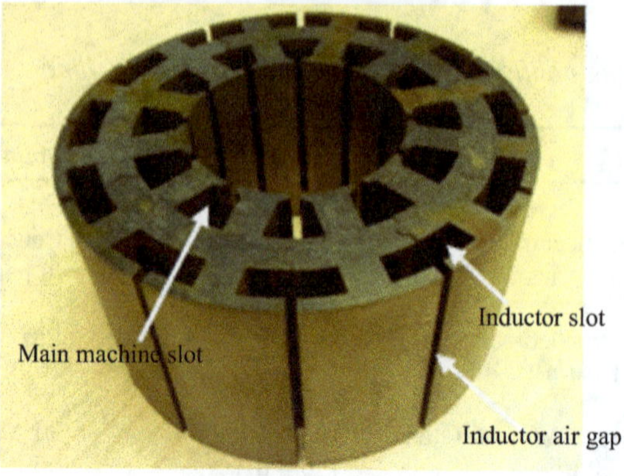

Figure 8.6 Pre-assembly of the double slot stator

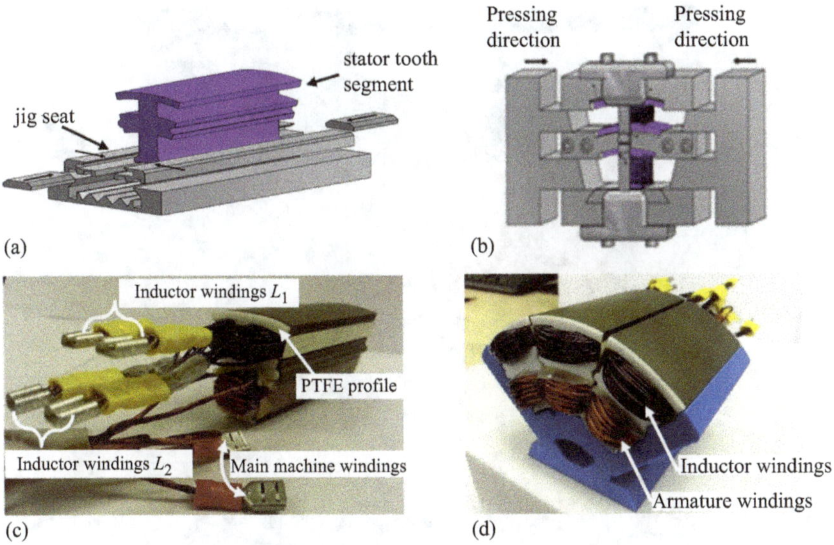

Figure 8.7 Manufacturing processes of the compressed coils: (a) winding jig seat, (b) on-tooth coil pressing, (c) fully assembled single tooth, and (d) three-tooth fully assembled motorette

Table 8.5 AC inductors comparison

	Integrated inductors	**Industrial inductors**
Drive rated power	38 kW	38 kW
Rated peak current	75 A	75 A
Frequency	50 Hz	50 Hz
Measured drive-side inductance/phase	182.6 μH	160.0 μH
Measured grid-side inductance/phase	5.1 μH	5.0 μH
Current density	12.0 A/mm^2	3.3 A/mm^2
Winding fill factor	50%	L_1: 40% and L_2: 60%
The volume of grid-side three-phase inductors	424 cm^3 $L_1 + L_2$	990 cm^3
The volume of drive-side three-phase inductors	424 cm^3 $L_1 + L_2$	2430 cm^3

density of 12.0 A/mm^2. The industrial inductors of the input LCL filter have the same voltage and current rating as the integrated LCL filter inductors.

Figure 8.8 shows the volume reduction achieved by integrating the three-phase filter inductors (L_1+L_2) into the base PM machine. The total volume of the integrated filters is 87.6% smaller than the total volume of the conventional discrete LCL filter inductors. It is worth noting that the integrated filters share the same forced cooling as the base machine, while the industrial inductors are air-cooled. The manufacture of extra slots and compressed coils for the integrated filters is also more difficult and expensive than the manufacture of the industrial inductors.

Figure 8.8 Volume comparison between conventional discrete inductors and the assembled stator of the IMD

The most important reason for the volume reduction of the inductors arises from improved cooling. By improving the cooling, the current density in the windings can be greatly increased. This, in turn, results in a greater per-unit resistance, so the winding losses in the inductors are actually increased in the integrated motor drive. Of course, there are now very short connections between the filters, inverter and motor, which act to reduce other components of loss. Overall efficiency comparisons with a conventional drive are very specific to the drive application, making it difficult to produce generalized statements.

8.1.5 Winding connection considerations

By integrating the input filters into the machine, the number of winding terminations is significantly increased compared to conventional three-phase PM machines.

Figure 8.9 shows the winding connections of the proposed double-slot integrated machine. There are 12 main phase windings in the stator. Every phase has 4 concentrated coils which need to be connected in parallel, giving 8 terminations per phase. The magnetic part of the LCL filters consists of 24 concentrated coils, which give 48 terminations.

The first challenge after the machine design of this IMD is the connection and routing of all the 72 terminations in a very limited space in the machine housing, without breaking any of the rules listed below:

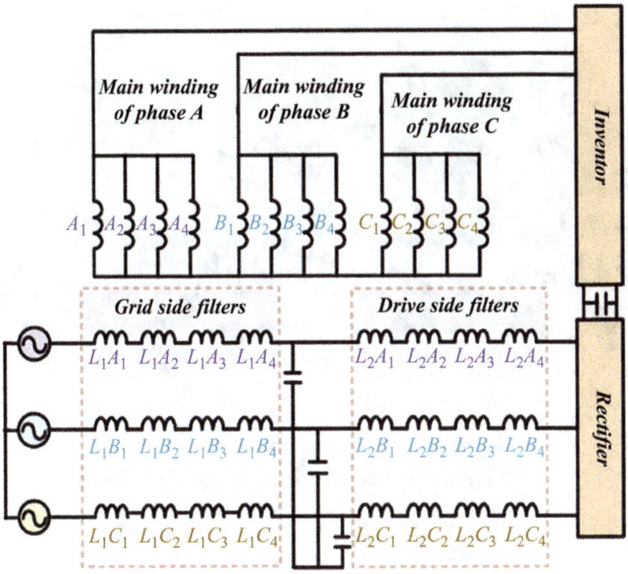

Figure 8.9 Winding topology of the IMD

- radial extension beyond the motor outside diameter must be avoided;
- axial extension must be kept as short as possible to maintain the power density;
- flying leads cannot be employed, as they are susceptible to failure.

PCB solutions were initially considered [11], because of their simple assembly, but due to copper thickness restrictions in the PCB manufacture and the radial length limitation in the machine housing, all the PCB designs were rejected. An integrated copper bus bar design was subsequently developed.

As shown in Figure 8.10, the integrated bus bar design consists of four layer bus bars, with all bus bars embedded into 3.0 mm deep grooves in 4.0 mm PTFE boards for insulation and mechanical fixing. The dielectric strength of this plastic material is 80 MV/m.

In order to get the breakdown voltage of the insulation to greatly exceed the required operating voltage of the drive, the minimum width of the barrier between two grooves in the same layer is selected to be 1.0 mm. The minimum thickness between two PTFE boards is also 1.0 mm, in accordance with the manufacturer's guidance. Considering the skin effect of conductors, the skin depth of all bus bars is calculated by (8.1)

$$\delta = \sqrt{\frac{\rho}{\pi \mu_0 f}} \tag{8.1}$$

where δ is the skin depth in m, ρ is the conductor resistivity in Ω/m, μ_0 is the free space permeability, and f is the frequency in Hz. The function and dimensions of each layer

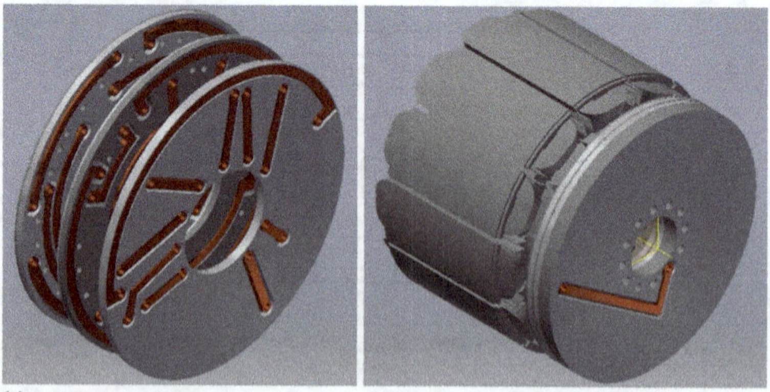

(a) (b)

Figure 8.10 3D representation of the integrated bus bar design: (a) first to third layer bus bars. (b) Bus bars assembled with machine

Table 8.6 Bus bar functions and dimension description

	Function	**Bus bar cross section (mm^2)**
Layer 1	Grid-side inductors connections	3 (thickness) × 4 (width)
	Three-phase machine windings connections	3 (thickness) × 4 (width)
Layer 2	Neutral point connections	3 (thickness) × 4 (width)
	Drive-side inductors connections	3 (thickness) × 4 (width)
Layers 3 and 4	Connections between machine and power electronics unit	3 (thickness) × 4 (width)

of bus bars are summarized in Table 8.6. Layers 1 and 2 are next to the machine and provide the internal connections between all the distributed coils of the filters and main windings. Layers 3 and 4 provide all the internal connections between the machine housing and the grid, three filter capacitors and two power converters.

The manufacture of the 3.0 mm deep grooves in the PTFE boards is shown in Figure 8.11. The layer-by-layer assembly processes and fully assembled bus bars are shown in Figure 8.12.

8.1.6 Power electronics integration

There are six main components in the power electronic unit of the IMD: a power board consisting of converters and power connections, a transducer printed circuit board (PCB) consisting of all the current/voltage sensors, a gate drive PCB, a DC bus capacitors PCB, an AC filter capacitors PCB and a position sensor for rotor position measurement.

Figure 8.11 Manufacture of grooves in insulation PTFE boards

All six individual components are illustrated in Figure 8.13. Details of each component employed in the IMD system are listed in Table 8.7.

The first PCB is the power board, which consists of 24 SiC MOSFETs distributed evenly around the periphery. These SiC MOSFETs are placed on the cooling jacket, which forms a good thermal heatsink. They are connected in 12 parallel pairs (six for the active rectifier and six for the motor drive inverter) providing 120.0 A drain current-carrying capability. The demonstrator machine has an electrical frequency of 1.67 kHz, which requires a PWM frequency in excess of 33.4 kHz in order to minimize PWM harmonics and give high quality current control. This frequency is generally beyond the capability of silicon based IGBTs and so SiC devices are required. The turn-on and turn-off dv/dt are 15.4 and 23.5 kV/μs, respectively. The 72 legs of the 24 switches also provide a strong mechanical support for the entire power electronics unit.

All the voltage and current sensors are arranged on the second PCB. Five current sensor circuits are designed on the second board: three for the three-phase current flowing into the active rectifier and two for the input currents of the star-connected PM machine. A high-speed input, wide bandwidth isolation amplifier is used to measure DC link voltage. Three voltage sensor circuits are placed on this board to get an accurate measurement of the phase of the input grid voltages for the active rectifier.

The third board is a 24-channel gate drive PCB. Just as the name suggests, there are 24 gate drive circuits designed on the board. Every single gate drive circuit occupies a 15 sector. 24 channel PWM signals generated by a DSP-based control unit flow from the periphery to the center of the PCB. The outputs of the gate drive board are fed to the SiC MOSFETs.

The final two boards contain the grid connected filter capacitor and the DC bus capacitor PCBs. The total length of the fully assembled drive is approximately 150.0 mm. These two populated PCBs occupy approximately half the length of the entire drive.

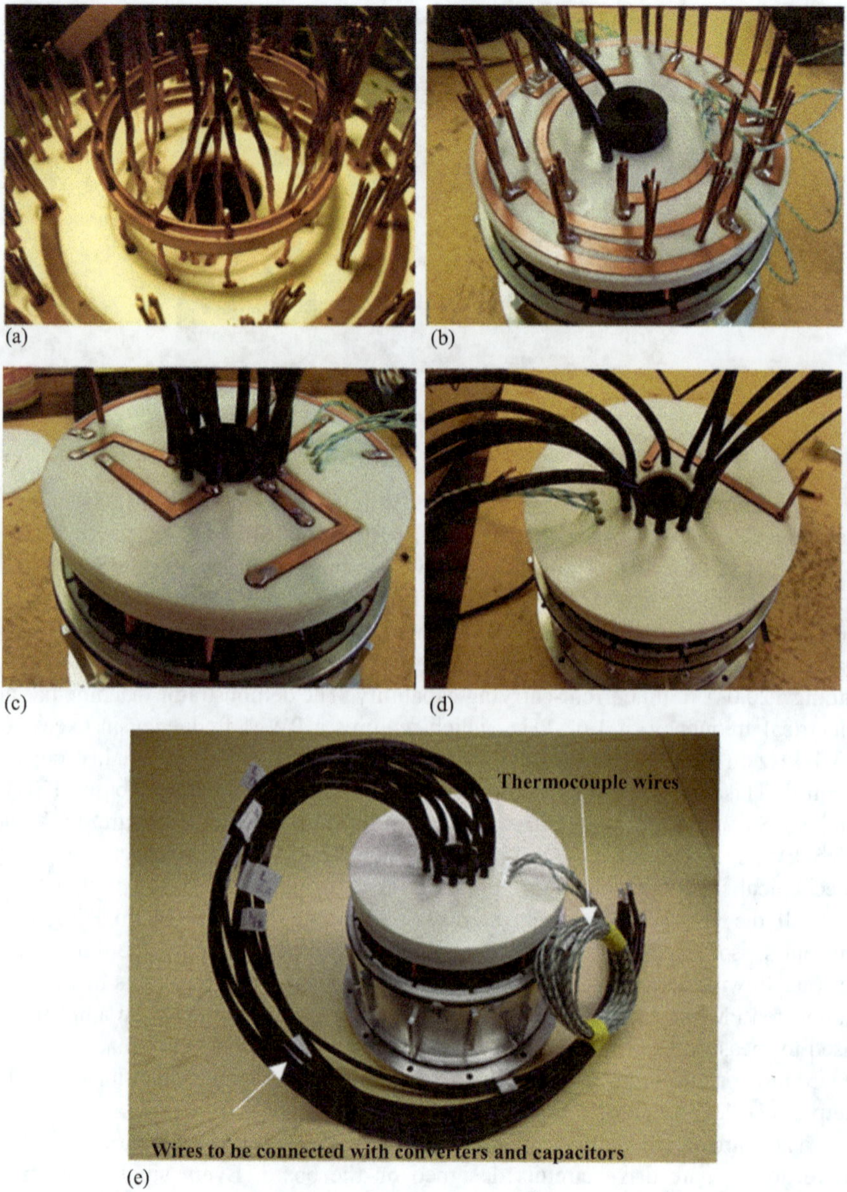

Figure 8.12 Bus bar solution manufacture and assembly: (a) layer 1, (b) layer 2, (c) layer 3, (d) layer 4, and (e) full assembly

8.1.7 Conclusion

In this section, development of a high-speed, PM, SiC-based drive with integrated input filters has been presented. The LCL filter inductors of AC motor drives have

Figure 8.13 Structure illustration of the power electronics design in the IMD

Table 8.7 Component employed in the IMD power electronics unit

	PCB1		
SiC MOSFETs	C2M0040120D		
	PCB2		
Current sensors	LEM LA 100-P	Voltage sensors	LEM LV 25-P
Isolation amplifier	AD215AY		
	PCB3		
Opto-coupler	ACPL-P480	DC:DC converter	MGJ2D152005SC
MOSFET driver	IXDN609SI		
	PCB4		
DC bus capacitor	B32776T1355		
	PCB5		
LCL filter capacitors	B32926H3156	Position sensor	TS2225N1112E102

been successfully integrated into a PM machine, manufactured and tested. The design of this integrated double-slot machine contains the novel feature of an integrated double-slot stator, pole number combinations of windings, segmental stator design and a direct on-tooth coil compressing technology.

Since the integrated filters and the base machine share parts of their magnetic circuits, a number of different pole combinations have been investigated to avoid magnetic coupling. The main machine windings are selected to be 8 pole. Since 16 pole (L_1) and 10 pole (L_2) combinations for the filter inductors do not couple with each other, this combination has been selected as the optimum design. This special double-slot stator geometry poses some manufacturing challenges. Segmental stator and direct on-tooth compressed coils are employed to increase manufacturability of the windings and increase the machine winding fill factor and filter inductor L_2 winding fill factor to 50% and 60%, respectively.

This integrated machine achieves significant volume reduction by integrating the three-phase filter inductors (L_1+L_2) into the base PM machine. The total volume of the integrated filters is 87.6% smaller than the total volume of the conventional discrete LCL filter inductors. The prototype offers far superior functionality and a much more compact envelope.

In order to accommodate a large number of winding terminals, a novel copper bus bar solution is employed. PTFE boards are designed and manufactured for insulation purposes. A 40.0 kHz 34.0 kW SiC-based drive is designed and built for the IMD. A series of static and dynamic tests are carried out with the fully assembled prototype. The static results show that the prototype gives filter inductances (L_1 and L_2) 2.0% greater and 6.8% smaller than the FEA predictions respectively. Dynamic results validate that the LCL filter can work effectively and independently. The THD has been effectively reduced from 5.2% to 2.1% at rated torque.

8.2 Research prototype 2: wide bandgap-based current source inverter integrated motor drive*

8.2.1 Introduction

The physical integration of power electronics and electric machines offers several benefits including reduced cost, mass, and volume, as well as higher efficiency [12–14]. The IMD concept is now widely accepted in electrified transportation applications [15–17] due to their inherently stringent volume and mass requirements. There has been a growing interest in the adoption of wide bandgap (WBG) power semiconductor devices for these IMDs. The WBG devices provide an order of magnitude faster switching speed and lower on-state resistance compared to the conventional Si-based devices, enabling efficient operation at high frequencies [18,19]. High switching frequency operation helps to reduce the size of passive components such as inductors and capacitors to improve the overall system power density [20]. The reduced power dissipation from the switching devices also enables compact cooling system design, which contributes to raising the power density of the power electronics [21].

However, the conventional voltage–source inverter (VSI) has several inherent limitations for IMD applications that were discussed in Chapter 4, and the presence of a bulky dc-link capacitor is the source of some of the most significant complications. The maximum operating temperature of the dc-link capacitors, which typically are implemented using electrolytic and film capacitors, is generally limited to 125°C [22]. Recognizing that the maximum temperature of WBG devices, ceramic capacitors, and inductors can be over 150°C [23], the presence of the dc-link capacitor ultimately limits the maximum operating temperature of the entire power electronics system. Since the power electronics in IMDs can be exposed to

*A significant portion of the text in Section 8.2 comes from the IEEE paper: W. Lee, R. A. Torres, H. Dai, T. M. Jahns and B. Sarlioglu, "Integration and cooling strategies for WBG-based current-source inverters-based motor drives," in *2021 IEEE Energy Conversion Congress and Exposition (ECCE)*, Vancouver, BC, Canada, 2021, pp. 5225-5232, doi: 10.1109/ECCE47101.2021.9595747.

high ambient temperature (>80°C) due to the electric machine's heat dissipation, high-temperature operation of the power electronics is of paramount importance.

Another challenge of the WBG-based VSI-IMD is the design of distributed dc-link capacitors [24]. It is a well-known fact that WBG devices are extremely sensitive to parasitic inductances in a power circuit loop due to their fast switching speed. The parasitic inductance between the dc-link capacitor and each phase-leg must be maintained as low as possible to minimize the voltage overshoot and ringing. This is particularly problematic in VSI-IMDs when the power electronics is spread out over the outer surface of the machine. Distributed dc-link capacitor configurations have been proposed and investigated in [25,26], but sharing the ripple current equally among paralleled capacitors can become a significant challenge due to parasitic inductances and resistances [27,28].

To overcome the drawbacks of the conventional VSI-IMD, WBG-based current–source power converters have been selected for this IMD application which do not require any bulky dc-link capacitors nor complicated distributed dc-link design. Special attention is devoted to the power density and thermal performance of the WBG-based current source inverter (CSI) developed for this IMD. This IMD prototype discussion is organized as follows: Section 8.2.2 introduces the design of a 3 kW WBG-based CSI tailored for an IMD configuration, including both its volumetric and specific power densities. The thermal performance of the electric machine and the integrated WBG-based CSI are experimentally investigated in Section 8.2.3. Conclusions are presented in Section 8.2.4.

8.2.2 Design of WBG-based CSI for IMD applications

One of the distinct requirements of CSIs compared to the conventional VSIs is reverse-blocking (RB) switches, which are generally implemented using a single active switch with a series-connected diode (FET+diode) or two active switches connected in anti-series (dual FET), as shown in Figure. 8.14. There is also growing interest in the development of monolithic RB bidirectional switches using gallium

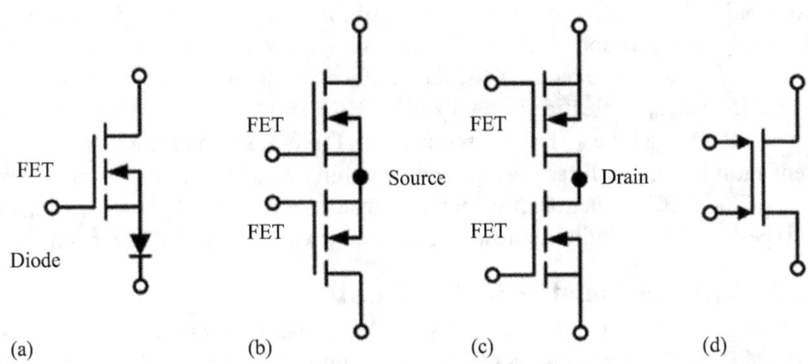

Figure 8.14 Reverse blocking (RB) switch configurations for CSI (a) FET+diode, (b) dual FET (common-source), (c) dual FET (common-drain), and (d) monolithic RB bidirectional switch

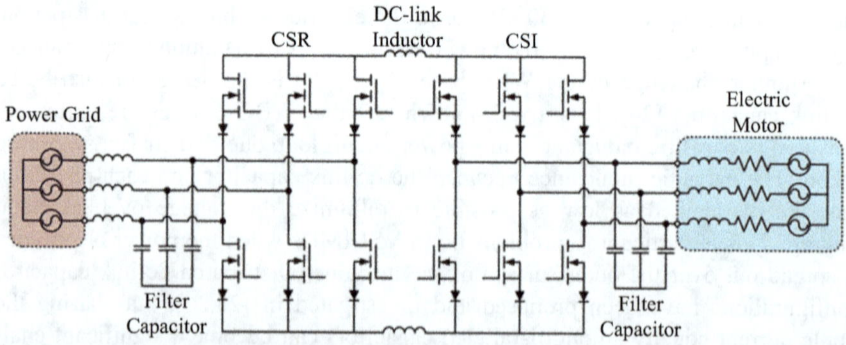

Figure 8.15 Schematic of WBG-based current–source inverter and rectifier interfaced with power grid

Table 8.8 Specifications of WBG-based CSI and Electric Machine

	Parameters	Value	Units
Power Electronics	Output phase voltage (rms)	131	Vrms
	Output phase current (rms)	8.4	Arms
	Output power	3.0	kW
	DC-link current (avg)	12	A
	DC-link inductor	320	μH
	Output filter capacitor	2.2	μF
	Switching frequency	125	kHz
SPM motor	Rated speed	1644	rpm
	Number of poles	10	-
	Phase resistance	0.19	Ω
	Phase inductance	2.9	mH

nitride (GaN) (Figure 8.14(d)), but the focus of this work is on FET+diode and dual FET switch configurations [29,30]. A schematic of a WBG-based CSI including a front-end current–source rectifier (CSR) interfaced with the power grid is shown in Figure 8.15. The specifications of a 3 kW WBG-based CSI and surface permanent magnet (SPM) machine [31] are presented in Table 8.8, which includes voltage, current ratings, output filter design, and switching frequency. During an investigation of the CSI's functionality, power density, and thermal performance, two prototypes have been built and tested: a benchtop version and IMD version.

8.2.2.1 IMD version of WBG-based CSI

The IMD shown in Figure 8.16(a) is designed with a compact power printed circuit board (PCB) and six phase-leg PCBs that fit into the limited volume inside the electric machine's end chamber (188 mm outer diameter with less than 64 mm in height). The size of the power board has been reduced to 0.025 m^2. Two versions of the phase-leg printed circuit boards (PCBs): Ver. 2.0 using the FET+diode reverse-blocking switch and Ver. 2.1 with the dual FET configuration, have been developed as shown in Figure 8.16(b) and 8.16(c), respectively, using arc-shaped boards that fit within 66.0 mm × 55.9 mm rectangular envelopes.

As depicted in Figure 8.17, the CSI-IMD unit consists of three layers of printed circuit boards (PCBs) stacked axially: a power and sensor PCB, 6 adjacent phase-leg PCBs, and a controller PCB. The power and sensor PCB closest to the machine includes the input and output power connectors, dc-link inductors, filter capacitors, and voltage and current sensors. The power phase-leg PCBs that are next in the stack include the power switching devices, isolated power supplies, and gate drivers. The controller PCB which is furthest from the machine is occupied by a digital signal processor (TMS320F28379D). Figure 8.18 shows the assembled 3 kW prototype CSI-IMD unit mounted on a dynamometer.

The achieved volumetric and specific power density values of the prototype CSI-IMD are provided in Table 8.9 for comparison with the target values of this project, showing that both targets were exceeded. In fact, volumetric power density target was exceeded by over 80% in the prototype IMD.

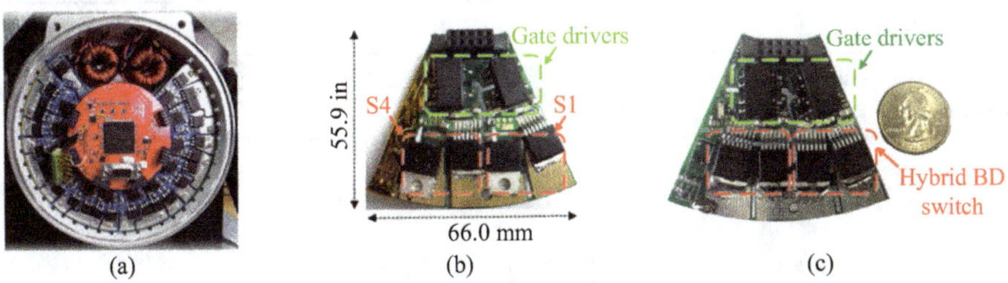

Figure 8.16 Views of WBG-based CSI in IMD configuration: (a) WBG-based CSI mounted in motor housing end chamber; (b) FET+diode phase-leg – Ver. 2.0; (c) dual FET phase-leg – Ver. 2.1

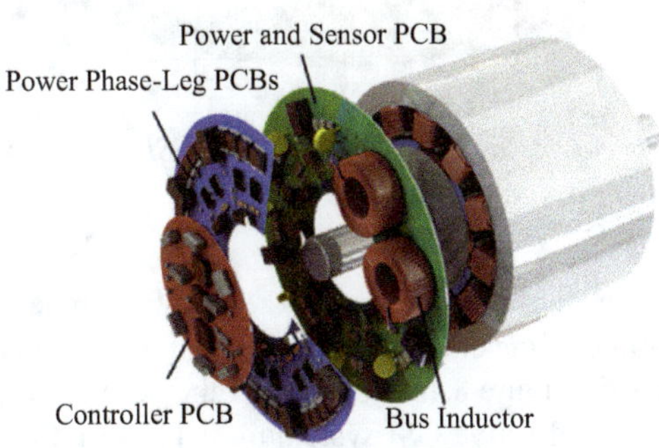

Figure 8.17 Exploded rendered drawing of IMD version WBG-based CSI with an electric machine without cooling components and housing

Figure 8.18 Prototype 3 kW IMD version of the WBG-based CSI mounted on a dynamometer

Table 8.9 Targets and estimated volumetric and specific power densities of benchtop and IMD version WBG-based CSIs (w/o controller)

Parameters	Targets	Benchtop	IMD	Units
Volumetric power density	5	2.97	9.14	kW/L
Specific power density	8	5.98	8.7	kW/kg

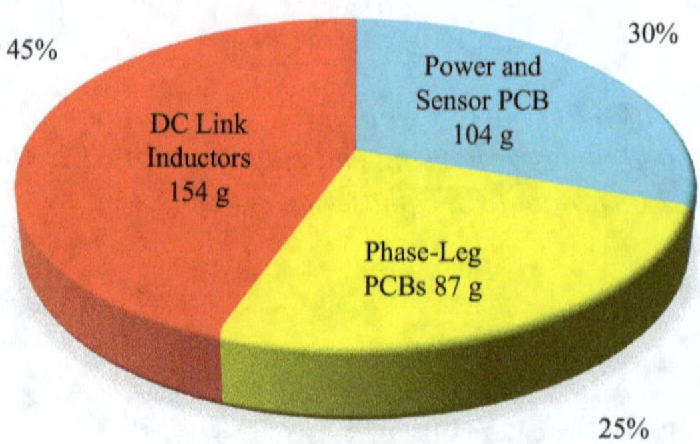

Figure 8.19 Mass breakdown of the WBG-based CSI in the prototype CSI-IMD unit, with a total active mass of 345 g

The specific power density target was exceeded by a much smaller margin of 8.8%, highlighting some of the challenges of reducing the CSI's mass. As shown in the CSI mass pie chart in Figure 8.19, the mass of the dc-link inductors corresponds to approx. 45% of the total active CSI system mass (154 g). Stated succinctly, it is critical to minimize the mass of the dc-link inductor (discussed in more detail in [23]) to achieve high specific power density in the CSI.

It is also important to consider the number of layers, thickness, and density of the copper in the PCBs since these factors also affect the mass of the boards. These results confirm that the expected mass reduction associated with smaller PCB area can be overwhelmed by increases in the copper fill factor, number of layers, and thickness. Therefore, these PCB design parameters need to be optimally selected considering the phase-leg's target mass power density and thermal performance.

8.2.3 Thermal performance analysis of electric machine and IMD

8.2.3.1 Measured machine end chamber temperatures

Since the IMD version of the WBG-based CSI shown in Figure 8.16(a) is installed in close proximity to the electric machine, as shown in Figure 8.18, it is critical to evaluate the thermal impact of this integrated configuration. The heat dissipated from the stator and the rotor of the electric machine expose the power electronics to an elevated ambient temperature (>25°C). The temperature of the electric machine is measured at five measurement points using type K thermocouples until the electric machine reaches its thermal steady-state condition which takes approx. 4 hours, as shown in Figures 8.20 and 8.21. The five measurement points are shown in Figure 8.22, where three measurement points are located on the inner wall and side wall of the end chamber where the bearing and encoder are also mounted. The other two measurements points are on top of the electric machine and on the endplate that covers the end chamber. There is no forced-air cooling during these tests.

The measurement results reveal that the middle of the end chamber inner wall exhibits the highest steady-state temperature during both half-load and full-load conditions with readings of 47.9°C and 56.5°C, respectively. The machine end windings located on the other side of the end chamber inner wall contribute to these high temperature readings. The temperatures near the center of the end chamber inner wall and the top of the electric machine are approx. 3°C lower than that at the middle of the end chamber inner wall. The side wall of the end chamber and the endplate cover are up to 10°C cooler than the hot spot at the middle of the end chamber inner wall.

8.2.3.2 Thermal performance evaluation of IMD

When a power electronics system is installed inside the electric machine housing, it is inevitable that the ambient temperature for the power electronics components is elevated. Generally, the power switching devices are the major sources of heat generation and tend to experience large temperature variations depending on the power losses. To evaluate the impact of the elevated temperature and PCB design on the thermal performance of SiC MOSFETs, a thermal equivalent circuit diagram and the corresponding thermal impedance values are presented in Figure 8.23(a) and Table 8.10, respectively. The SiC MOSFET devices used in the prototype motor drive are mounted in TO-220 packages, as shown in Figure 8.23(b).

The heat generated at the SiC MOSFET junction is dissipated to the ambient environment through two channels: junction-to-case top and junction-to-case/PCB.

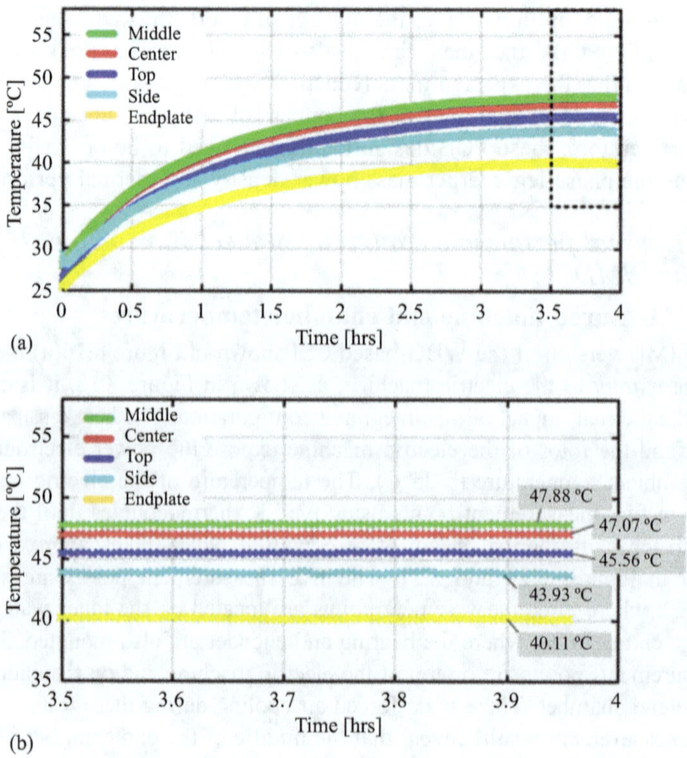

Figure 8.20 Measured transient electric machine temperatures for the sudden application of half-load (50%) operating conditions (1.5 kW at 1,250 rpm): (a) full measurement cycle from 0 to 4 hours; and (b) expanded view of measurement cycle from 3.5 to 4 hours

Therefore, the junction temperature of the device can be estimated using

$$T_J = P_{top}(\theta_{JCtop} + \theta_{CtopA}) + T_A = P_{bottom}(\theta_{JC} + \theta_{CA}) + T_A$$
$$P_{loss} = P_{top} + P_{bottom} \qquad (8.2)$$

where T_J is the junction temperature of the device, P_{top} is the power loss dissipated through the top of the device case, P_{bottom} is the power loss dissipated through the base of the device, T_A is the ambient temperature, and P_{loss} is the total power loss from the device. The thermal resistances are defined in Table 8.3. However, as shown in the table entries and Figure 8.23, the thermal impedance to ambient from the top of the device case (>90°C/W) is significantly higher than the thermal impedance to ambient from the device base (<40°C/W). As a result, the heat dissipation from the top of the device is ignored so that (8.2) can be rewritten in simplified form as

$$T_J = P_{loss}(\theta_{JC} + \theta_{CA}) + T_A \qquad (8.3)$$

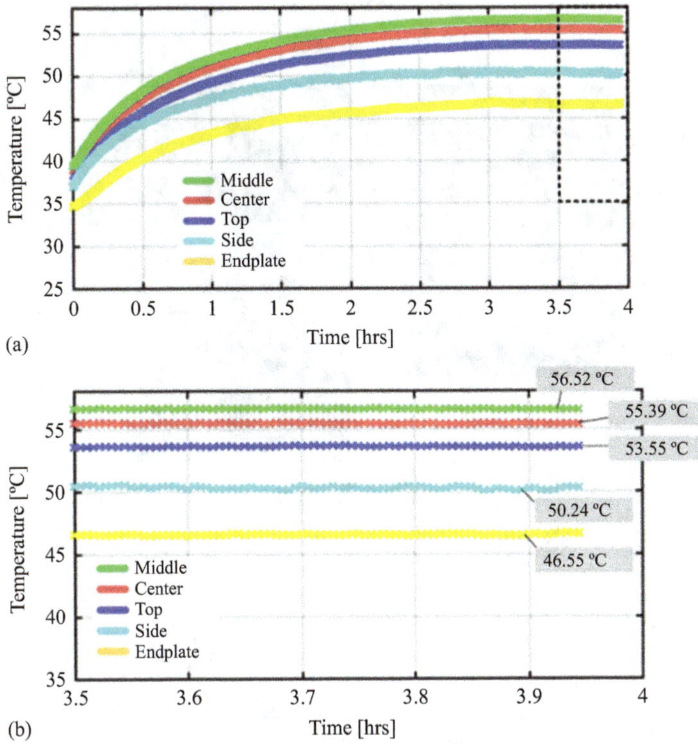

Figure 8.21 Measured transient electric machine temperatures for the sudden application of full-load (100%) operating conditions (3.0 kW at 1,500 rpm): (a) full measurement cycle from 0 to 4 hours; and (b) expanded view of measurement cycle from 3.5 to 4 hours

assuming

$$P_{bottom} \approx P_{loss} \tag{8.4}$$

$$P_{top} \approx 0 \tag{8.5}$$

To reduce the junction temperature of the devices in the IMD version of the WBG-based CSI, there are three design variables to be considered: device power loss P_{loss}, the case/PCB-to-ambient thermal impedance θ_{CA}, and ambient temperature T_A. The junction-to-case thermal impedance θ_{JC} is determined by the SiC MOSFET package type (TO-220).

First, the device power loss can be reduced by using the dual FET configuration in Figure 8.14 in place of the FET+diode configuration because the voltage drop across the second MOSFET is lower than the voltage drop across the SiC Schottky diode for the same current. The thermal performance can be further enhanced using low on-state resistance devices or monolithic bidirectional switches, but both options tend to increase the total system cost.

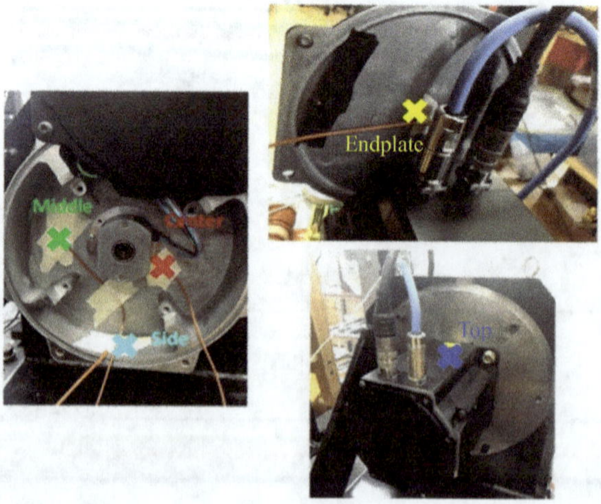

Figure 8.22 Five temperature measurement points on the electric machine using type K thermocouples

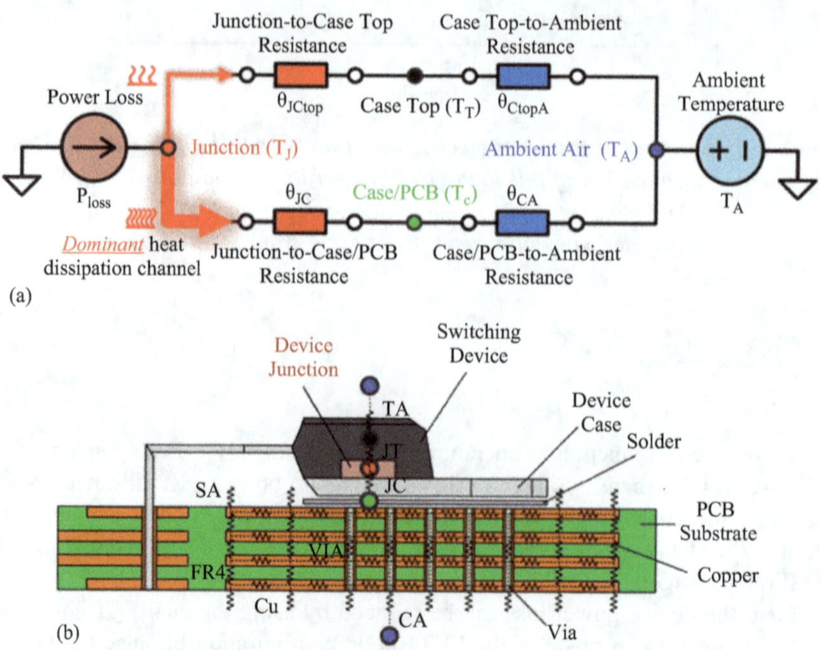

Figure 8.23 SiC MOSFET thermal model and cooling configuration: (a) thermal equivalent circuit diagram, using a model with 4 thermal impedances; and (b) SiC MOSFET in TO-220 package mounted on the phase-leg PCB with multiple copper layers

Table 8.10 Typical Thermal Impedance Values of SiC MOSFET in TO-220 Package and PCB

Parameter	Value	Units
Junction-to-Case Top (θ_{JCtop})	>40	°C/W
Case Top-to-Ambient (θ_{CtopA})	>50	°C/W
Junction-to-Case/PCB (θ_{JC})	0.5 - 4	°C/W
Case/PCB-to-Ambient (θ_{CA})	15 - 30	°C/W
Junction-to-Ambient(w/o PCB or heatsink)	50 - 65	°C/W
Junction-to-Ambient(w/ 4-layer JC51 Standard Board)	20 - 40	°C/W

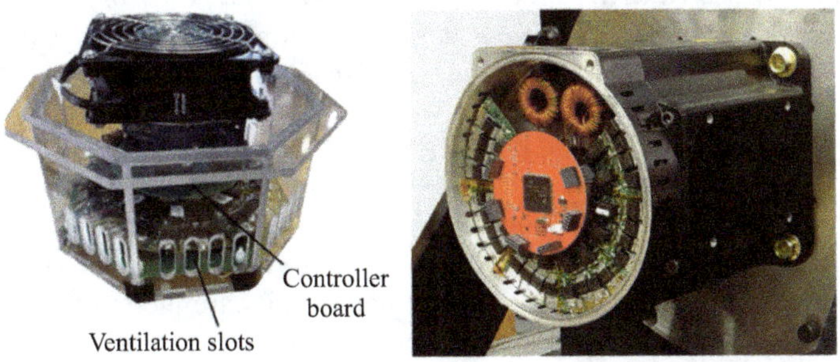

Figure 8.24 (Left) Benchtop version of the WBG-based CSI for thermal performance evaluation with fan; and (right) IMD version of the WBG-based CSI using IMD #1 configuration in Figure 8.26

Second, the case/PCB-to-ambient thermal impedance (θ_{CA}) is highly dependent on the PCB design parameters such as copper thickness, board size, copper fill factor, and the number of copper layers. However, increasing these parameters increases the mass of the system, leading to the lower specific power density.

Finally, it is desirable to find an optimal location for the power electronics providing the lowest ambient temperature inside the electric machine, as shown in Figures 8.20 and 8.21. However, it is nearly impossible to avoid having an elevated ambient temperature in the IMD applications, with values often exceeding 45°C. The increased ambient temperature directly raises the junction temperature, as shown in (8.3).

To maintain the device junction temperature lower than the maximum limit of 150°C, forced air convection using a 12 Vdc fan was chosen for the CSI-IMD design. To investigate the even distribution of the air flow around the 24 switching devices placed at the outer periphery of the power electronics, an octagonal polycarbonate housing with 32 ventilation slots was fabricated for the benchtop version of the WBG-based CSI, as shown in Figure 8.24. The dc fan was mounted on top of the housing to draw air into the housing through the side slots. This polycarbonate housing design emulates the power electronics housing design adopted for the IMD version.

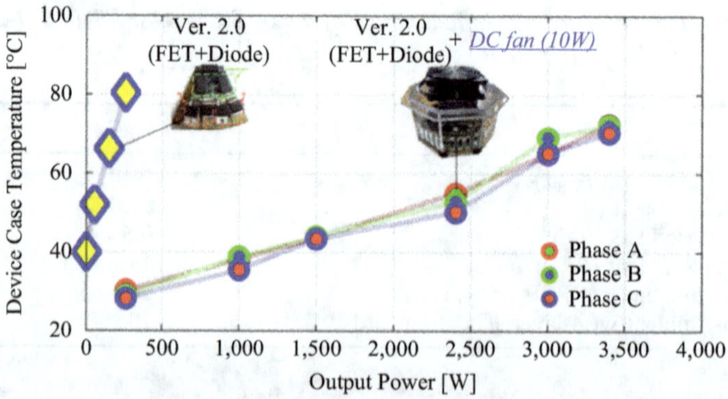

Figure 8.25 Measured device case temperature vs. output power of phase-leg PCB Ver. 2.0 with and without 10 W dc fan drawing air through polycarbonate housing (air speed = 8 m/s) at an ambient temperature of 25°C

Both fan orientations that either blow or draw air through the housing were tested. Drawing air through the housing was shown to be more efficient due to less air leakage and more constant air flow over the switching devices (i.e., less turbulence). The device temperature measurement results are plotted in Figure 8.25, where the effectiveness of the proposed housing and dc fan installation is verified. The case temperatures of all of the devices are maintained lower than 80°C during full-load operation with the fan. Even with an elevated ambient temperature of 45°C inside the machine housing for the IMD version, the device case temperatures are maintained <100°C.

8.2.3.3 Proposed IMD configurations

Based on the power density and thermal performance evaluation results, six IMD configurations with different air-cooled thermal management systems are proposed. The first three configurations, shown in Figure 8.26, share the same three-layer power electronics design shown in Figure 8.17, and the power electronics system is installed in the end chamber adjacent to the electric machine housing. IMD #1 is a ventilated enclosure, blower cooled (VEBC) configuration that includes a separately-powered blower installed on the endplate either blowing or drawing air through the end chamber as shown in Figure 8.26(a). An exploded 3D view of IMD #1 is provided in Figure 8.28 and a hardware implementation of the WBG-based CSI in this configuration is provided in Figure 8.24. IMD #2 is a totally enclosed, non-ventilated (TENV) configuration that mounts a heatsink on the outside of the endplate to dissipate the heat, providing a totally enclosed housing for the IMD. This configuration is advantageous in applications where exposing the power electronics to the external environment is not acceptable due to particulate contamination or precipitation. IMD#3 shown in Figure 8.26(c) is a totally enclosed, blower cooled configuration that combines IMD #1 and IMD #2 by adding a separately-excited blower to force air through the fins of the externally

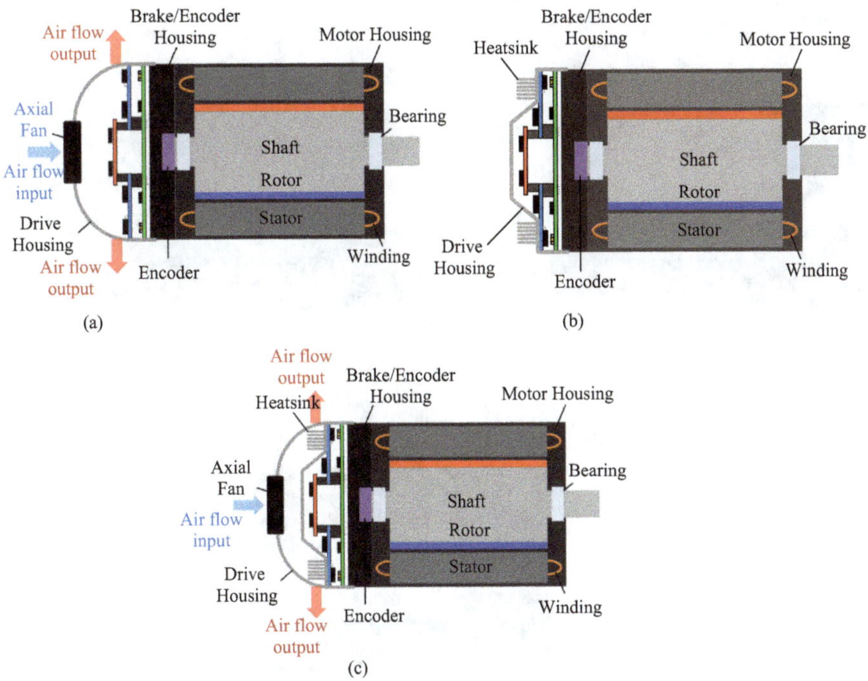

Figure 8.26 Alternative air-cooled IMD configurations with the power electronics axially mounted in the end chamber of the machine housing. (a) IMD #1: ventilated enclosure, blower cooled (VEBC); (b) IMD #2: totally enclosed, non-ventilated (TENV); (c) IMD#3: totally enclosed, blower cooled (TEBC)

mounted heatsink. This configuration provides the combined advantages of high thermal performance and a totally enclosed IMD housing.

The other three configurations shown in Figure 8.27 include a two-layer power electronics design in which the controller board is integrated into the power and sensor PCB. Importantly, the power electronics is installed inside the electric machine housing, resulting in the highest level of IMD integration and power density. A fan blade is mounted on the rotating machine shaft to provide forced air convection cooling of the power electronics and the electric machine using external air drawn into the housing through housing vents, a configuration referred to as ventilated enclosure, shaft fan cooled (VESFC). IMD #4, referred to as VESFC-Ver1, draws the air through the power electronics and the machine airgap using the shaft-mounted fan before exhausting the air at the opposite end of the machine housing as shown in Figure 8.27(a). A conceptual 3D view of IMD #4 is provided in Figure 8.29 that shows the design details more clearly, including the shaft-mounted blower fan. However, the power electronics is more directly exposed to the heat dissipated by the machine compared to the IMD

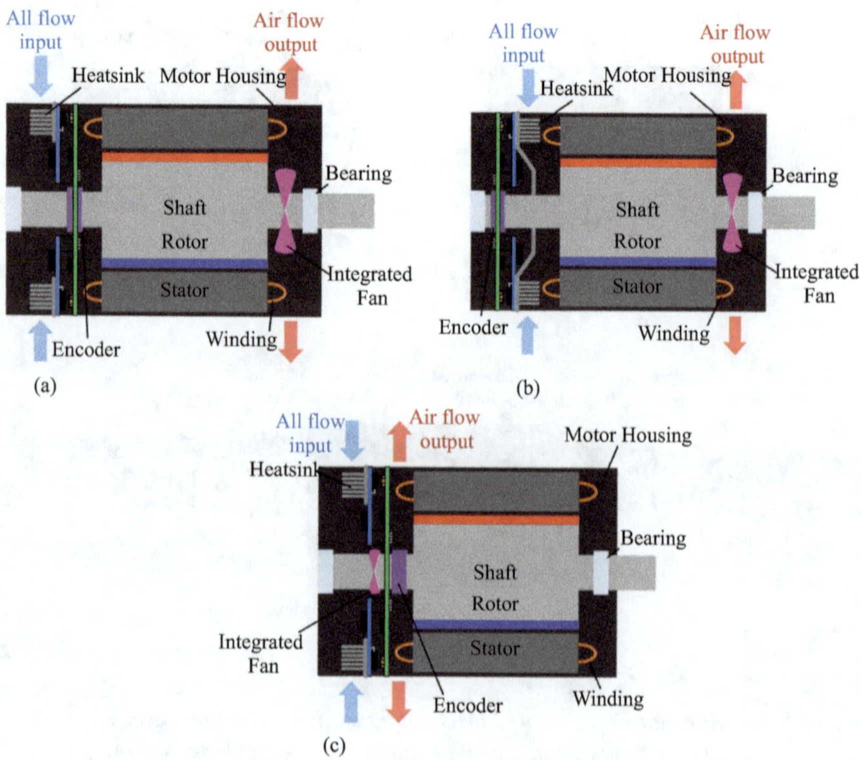

Figure 8.27 Alternative IMD configurations with the power electronics axially mounted inside the electric machine housing. (a) IMD #4: ventilated enclosure, shaft fan cooled, Version 1 (VESFC-Ver1); b) IMD #5: ventilated enclosure, shaft fan cooled, Version 2 (VESFC-Ver2); and (c) IMD #6: ventilated enclosure, shaft fan cooled, version 3 (VESFC-Ver3)

configurations in Figure 8.26 that mounts the inverter in the end chamber, leading to higher ambient temperature for the inverter in the IMD #4 to #6 configurations.

To reduce the impact of the additional heat from the electric machine, IMD #5 in Figure 8.27(b) moves the internal heatsink to an axial position between the power electronics and electric machine where it acts as a thermal barrier. The air drawn from the outside of the electric machine by the internal blower fan forces air through the heatsink to cool the power electronics. To minimize the impact of the heat dissipated by the electric machine, IMD #6 shown in Figure 8.27(c) re-positions the shaft-mounted blower fan closer to maximize air flow and cooling of the power electronics. More specifically, air drawn from the outside flows through the heatsink fins of the power electronics and then passes across the other side of the power electronics to provide double-sided cooling for improved thermal performance. It should be noted that air flowing between the power electronics and

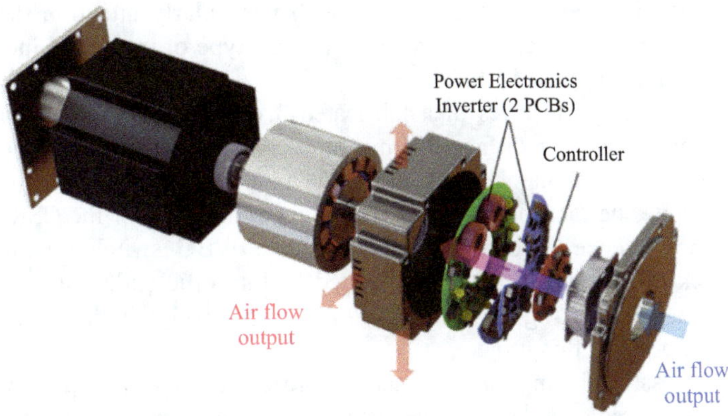

*Figure 8.28 Exploded rendered drawing of the IMD #1 (VEBC) configuration with the power electronics mounted inside an end chamber (*note: *The airflow direction can be reversed)*

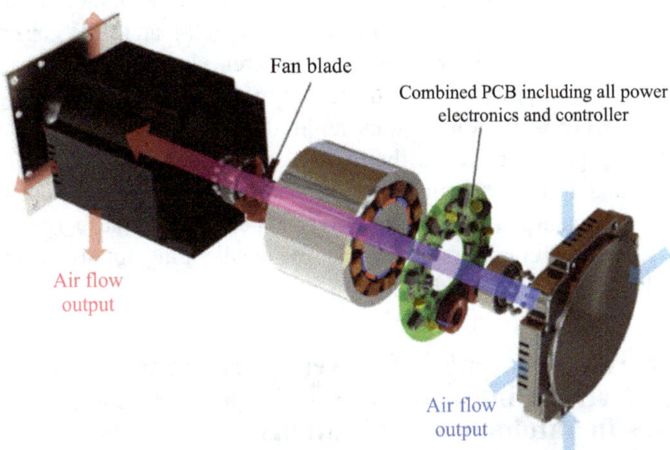

Figure 8.29 Exploded rendered drawing of the IMD #4 (VESFC-Ver1) configuration with the power electronics mounted in the same housing with the electric machine

the electric machine functions as an air curtain that suppresses heat convection from the electric machine to the power electronics.

8.2.4 Conclusions

A new class of integrated motor drives has been presented that incorporates wide-bandgap power semiconductor devices, a current-source power converter consisting of back-to-back CSR and CSI stages, and a surface permanent magnet (SPM) machine. Several issues associated with the integration of the power electronics inside the machine have been discussed, with particularly attention devoted to thermal management and cooling strategies for this type of integrated motor drive incorporating a WBG-based CSI.

The major design features of a fabricated 3 kW CSI-based IMD have been presented, consisting of a surface PM machine combined with the power electronics mounted in an end chamber attached to the axial end of the aluminum machine housing. Despite the fact that hybrid switches were required for the CSR and CSI stages in the absence of commercially available WBG switches with reverse voltage-blocking capability, the demonstrated CSI specific and volumetric power density values (8.7 kW/kg and 9.14 kW/liter, respectively) both exceeded their aggressive targets for this project.

The successful thermal management approach for cooling the SiC power MOSFETs was based on a three-pronged strategy that consisted of: (a) minimizing the switch losses by careful selection of the MOSFET devices and their switch topology; (b) mounting of the devices on PCBs that were designed with a matrix of multiple copper layers and large numbers of vertical vias to maximize the heat spreading; and (c) adoption of a small separately excited fan that blows air over the surface of the PCBs and power devices. The maximum device case temperatures were all limited to values less than 100°C in the final IMD configuration during full-load (3 kW) operation.

Finally, six IMD configurations with different integration and air-cooling approaches have been presented and discussed in the last section of this prototype IMD discussion. These configurations highlight the wide range of alternatives available to designers to tailor their IMD designs for specific applications. Promising results from subsequent experimental testing have demonstrated that WBG-based CSIs can be designed to deliver appealing thermal performance in a variety of air-cooled configurations including challenging totally-enclosed non-ventilated (TENV) enclosures [32].

8.3 Research prototype 3: a novel approach to thermal and electrical integration of high power density drives in automotive applications

8.3.1 Introduction

Recent developments in energy storage technologies together with national and international policy changes and the desire to progress to a low carbon economy, have made the all-electric vehicle concept, not only more attractive, but a critical path for the automotive industry [33]. Higher energy density batteries combined with highly efficient variable speed motor drives offer a realistic prospect to create higher performance electric vehicles with extended ranges that are acceptable to the general public [34]. Electric propulsion systems implemented in the all-electric vehicles have advanced in recent years but still need to achieve a higher degree of power density, efficiency, and lower system cost to meet the requirements of this industry. Increased power density and efficiency leads to lower weight, reduced energy storage requirements and longer range. The integration of the motor with the power electronics, in itself, reduces system weight by reducing the number of connectors and heavy cables within a system [35] but system power density can be increased further by careful design of the integrated package. Direct liquid cooling of power devices can reduce the power converter size but one of the paths to be able to attain the goal of increased system power density is the use of direct liquid cooling of the stator windings [36,37]. This allows higher current densities to be used in the windings and therefore can push the present boundaries of achievable power densities and reduce the quantity of material used in the production and therefore reduce the manufacturing energy needed and total carbon footprint of the drive system. This would seem to be a reasonable methodology within the automotive sector due to the common availability of liquid cooling loops in direct contrast to an industrial application where liquid cooling is often unavailable. In this regard, thermal models have been previously developed in order to optimize the design with respect to power density and efficiency, and explore the maximum current density in the stator windings [38,39].

This technique of integrating the power converter with the motor to create an integrated drive can help to reduce EMI problems and reduce the effect of switching noise on other parts of the electric vehicle while reducing the need for connectors, cables, and housing material since the housing would be shared in an integrated drive. The integrated drive concept creates new challenges for various design aspects. First, the proximity of the switching devices to the motor stator. This implies that the thermal aspects of the two systems are tightly coupled to each other. Second, to maximize efficiency and performance, the use of wide bandgap semiconductors is also being investigated. Since these are optimized to operate at high switching frequencies, proper design of the integrated switching circuit becomes a determining factor in their successful implementation. An optimized design can enable the exploration of the advantages of high-speed devices and increase the power density of the integrated drive. The use of high-speed wide bandgap devices has implications for the motor system. Winding insulation, waveform reflections, and voltage overshoot at the terminals are all important factors which affect motors which are driven from pulse width modulated power electronic converters of any kind. The close proximity of the converter to the motor in an integrated drive is a significant benefit in terms of motor terminal voltage overshoot and radiated EMI. The typical issues relating to motor insulation degradation under high dv/dt stress is mitigated by using the thermal management system presented in this paper due to the controlled proximity of each winding needed to facilitate effective coolant flow. For the above reasons, the approach presented herein facilitates the use of wide bandgap semiconductors.

Finally, new ways of thinking are required when designing the mechanical aspects, the electrical connections, and the assembly of the whole system so that the requirements of the individual components can be met. This prototype presents a novel method of structural and functional integration of a power electronic converter into a permanent magnet motor. Within this approach, a single liquid cooling loop is utilized for the thermal management of both the power modules and stator windings. The power modules are directly assembled into the motor housing and are connected to the stator windings, through a novel multilayer PCB structure for space saving. Also, control hardware, DC-bus decoupling, and gate driver circuits are also integrated into the same space, close to the power modules. This paper will focus the attention on the design and the implementation of the integrated power electronics and its control.

8.3.2 Overview of the system

The system comprises a 9-phase salient permanent magnet motor which is configured as 3 separate, 3 phase sub-machines. The nominal power of the designed motor is 80 kW at 30,000 rpm. At the nominal power, the rated current is 105 Arms. In the designed stator, this current level is equivalent to a current density of 28 A/mm^2 within the windings. This high level of current is only possible using a direct liquid cooling system for the windings and a careful design of the thermal management system. The power electronic parts can be considered to be more susceptible to high temperatures as the maximum operating temperatures of the semiconductor devices and their interconnecting systems are lower than the copper and iron of the stator. The same coolant, in this case de-ionized water, flows first through the power module cooling system. This will be explained later. Each motor winding is connected to one power module through an innovative multi-layer PCB structure, designed to minimize space required for the end-windings. A general overview of this integrated drive system is presented in Figure 8.30.

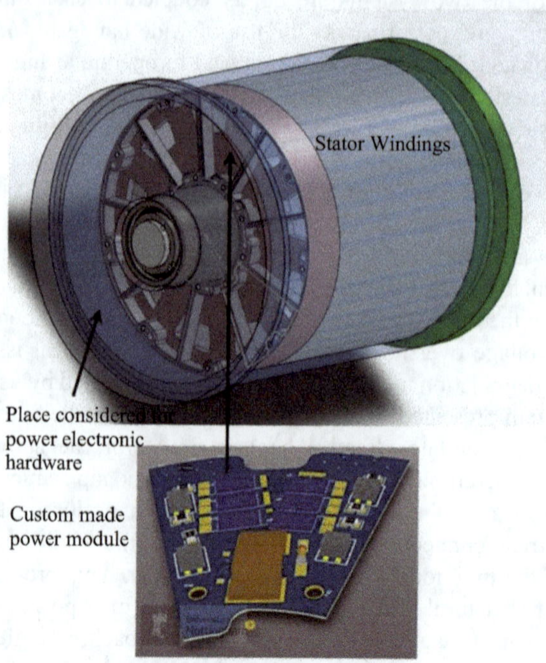

Figure 8.30 Representation of the integrated motor drive together with custom power tile

This figure shows a cut-away view of the drive and the interface between the stator winding areas. The power converter section can also be seen. Each power device "tile" is mounted at the end of the stator windings. The only thing that separates the power tiles from the windings is a coolant guide to ensure effective thermal management of the system.

8.3.2.1 Control hardware

Each sub-machine has its own control hardware to manage the high-speed, low level control. Figure 8.31 shows a block diagram of the control system. The low level control is responsible for the current control of each sub-machine. Functions include the generation of PWM vectors, fast protection functions, DQ transformations, and current control. A higher-level supervisory system provides the speed and torque control for the whole drive together with other supervisory and communications functions. The master and auxiliary control units are connected together using high-speed, differential serial peripheral interface (SPI).

The auxiliary controllers cannot perform the high-level control of the overall motor function and as such need information from the master controller such as demands for flux and torque, and information on the rotational position of the rotor for Park-Clark transformations. The master controller also requires data from the auxiliary controllers in order to perform the high level supervisory and control tasks such as motor line currents, demanded output voltages, and temperatures. The control PCBs which were implemented for the drive are shown in Figure 8.32.

Shown in the smaller insert picture, the first layer, at the top is the master controller board consisting of a Texas Instrument TMS320F28379D microcontroller and a Microsemi iGLOO2 FPGA. The master board is connected to three local controller boards (auxiliary board) through a 22 MHz customized SPI connection. The information sent to each of the local auxiliary controllers from the

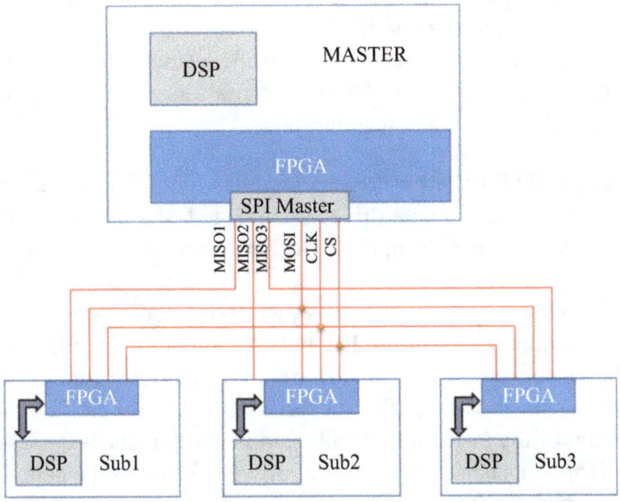

Figure 8.31 Schematic to show the control structure of the system

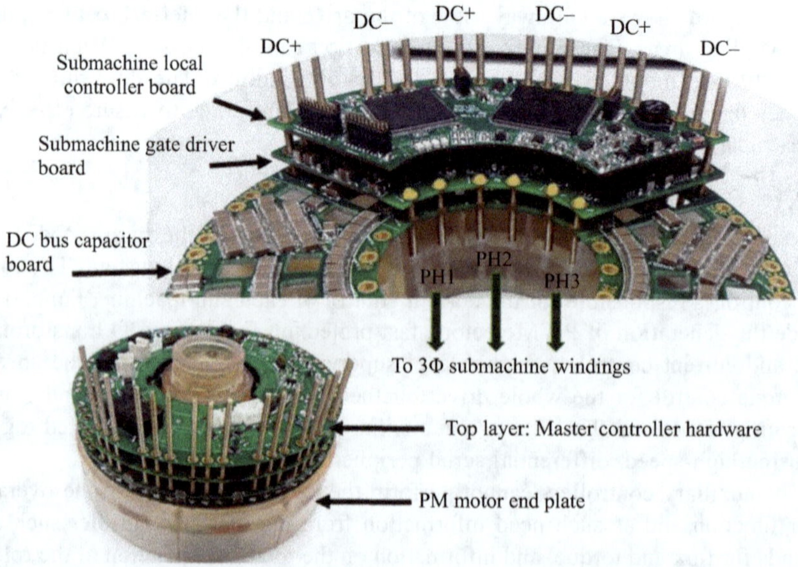

Figure 8.32 Power electronics and control hardware integrated into the PM machine

master: rotor speed, electrical angle, stator frame current will be available to the local controllers at the beginning of each switching cycle with an update rate of 60 kHz. The second boards from the top are auxiliary controller boards which use the same microcontroller-FPGA structure as the master board. Four integrated analogue to digital converters are used to make a synchronous measurement of the three local motor phase currents and the local DC bus voltage. This 16-bit data will be sent once per switching cycle to the Master controller. A space vector calculator is implemented in each of the three local-controllers (for each sub-machine). PWM waveforms are then generated by the FPGAs and sent to the gated driver boards through high-speed differential signaling lines.

8.3.2.2 Gate driver integration

The final two boards are the gate driver and DC-link decoupling section. A cross section of the different layers from master controller board to the stator windings is represented in Figure 8.33.

The DC bus capacitor board is placed just next to the power modules and gate driver board. It is important that the DC link decoupling capacitors are as close as possible to the power sections in order to minimize any parasitic inductance in the switching cell commutation loop. Placing a distributed, 37 μF, X7R capacitor board within a few mm of the switching module considerably reduces the EMI noise and commutation loop inductance. The physical distance between the gate driver boards and the power modules is less than 8 mm. This short distance will prevent oscillations on the gate signals which can occur if significant gate loop inductance is present

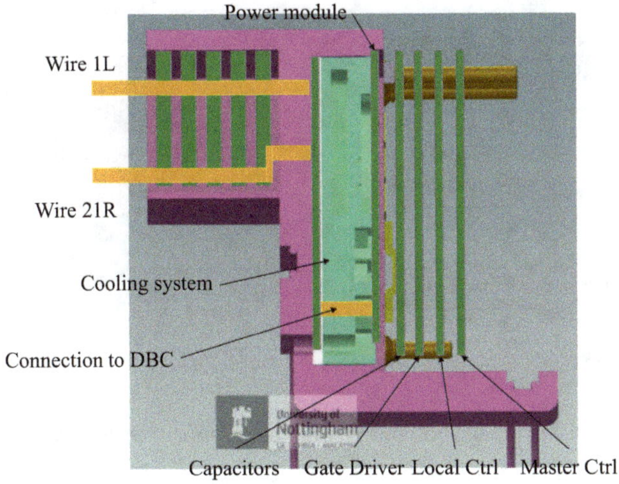

Figure 8.33 Schematic to show the connection of the power modules to the rest of the system including the stack-up of control boards

where distances cannot be minimized. A 3 W custom made gate driver power supply, with low capacitive coupling has been designed for each phase to switch three parallel silicon carbide transistors at 60 kHz. The connections between the external DC source and capacitor board are provided through vertical brass bars.

8.3.3 Integrated power module design

Each power module consists of six SiC MOSFET (2 × 3 in parallel) transistors arranged as an inverter output leg (1.2 kV) which are sintered onto an aluminum nitride (AlN) substrate. Gate resistors, current measurement shunt resistor as well as a local decoupling capacitor are also included on this substrate. The lower inset in Figure 8.1 shows this in more detail. The source and gate contacts are on the top side of the 7.3 × 4.35 mm transistor die and the drain contact is on the bottom side of the die. The transistors are first sintered onto a flex PCB using a nano-silver die paste to provide electrical connection to the gate and source contacts. The transistors together with the flex PCB are then sintered back onto the AlN substrate as a single unit to create the desired drain contact electrical connections, thermal interface connections and any necessary connections between the flex PCB and the substrate. To complete the module connections, inter-board connections are then soldered into place to allow the DC bus connections and the gate drive connections to be made to the subsequent boards. The DC bus connections are brass bus bars, as described in the previous section. The brass bars are directly soldered onto the AlN substrate and then connected to the decoupling capacitor board and main DC source (Figures 8.33 and 8.34). This method is an enhanced version of the one presented in [40] to improve the reliability of the contacts. The cooling system is then placed on the back side of the AlN substrate, as shown in Figure 8.33. Connection to the windings is provided via a multi-layer PCB design. U-shape stator windings are

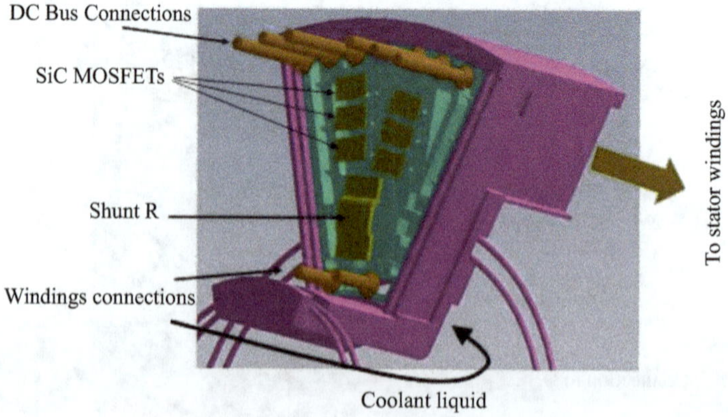

Figure 8.34 Detail of the power module positioned in the stator housing

closed by a special arrangement of PCBs, placed at the end of the u-shape wires. These end termination boards not only connect the ends of the windings to each other in the correct way to create an electrical coil system but also allow fluid to flow between them to aid cooling. The first and the last turns of 21 stator winding turns will pass through all of the PCB layers, through the cooling system section, underneath the DBCs, and finally be terminated onto the power module via two small brass bars. The power section is then encapsulated using silicone to enhance the voltage breakdown capability of the system.

8.3.4 Thermal management

The design of the thermal management system is critical for the effective operation of the integrated drive. The ability to create a compact drive system relies on the effectiveness of the cooling system to enable an efficient removal of heat and to keep the operating temperature of the critical parts to acceptable levels. In general, the switching devices have a lower maximum operating temperature and many more interfaces through the converter structure before the thermal energy reaches the coolant medium. They are more susceptible to elevated and/or rapidly changing temperatures and it is for this reason that the coolant first passes through the power section. The coolant then flows through the stator windings as shown in Figure 8.34. The stator windings are separated using a series of plastic pre-form guides which not only keep the windings from bunching up, but guide the fluid so that an even cooling is ensured for the entire length of the stator winding.

The coolant liquid, flowing through the outer circumference of the rotor shaft, is sprayed onto the DBC, directly underneath each MOSFET Die. This solution is based on the direct substrate, jet impingement cooling method which can provide a very low specific thermal resistance for power devices [41]. The liquid is guided via the coolant channels through a 3D printed structure and is sprayed through holes placed behind each MOSFET die. The liquid is then collected before it is further guided to the end of the stator windings. It circulates around and through the end winding connection PCBs and finally flows through the stator windings. Figure 8.35 shows the liquid flow path through the jet impingement section that directly cools the back side substrate, under each switching device. Figure 8.36 shows a computational fluid dynamic (CFD) simulation of the coolant path through the system and Figure 8.37 shows a simulation of the device temperatures for a particular operating point of the system. Simulation and analysis of this kind can help to determine the feasibility of different approaches to the challenges faced by the structural and functional integration of power converter, motor, thermal

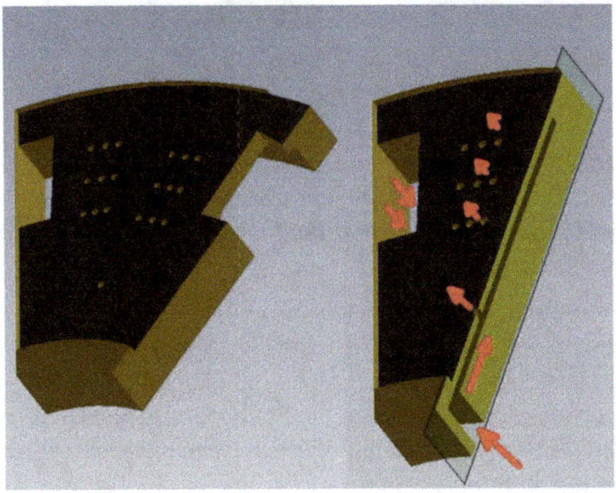

Figure 8.35 Liquid coolant path through the power module coolant guide section

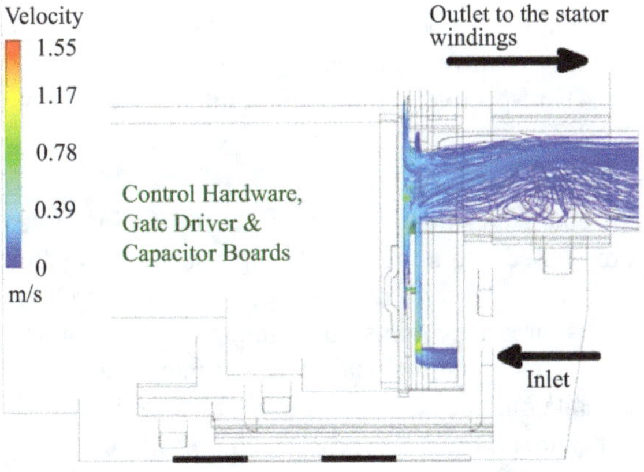

Figure 8.36 This figure shows a CFD simulation of the coolant path through a cross section of the power module coolant guide and then on to the motor stator

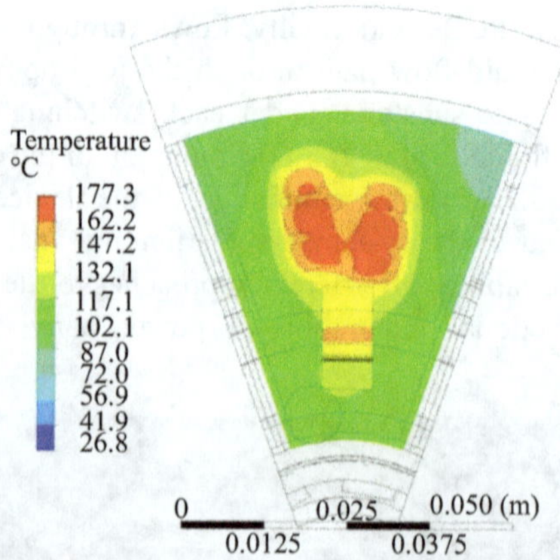

Figure 8.37 This figure shows a simulation of the device temperatures when operated at full load current where the coolant is a mixture of water and glycol (50/50) with a fluid flow rate of 50 ml/s and coolant inlet temperature of 50°C

management system and control and driving electronics. Variables such as coolant temperature, viscosity, specific heat capacity, jet diameter, flow rate, and pressures can all have a very significant effect on the final operating junction temperatures of the devices and as such, integrated thermal management simulations are an essential part of the design process.

8.3.5 Results

Initial results of the commissioning and testing of the power section of the integrated drive are presented in Figures 8.38 and 8.39. These show the output waveforms of the converter when operating from a 300 V battery voltage with 1.5 A load current. The device waveforms during turn on and turn off are shown in Figures 8.40 and 8.41, respectively, to highlight the dv/dt of the switching devices.

8.3.6 Conclusions

In this section, the challenges met in the next-generation integrated drive research work have been discussed and an innovative approach to the structural and functional integration of power converter, motor, control, and thermal management has been proposed. This prototype aims to highlight the importance of providing solutions to both domains of thermal and electrical integration of a drive system. In terms of electrical integration, several compact PCB boards have been designed and built to not only incorporate the power components but also the control and monitoring hardware and the gate driver modules as close to the stator windings as physically possible given the design constraints. A switching module has been designed to fit into the endplate of a compact permanent magnet motor and to be integrated onto the stator windings. This highly compact electrical design made it

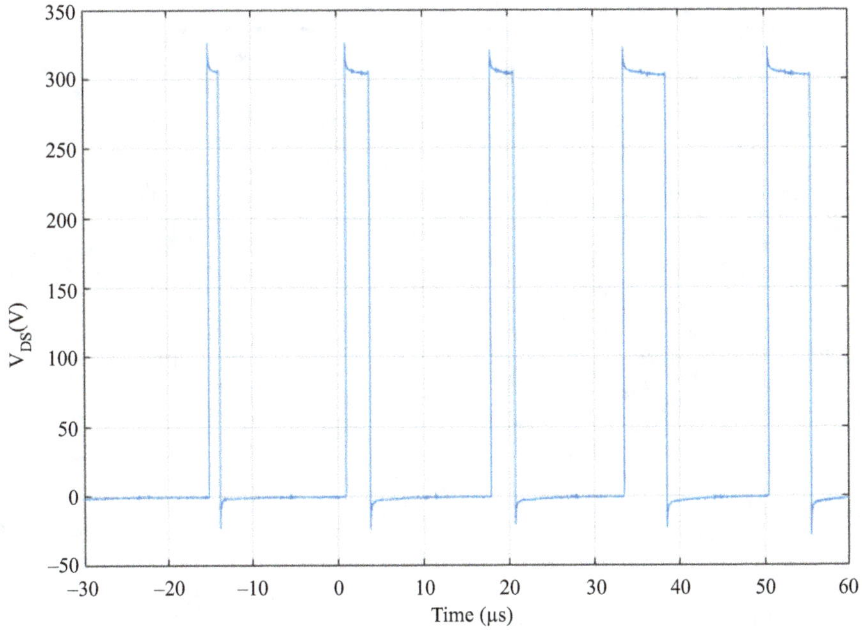

Figure 8.38 Commissioning results of the switching waveform of the power converter fed from 300 V DC

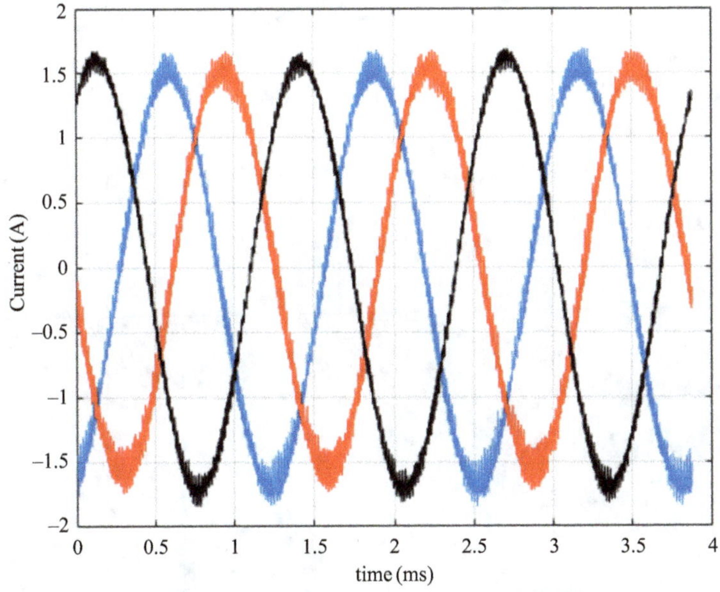

Figure 8.39 Commissioning results of the power converter fed from 300 V DC 1.5 A out

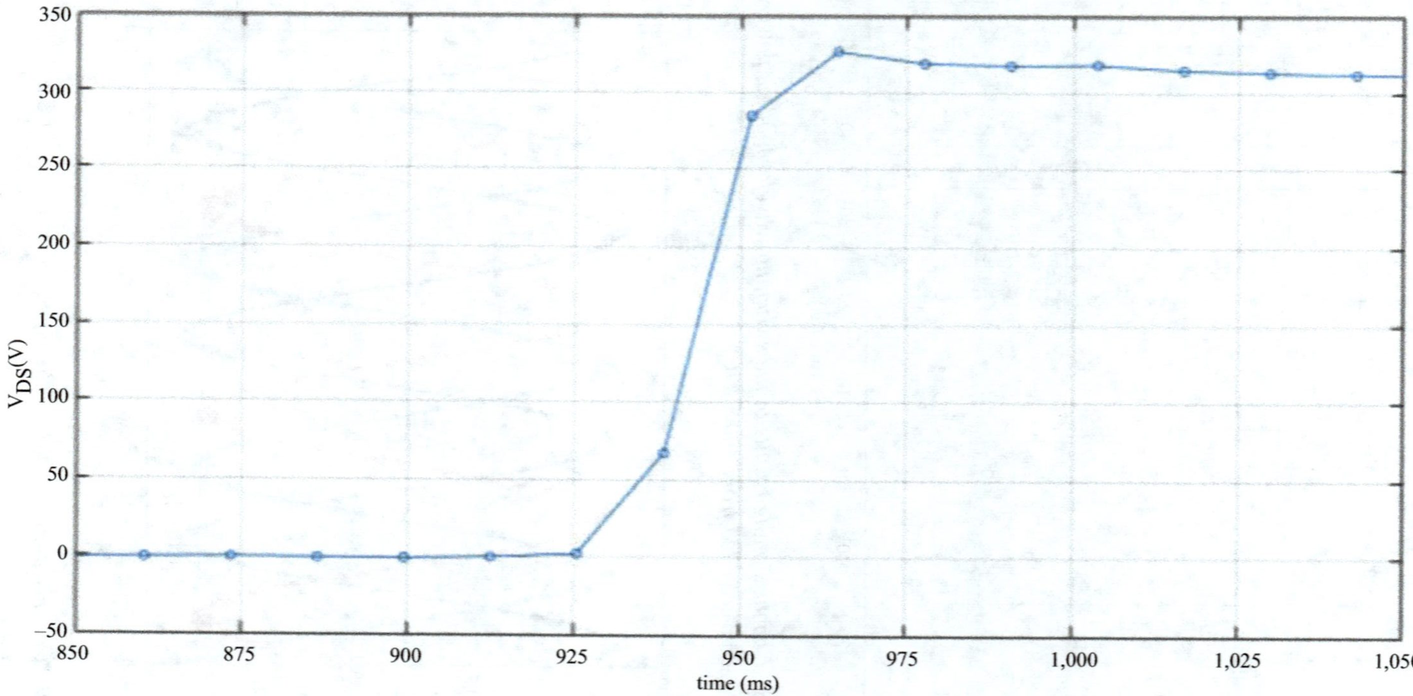

Figure 8.40 Commissioning results of the turn-off of the SiC MOSFETS

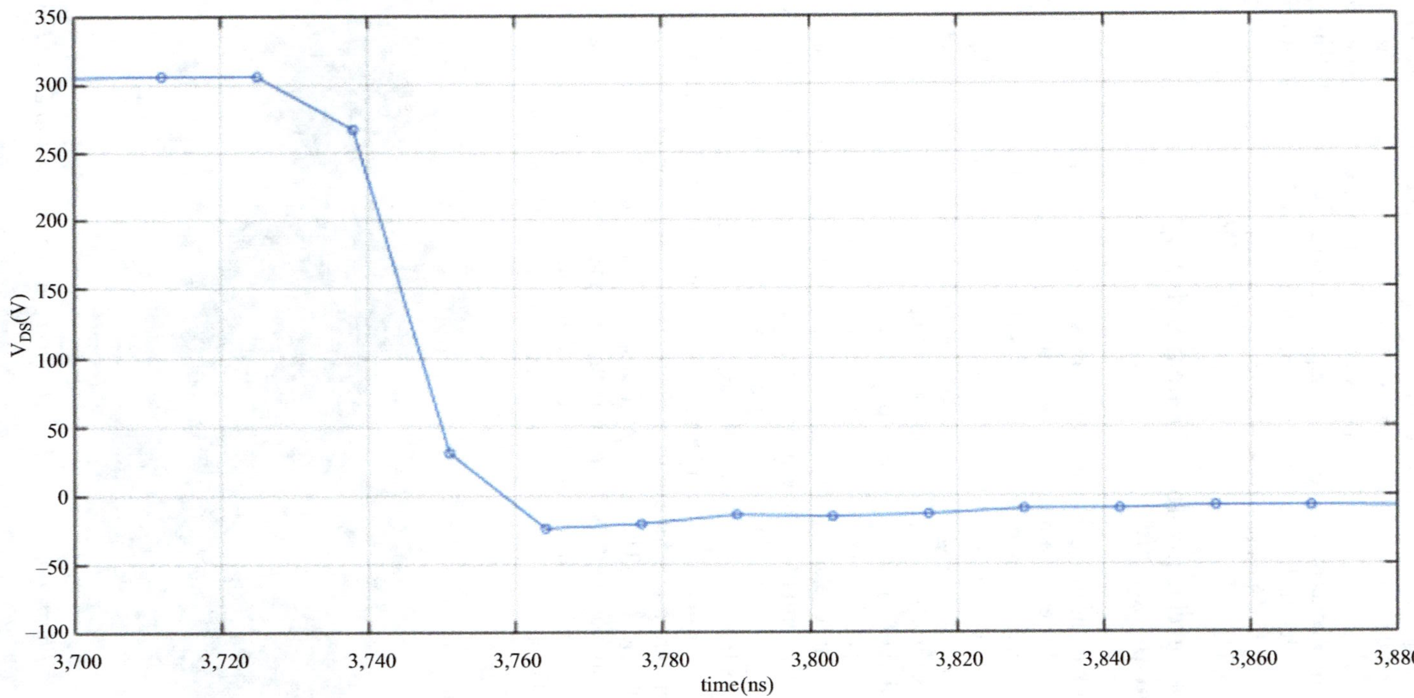

Figure 8.41 Commissioning results of the turn-on of the SiC MOSFETS

essential to implement a new thermal design which incorporates both the stator and the power module cooling systems utilizing a single coolant source. Finite-element simulations of the design of the drive system show that a viable solution for the integrated drive concept in automotive applications can be achieved. Initial commissioning results of the experimental implementation and validation of the analysis performed in simulation have also been presented.

8.4 Commercial product: ProteanDrive by Protean Electric

ProteanDrive is an in-wheel electric motor system for passenger cars and light commercial vehicles. The motors can be used in two- or four-wheel-drive configurations in electric and hybrid vehicles with 400 V dc or 800 V dc systems. A pair of in-wheel motors represents an alternative to a more conventional electric axle, in which a motor and inverter are mounted on the vehicle body and drive is transferred to the wheels through a gear, differential and half-shafts.

Illustration of a vehicle platform powered by ProteanDrive in-wheel motors is shown in Figure 8.42. The attraction of in-wheel motors to provide the drive for a vehicle is primarily related to the opportunities in vehicle design. Without the need to package a powertrain on the vehicle body, greater and more flexible space can be created on a given footprint and better aerodynamics can be achieved. Secondary advantages include decreased overall mass and improved vehicle dynamic control which improves handling and active safety functions. Against these advantages, in-wheel motors are a less mature technology with higher cost compared to a conventional electric axle. The addition of unsprung mass, that is the mass between suspension and road, negatively impacts the secondary ride of the vehicle.

Protean's in-wheel motors are direct drive, meaning that they are not geared, and are packaged to fit entirely within a wheel rim accompanied by a largely conventional

Figure 8.42 Illustration of a vehicle platform powered by ProteanDrive in-wheel motors

friction brake. The electric machine is a permanent magnet synchronous motor (PMSM) with an outer rotor and the inverter is integrated in the same package, sharing a heat-sink with the motor. The assembly is shown in an exploded view in Figure 8.43. There are alternative in-wheel drive concepts from other companies, some of which are geared and some of which occupy space outside of the wheel.

Packaging the motor into the space available in a wheel rim presents a challenge. The goal is to minimize disruption to other systems at the wheel-end, particularly the wheel-bearing, suspension, steering, and friction brake. While the large diameter annular volume available lends itself to a high radius, axially short, high torque motor, it is not ideal for packaging an inverter. ProteanDrive is designed such that standard wheel rim and bearing, suspension and steering systems can be adopted but the brake disc has to be inverted compared to conventional solutions.

The in-wheel environment necessitates a design that is robust against vibration and shock. Vibration testing is carried out over a prolonged period at an average of 11 g rms acceleration, and shock accelerations up to 100 g need to be tolerated. Sealing is also more challenging than for the motor and inverter in a conventional vehicle. The system may need to stand in water and is subjected to dust, grit and other materials. Furthermore the wheel-bearing, which forms the bearing of the ProteanDrive system, deflects in cornering which in turn means that the dynamic seal needs to tolerate relative movement between the rotor and stator.

Heat needs to be managed in the system. The motor, inverter, and friction brake are all generators of heat but at the same time they are tolerant of very different temperatures. Typically the motor windings can be operated up to 180°C while most electronic components are rated up to 125°C and the capacitor elements up to 100°C. The brake disc may reach temperatures of the order of 500 and, in the

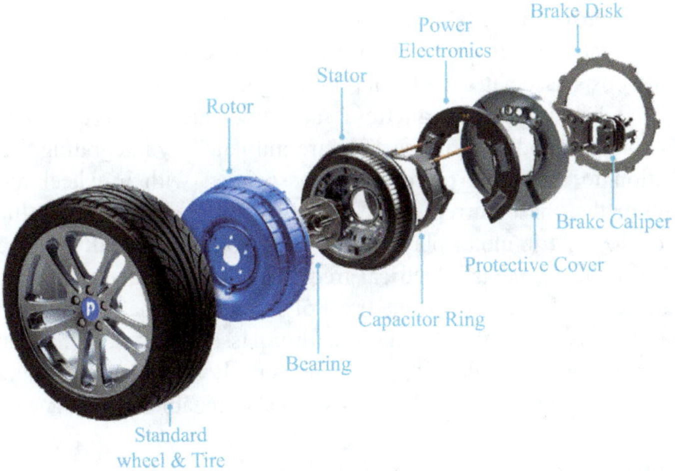

Figure 8.43 Exploded view of the ProteanDrive in-wheel motor system including the integrated inverter

ProteanDrive system, is attached to the rotor, meaning that heat conduction to the NdFeB magnets needs to be accounted for.

Noise, vibration and harshness (NVH) have to be managed. The large diameter outer rotor format permits unacceptable levels of noise production if the excitation of the rotor by electromagnetic performance is not constrained by design. In particular, the slot-pole configuration is important in determining the modes that are excited. Additionally, structure-borne noise is generated by torque ripple from the in-wheel motor which therefore needs to be minimized, particularly at frequencies below about 100 Hz which transmit well through the vehicle structures.

8.4.1 Motivation for integration

Integrating the inverter with the electric machine in an in-wheel motor increases the design challenges around packaging, shock and vibration and thermal management. Against this, there are advantages for the vehicle designers and manufacturers that make it an attractive goal.

First, the primary benefit of in-wheel motors in relation to more conventional electric drive systems is that powertrain components do not need to be housed on the vehicle body, so creating space and design freedoms. Where the inverter is not integrated with the in-wheel motor, a space will need to be found for it on the vehicle body which dilutes the advantage.

Second, a cost advantage can be achieved. Compared to an inverter that is integrated with the in-wheel motor package, a separate inverter has extra cost and material, associated with the housing, including cooling jacket, and connectors, or alternative forms of connections to external services: six additional phase connectors, two additional coolant hose connectors and two additional low voltage/control/sensor signal connectors. Connectors and other forms of connections are typically amongst the least reliable components of a system so reducing the number of them will improve the reliability of the product in service.

Third, the vehicle manufacturer has fewer components to handle which reduces work and costs associated with sourcing, logistics, inspection and assembly.

Finally, radiated electromagnetic issues associated with the phase cables between the inverter and electric machine are mitigated by integrating the inverter.

In relation to the NVH performance associated with in-wheel motors, the ability to tailor the phase current wave-forms to complement the non-linear characteristics of the electric motor plays a significant role. Specifically, the addition of harmonics of the fundamental electrical frequency at the right amplitude and phase can reduce torque ripple such that the level of structure-borne noise experienced in the vehicle cabin is acceptable for high quality passenger cars. This is achievable with an inverter that is not integrated, so is not strictly a motivation integration, but it does require the inverter control code to be customizable.

8.4.2 Description of integrated approach

Having established a motivation for integrating the inverter in a single package with the electric machine based on the advantages for customers, the engineering challenge is how to realize that goal without compromising the product in some other way. In the case of in-wheel motors that are intended to be integrated in vehicles with conventional chassis components, it is the space available to package the inverter that drives design decisions that are different from the usual inverter-in-a-box design. The volume available to house the electric machine and inverter is annular. The outer diameter is constrained by the wheel-rim diameter, the inner diameter by the need to keep a center-bore available for a wheel-bearing, and the total axial length is constrained by the wheel-rim width, allowing space for the friction brake. For example, a ProteanDrive motor-inverter designed to fit on passenger cars with 18 wheel rims has an outer diameter of approximately 400 mm, an inner diameter of approximately 200 mm, and a total axial extent of about 140 mm.

For this product, the power modules and PCB assemblies associated with the inverter are distributed on the in-board (vehicle side) face of the assembly. The capacitor is formed as a ring of lower radius than the electric machine, inside the electric machine. The chassis of the stator is a machined aluminum casting with internal cooling channels through which a water/glycol mix flows. The arrangement creates heat-sinks that thermally separate the electronics from the electrical machine from the capacitor. The arrangement is shown in Figure 8.44.

At the time of design, commercial off-the-shelf power modules were not available that were sufficiently compact to fit in the volume and therefore custom modules are adopted. The power transistors sit on a cooling plate with pin-fins directly in the cooling channel. This was found to provide superior thermal isolation from the electric machine compared to an indirect cooling arrangement in which a flat cooling plate is fixed to the heat-sink with a thermal interface material between the two. To meet the packaging constraints and improve cooling of the power modules, there are four independently controlled 3-phase power modules distributed around the circumference, with each driving two parallel 3-phase motors. Each pair of power modules is housed in a single module along with the associated control electronics. The arrangement is shown schematically in Figure 8.45. This distributed architecture also reduces the impact of faults but has

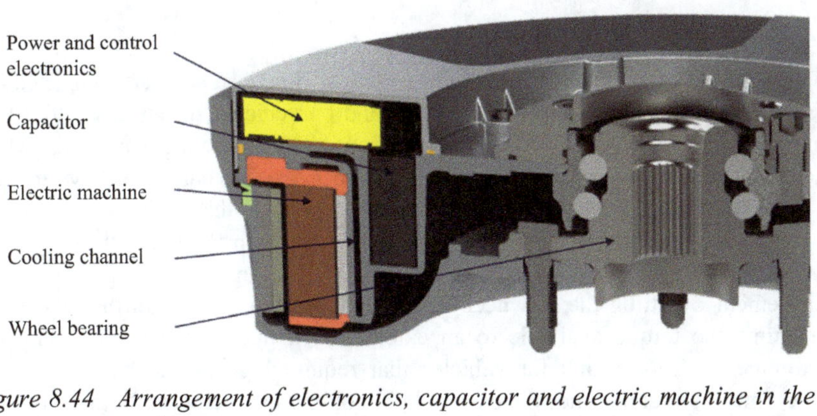

Figure 8.44 Arrangement of electronics, capacitor and electric machine in the package

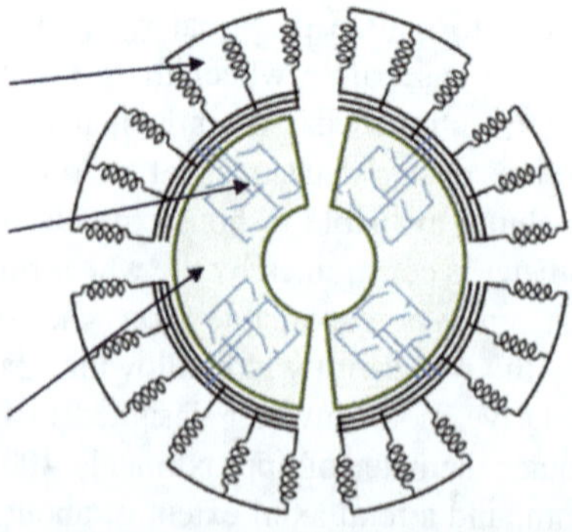

Figure 8.45 Schematic representation of the distributed drive architecture

the disadvantage of increasing cost as a result of the increase in the number of drivers and current sensors and the generally increased complexity.

The capacitor is formed as a single ring shaped assembly together with the DC bus bars. The DC bus bar distributes the supply to the four inverters. Thin film capacitor elements are arranged around the ring to create the desired total X and Y capacitance. The arrangement ensures that each inverter has capacitance in close proximity.

In order to avoid failures due to vibration and shock loads, the use of heavy components on PCBs is avoided where possible, or they are supported mechanically. The electronics modules are filled with a silicone based gel which also reduces stress on components and their connections under vibration. Attention in design is also paid to differential movement between different PCBs and between power modules and PCBs across which connections are made. Such connections are subject to continuous stress due to vibration of the unit. This is reduced in design by minimizing the relative movement and by supporting the cables that carry signals between PCBs. For example, ribbon cables should be kept short or supported along their length.

8.4.3 Operational experience

At the time of writing the in-wheel motor system as described here has completed full life-time durability testing in the laboratory and on vehicles. In this case, "full life-time" is equivalent to 300,000 km, 15 years life and 8,000 operating hours in the wheel of a passenger car. No servicing is required. As such, the feasibility of the system, and of the integration of the inverter in particular, is proven for this application.

The integration of the inverter achieved the goals in terms of advantages for the customer in most of the applications of interest. In some cases though, the requirement of fitting the in-wheel motor package entirely within the wheel-rim constrains the torque available to an extent that it does not meet the vehicle's requirements. This occurs for vehicles that require high torque but have small wheel rims, as is sometimes the case in light commercial vehicles. In this case, the advantages of having an integrated inverter may be out-weighed by the disadvantage of taking up space that could be used to increase the torque output of the motor by increasing its stack length.

The integration of the inverter into an in-wheel has also been seen to present particular considerations in the event of a crash. In a crash, the high-voltage supply and other services to the motor may be severed and conductors exposed. With an integrated inverter the DC link capacitor retains its charge. Contact with the exposed conductors attached to the in-wheel motor is dangerous and therefore the capacitor needs to be discharged rapidly. To achieve this, there are typically two mechanisms. First, an active fast discharge switches current through the coils or through a discharge resistor, typically discharging the capacitor to safe voltage levels within less than 5 sec. Second, a passive discharge through a resistor that is permanently connected in parallel with the capacitor and results in discharge in about 2 min. Preferably the active mechanism draws power off the HV DC line so that it functions even if the low voltage supply is removed. The passive mechanism is merely a back-up in case all other measures fail.

8.4.4 Future integration possibilities

In the short term, the greatest impact on inverter integration in in-wheel motors results from the increased availability and decreased cost of WBG, silicon carbide (SiC) or gallium nitride (GaN), power transistors. Critically for in-wheel applications, they allow the total volume of the electronics to be reduced for a given power output. Power modules themselves can be smaller, but also the opportunity for higher switching frequencies allow the DC link capacitance to be reduced. With modern power modules, it is no longer necessary to configure in-wheel motors with multiple, custom built power modules, as described above, and so cost can be reduced significantly.

In the longer term, there are opportunities for integrating further control functions into the in-wheel motor electronics. As in-wheel motors are more widely adopted the corner of the vehicle will become more adapted and brake-by-wire and steer-by-wire systems that work together with the in-wheel motor may become available. In this case, there are savings in cost and complexity to be achieved by combining the control of the various components using the processor set integrated within the in-wheel motor. Furthermore, fully coordinated functionality incorporating the in-wheel motor, brake, and steering provides opportunities at the vehicle level in terms of controllability, active safety such as stopping distances and object avoidance, and power consumption.

References

[1] M. R. Khowja, C. Gerada, G. Vakil, P. Wheeler, and C. Patel, "Integrated output filter inductor for permanent magnet motor drives," in *IECON 2016 – 42nd Annual Conference of the IEEE Industrial Electronics Society*, 23–26 Oct. 2016 2016, pp. 2827–2832, doi:10.1109/IECON.2016.7793027.

[2] M. R. Barzegaran, T. A. Youssef, A. Berzoy, and O. A. Mohammed, "Electric machine drive design improvements through control and digital

signal processing techniques," *IEEE Transactions on Energy Conversion*, vol. 30, no. 3, pp. 1255–1264, 2015, doi:10.1109/TEC.2014.2375382.
[3] R. M. Tallam, G. L. Skibinski, T. A. Shudarek, and R. A. Lukaszewski, "Integrated differential-mode and common-mode filter to mitigate the effects of long motor leads on AC drives," in *2010 IEEE Energy Conversion Congress and Exposition*, 12–16 Sept. 2010, 2010, pp. 838–845, doi:10.1109/ECCE.2010.5617910.
[4] M. A. Mohamed, S. M. Lambert, B. C. Mecrow, D. J. Atkinson, and A. C. Smith, "An electrical machine with integrated drive LCL filter components," in *8th IET International Conference on Power Electronics, Machines and Drives (PEMD 2016)*, 19–21 April 2016 2016, pp. 1–6, doi:10.1049/cp.2016.0222.
[5] M. A. S. Mohamed, S. M. Lambert, B. C. Mecrow, X. Deng, S. Ullah, and A. C. Smith, "Integrating the magnetics of an LCL filter into a high speed machine with pre-compressed coils," in *2017 IEEE International Electric Machines and Drives Conference (IEMDC)*, 21–24 May 2017, 2017, pp. 1–7, doi:10.1109/IEMDC.2017.8002292.
[6] X. Deng, S. Lambert, B. Mecrow, M. A. S. Mohamed, and S. Ullah, "Winding connection solution for an integrated synchronous motor drive," in *2017 IEEE International Electric Machines and Drives Conference (IEMDC)*, 21–24 May 2017, 2017, pp. 1–7, doi:10.1109/IEMDC.2017.8002162.
[7] J. D. Ede, K. Atallah, G. W. Jewell, J. B. Wang, and D. Howe, "Effect of axial segmentation of permanent magnets on rotor loss in modular permanent-magnet brushless machines," *IEEE Transactions on Industry Applications*, vol. 43, no. 5, pp. 1207–1213, 2007, doi:10.1109/TIA.2007.904397.
[8] W. Huang, A. Bettayeb, R. Kaczmarek, and J. Vannier, "Optimization of magnet segmentation for reduction of eddy-current losses in permanent magnet synchronous machine," *IEEE Transactions on Energy Conversion*, vol. 25, no. 2, pp. 381–387, 2010, doi:10.1109/TEC.2009.2036250.
[9] X. Deng, M. A. S. Mohamed, S. Lambert, and B. Mecrow, "Development of a high-speed, permanent magnet, SiC-based drive with integrated input filters," *IEEE Transactions on Energy Conversion*, vol. 35, no. 2, pp. 863–874, 2020, doi:10.1109/TEC.2020.2973449.
[10] A. G. Jack, B. C. Mecrow, P. G. Dickinson, *et al.*, "Permanent-magnet machines with powdered iron cores and prepressed windings," *IEEE Transactions on Industry Applications*, vol. 36, no. 4, pp. 1077–1084, 2000, doi:10.1109/28.855963.
[11] X. Deng, S. Lambert, B. Mecrow, and M. A. S. Mohamed, "Design consideration of a high-speed integrated permanent magnet machine and its drive system," in *2018 XIII International Conference on Electrical Machines (ICEM)*, 3–6 Sept. 2018, 2018, pp. 1465–1470, doi:10.1109/ICELMACH.2018.8507155.
[12] T. M. Jahns and H. Dai, "The past, present, and future of power electronics integration technology in motor drives," *CPSS Transactions on Power Electronics and Applications*, vol. 2, no. 3, pp. 197–216, 2017.

[13] B. Cheong, P. Giangrande, M. Galea, P. Zanchetta, and P. Wheeler, "Integrated motor drive design for weight optimization," in *2017 IEEE Energy Conversion Congress and Exposition (ECCE)*, 2017, New York, NY: IEEE, pp. 816–823.

[14] W. Lee, S. Li, D. Han, B. Sarlioglu, T. A. Minav, and M. Pietola, "A review of integrated motor drive and wide-bandgap power electronics for high-performance electro-hydrostatic actuators," *IEEE Transactions on Transportation Electrification*, vol. 4, no. 3, pp. 684–693, 2018.

[15] S. Chowdhury, E. Gurpinar, G.-J. Su, T. Raminosoa, T. A. Burress, and B. Ozpineci, "Enabling technologies for compact integrated electric drives for automotive traction applications," in *2019 IEEE Transportation Electrification Conference and Expo (ITEC)*, 2019, New York, NY: IEEE, pp. 1–8.

[16] L. Tang and G.-J. Su, "High-performance control of two three-phase permanent-magnet synchronous machines in an integrated drive for automotive applications," *IEEE Transactions on Power Electronics*, vol. 23, no. 6, pp. 3047–3055, 2008.

[17] A. Tenconi, F. Profumo, S. E. Bauer, and M. D. Hennen, "Temperatures evaluation in an integrated motor drive for traction applications," *IEEE Transactions on Industrial Electronics*, vol. 55, no. 10, pp. 3619–3626, 2008.

[18] E. A. Jones, F. F. Wang, and D. Costinett, "Review of commercial GaN power devices and GaN-based converter design challenges," *IEEE Journal of Emerging and Selected Topics in Power Electronics*, vol. 4, no. 3, pp. 707–719, 2016.

[19] A. K. Morya, M. Gardner, B. Anvari, *et al.*, "Wide bandgap devices in AC electric drives: opportunities and challenges," *IEEE Transactions on Transportation Electrification*, vol. 5, no. 1, pp. 3–20, 2019.

[20] R. A. Torres, H. Dai, W. Lee, T. M. Jahns, and B. Sarlioglu, "Development of current-source-inverter-based integrated motor drives using wide-bandgap power switches," in *2019 IEEE 15th Brazilian Power Electronics Conference and 5th IEEE Southern Power Electronics Conference (COBEP/SPEC)*, 2019, New York, NY: IEEE, pp. 1–6.

[21] A. Marzoughi, R. Burgos, and D. Boroyevich, "Investigating impact of emerging medium-voltage SiC MOSFETs on medium-voltage high-power industrial motor drives," *IEEE Journal of Emerging and Selected Topics in Power Electronics*, vol. 7, no. 2, pp. 1371–1387, 2018.

[22] Film Capacitors for High Temperature Switches and Power Electronics Applications above 125°C. Available: https://www.psma.com/sites/default/files/uploads/tech-forums-capacitor/presentations/is96-high-operating-temperature-metalilized-film-capacitors-downhole-tools-oil-and-gas-industry.pdf

[23] Designing with Magnetic Cores at High Temperatures. Available: https://www.mag-inc.com/Design/Design-Guides/Designing-with-Magnetic-Cores-at-High-Temperatures

[24] J. Van Damme, L. Verkroost, H. Vansompel, F. De Belie, and P. Sergeant, "A holistic DC link architecture design method for multiphase integrated

modular motor drives," in *2019 IEEE International Electric Machines & Drives Conference (IEMDC)*, 2019, New York, NY: IEEE, pp. 1593–1598.

[25] L. Verkroost, J. Van Damme, D. V. Bozalakov, F. De Belie, P. Sergeant, and H. Vansompel, "Simultaneous DC-Link and stator current ripple reduction with interleaved carriers in multiphase controlled integrated modular motor drives," *IEEE Transactions on Industrial Electronics*, vol. 68, no. 7, pp. 5616–5625, 2020.

[26] A. Hopkins, B. Hopfensperger, and P. Mellor, "DC-link capacitor reduction in low voltage and high power integrated modular motor drives," in *2019 IEEE Energy Conversion Congress and Exposition (ECCE)*, 2019, New York, NY: IEEE, pp. 3208–3214.

[27] E. van Dijk, J. A. Ferreira, and P. Van Gelder, "Current sharing between diode clamps of polar capacitor bank," *IEEE Transactions on Industry Applications*, vol. 38, no. 6, pp. 1600–1605, 2002.

[28] N. Chen and M. Nawaz, "Current sharing design assessment of DC link capacitor module," in *2014 16th European Conference on Power Electronics and Applications*, 2014, New York, NY: IEEE, pp. 1–9.

[29] M. Guacci, D. Zhang, M. Tatic, *et al.*, "Three-phase two-third-PWM buck-boost current source inverter system employing dual-gate monolithic bidirectional GaN e-FETs," *CPSS Transactions on Power Electronics and Applications*, vol. 4, no. 4, pp. 339–354, 2019.

[30] H. Ueno, Y. Kinoshita, Y. Yamada, *et al.*, "A 3-phase T-type 3-level inverter using GaN bidirectional switch with very low on-state resistance," in *PCIM Europe 2019; International Exhibition and Conference for Power Electronics, Intelligent Motion, Renewable Energy and Energy Management*, 2019, New York, NY: VDE, pp. 1–4.

[31] AC Synchronous Servo Motor – AKMTM Series. Available: https://www.kollmorgen.com/en-us/products/motors/servo/akm-series/akm-series-ac-synchronous-motors/ac-synchronous-servo-motors/

[32] R. A. Torres, H. Dai, W. Lee, K. Saviers, T. M. Jahns, and B. Sarlioglu, "Investigation of cooling techniques and enclosure types for integrated motor drives," *2022 IEEE Energy Conversion Congress and Exposition (ECCE)*, Detroit, MI, USA, 2022, pp. 1–8, doi: 10.1109/ECCE50734.2022.9947618.

[33] T. Placke, R. Kloepsch, S. Dühnen, and M. Winter, "Lithium ion, lithium metal, and alternative rechargeable battery technologies: the odyssey for high energy density," *Journal of Solid State Electrochemistry*, vol. 21, no. 7, pp. 1939–1964, 2017.

[34] M. S. Kumar and S. T. Revankar, "Development scheme and key technology of an electric vehicle: an overview," *Renewable and Sustainable Energy Reviews*, vol. 70, pp. 1266–1285, 2017.

[35] M. N. Anwar, M. Teimor, R. E. Luken, and J. J. Brautigan, "Electric machine with integrated electronics in a circular/closed-loop arrangement," ed: Google Patents, 2007.

[36] P. Lindh, I. Petrov, A. Jaatinen-Varri, *et al.*, "Direct liquid cooling method verified with an axial-flux permanent-magnet traction machine prototype,"

IEEE Transactions on Industrial Electronics, vol. 64, no. 8, pp. 6086–6095, 2017.

[37] Z. Huang, S. Nategh, V. Lassila, M. Alaküla, and J. Yuan, "Direct oil cooling of traction motors in hybrid drives," in *2012 IEEE International Electric Vehicle Conference*, 2012, New York, NY: IEEE, pp. 1–8.

[38] R. Camilleri, D. A. Howey, and M. D. McCulloch, "Predicting the temperature and flow distribution in a direct oil-cooled electrical machine with segmented stator," *IEEE Transactions on Industrial Electronics*, vol. 63, no. 1, pp. 82–91, 2015.

[39] S. Nategh, Z. Huang, A. Krings, O. Wallmark, and M. Leksell, "Thermal modeling of directly cooled electric machines using lumped parameter and limited CFD analysis," *IEEE Transactions on Energy Conversion*, vol. 28, no. 4, pp. 979–990, 2013.

[40] J. Espina, B. Ahmadi, L. Empringham, L. De Lillo, and M. Johnson, "Highly-integrated power cell for high-power wide band-gap power converters," in *2017 IEEE 3rd International Future Energy Electronics Conference and ECCE Asia (IFEEC 2017-ECCE Asia)*, 2017, New York, NY: IEEE, pp. 146–150.

[41] C. M. Johnson, A. Castellazzi, R. Skuriat, P. Evans, J. Li, and P. Agyakwa, "Integrated high power modules," in *2012 7th International Conference on Integrated Power Electronics Systems (CIPS)*, 2012, New York, NY: IEEE, pp. 1–10.

Chapter 9
Emerging technologies and needs

Andrea Stratta[1], Lee Empringham[1], Liliana de Lillo[1] and Luca Rovere[2]

Literature and advanced industry applications are the main source of information when it comes to spot emerging technologies and new trends. This is valid in every research field, and integrated motor drives make no exception.

However, there are two main risks in producing a list of the most novel technologies currently published, or available on the market. First, this list would be an instantaneous frame destined to become outdated. Second, being a selection from a very broad range of examples, it is likely to be incomplete. An outdated and incomplete list of manufacturing processes would not make a great service to the reader and the community. A method, on the other hand, that can be used to understand weather a certain new concept can be beneficial for IMDs development, can be a pragmatic and profitable aim for this chapter.

To this end, the main challenges in the design, manufacturing, and operation of IMD are here recalled and discussed. Some of these challenges pose limitations in the application of IMD to higher power level or to production process scalability. A representative selection of emerging concepts is then considered, maintaining the focus on the challenge that they address. In this way, it is possible to include in the selection, concepts that still have not been applied to the IMD world but are eligible as technology enabler. These emerging concepts are presented following a sensible classification: design concepts and assembly technologies.

9.1 Summary IMDs' challenges

As discussed in previous chapters, IMD design presents several challenges, which are currently limiting its full potential and application. The challenges that will be addressed in this chapter are briefly recalled and summarized in this section.

There are two main reasons why these are so difficult to tackle. They belong to different field of expertise, and often they are placed at the boundary between different fields of expertise. Either they are multi-physic problems, or they arise from

[1]Power Electronics Machine and Control (PEMC) Group, University of Nottingham, UK
[2]The Thinking Pod Innovations Ltd, UK

non-strictly engineering problem, like the need to manufacture a product that respects the sustainability principles. If these challenges are addressed with a narrow perspective, not considering all the physical and social domains at the same time, the solution will necessary be a biased sub-optimum. In addition to being multidisciplinary, these challenges are heavily correlated and often contradictive. In a similar way as the approach to these challenge needs to be holistic (e.g., multi-physics approach), when possible, they need to be tackled simultaneously or solved iteratively that the correlations between them are well described and addresses. If this approach is not followed, in solving one challenge, other challenges may be exacerbated. Following systematic design approaches tradeoffs can be found, which can guarantee that the manufacturing technologies used to produce IMDs are addressing the following challenges:

- *Thermal management integration*: The PE converter losses are concentrated in a small surface, sharing this surface with other heat sources (e.g., EM stator or end winding) limits the maximum power that the converter can dissipate.
- *High-temperature electronics*: WBG devices are capable to withstand higher temperature than Si; however, exceedingly elevated temperatures can compromise the whole power module and converter reliability.
- *High dv/dt and di/dt*: Compared to Si devices, WBG offer the possibility to have extremely fast switching transitions. Exceedingly fast waveforms pose a problem for winding insulation and electro-magnetic compatibility.
- *Vibration and mechanical stability*: Mounting the converter on or in the housing of electrical machines exposes the converter to significant vibration. Reliability of attachment methods such as thermal-joint compounds and screws may be severely affected by high vibration levels.
- *Assembly process compatibility*: The latest interconnect and bonding packaging techniques require a surface free from any particle and contaminant. On the other hand, many of the EM manufacturing steps include the use of lubricants and resins.
- *Assembly cost reduction*: The process that brings together EM, PE converter and cooling system is often complex and difficult to automate. Assembly processes compatible with large scale and automatic manufacturing should be preferred.
- *Design for recycling*: A higher level of integration results in reduction of used material which meet the environmental sustainability requirements. However, the assembly process needs to be designed to be reversible and ease recycling and reuse of materials/components.

9.2 Emerging design concepts

In the following sections of this chapter concepts and manufacturing techniques are discussed to understand if they can be a potential technology enabler for IMDs. However, the only use of advanced manufacturing technologies does not guarantee an improvement in terms of drive performance.

Many of the challenges listed in Section 9.1.1 are strongly interdependent and often contraindicative. To give an example, the need to increase the drive power density poses the technical problem of operating the PE converter at higher temperatures. SiC-based devices can be operated at significantly higher temperature compared to Si devices, and they could potentially improve the drive performance. Nevertheless, this choice affects all the other component of the PE converter (e.g., power module package, gate drivers, power supplies, etc.), increasing their thermal cycling and reducing reliability and time life. This chain of considerations often leads the designer to adopt WBG semiconductor for all the others advantages they introduce, but without operating them at temperature limit to preserve the rest of the converter.

The complexity of IMD systems requires an integral approach to design and realization. The design tools need to facilitate the design and analysis of a heterogeneously integrated system, which involves combining device and system models and the modeling of PE modules, thermal and EMI analysis, EM performance and so forth. In the automotive sector, design, analysis, and verification, V-shape workflow (top-down and down-up) has become a standard. As shown in Figure 9.1, this method is based on two main phases modeling, from system to component, and validation from component to system. Given the complexity of IMD design, this workflow could be successfully followed. However, this process is not linear but several iterations, parallel and transverse to the main flow are needed.

Undeniably, it is just with the correct design process that IMD performance can be improved. A systematic design workflow that is of interest for IMDs design should start from the deep knowledge of every component of the system and all the manufacturing processes involved in the assembly. Second it should be capable of catering the interactions between heterogeneous components in a holistic-multi-physics fashion.

In this way, instead a linear hierarchical design, which might overlook components interaction, multi-variables, and multi-physics optimization can be carried out to pursue one or multiple objectives (e.g., maximize efficiency, power density, lifetime).

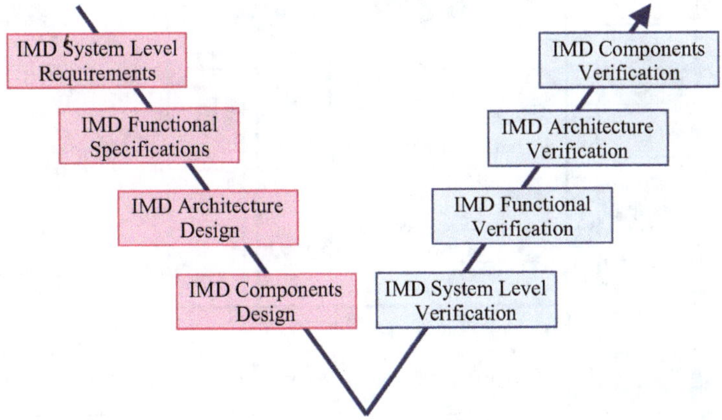

Figure 9.1 Top-down and bottom-up approach V-shape design workflow

9.2.1 Holistic design approach

Traditionally, each part of a motor drive is designed independently, and, as a result, individual parts are available on the market as products. This serial approach to design and manufacture leads to electrical motors, power modules, passive components, thermal management systems, gate drives and converters, that have been developed without full consideration of inter-component interactions.

This serial design approach has reached its limit and further performance advances will only be made avoiding hierarchical design workflow that might overlook the effect of a priori choices of a single sub-system has on other sub-systems, leading to sub-optimal solutions. In this section, the authors discuss the two main steps involved in a holistic design workflow. First, modeling of individual components and their interactions. Second, the automatized use of this comprehensive model to find the optimal set of design variables to meet the application requirements.

9.2.1.1 IMD modeling

As presented in Figure 9.2, IMDs are complex systems that include several heterogeneous components, each one belonging to multiple physical domains.

Together with filters, gate drivers, and power supplies, the PE modules need to be modeled in circuit simulation, thermal management, and electromagnetic interference (EMI) compatibility tool. On the other hand, the electric motor is often modeled on its own to study its electromagnetic and mechanical behavior. The two set of models are then simplified and merged to study the stability and performance of the sensors and control strategy.

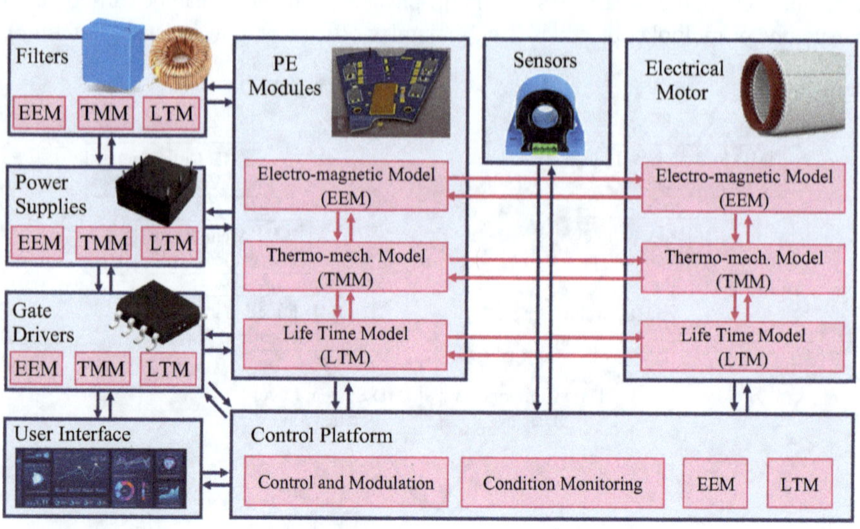

Figure 9.2 IMD model

For circuit simulation, Simulation Program with Integrated Circuit Emphasis (SPICE) models can be used when the focus is on detailed parts of the converter, while MATLAB/Simulink can be used for simulating the whole converter circuit/schematic with control. Finite-element models (FEM) are the standard for both converter EMI analysis and motor electromechanical analysis. Electro thermal models can be either FEM or computational fluid dynamic (CFD) depending on the complexity of the system. All these models can be classified as physical-based models, while statistical-based models are used when it comes to address reliability and availability problems.

Although this set of design tools have improved significantly over time, they have remained separate and stand-alone.

Consequently, when engineers convert specifications to electrical circuits and motors, and then implement them in circuit boards and 3D drawing for manufacturing, they employ multiple tools to ensure that the final product is performing satisfactorily with high reliability and low cost.

However, in the last decade or so, efforts have been made, especially in power electronics, to integrate these varied tools into one process flow for the design and realization of power systems. As a result, automation in design tools for power electronics circuits started to emerge but they are still not integrated with motors design tools.

For a model to be used within an automated design workflow, three requirements need to be met: accuracy, generality and computational efficiency. Accuracy is clearly the most important of the requirements, however, it is often difficult to establish how accurate a model needs to be. Models with percentage error of 10–15% are widely accepted and adopted in literature. It is important to consider that, when several models need to be used, like in IMD design, the accuracy of each model should be comparable. Having one model that is far more accurate than the others does not apport any advantage to the global accuracy, while it could affect the computational efficiency. On the other hand, it takes just an inaccurate model to affect the validity of the design. Indeed, it is the validity of the solution that establish if the models employed in the design are accurate enough, thus suggesting that one or more validation steps are required.

Developing an automated design workflow takes a significant amount of effort and resources, for this reason, one of the objectives is for it to be general and applicable to different scenarios. For this reason, the models employed need to be as more general as possible, without using simplifying assumptions that are dependent to applications, converter topology, machine structure, and control strategy.

Computational efficiency and accuracy are two contradictory principles. A good compromise is to perform more design stages with increasing accuracy and computational time. The first design attempts need to be fast, and they can be relatively inaccurate because they are used to narrow down the research space. The last design steps can be slower and more accurate because they focus just on fewer variables of the design space.

The digital twin is another approach to develop comprehensive and interconnected models of complex systems. In the last decade, this concept has been

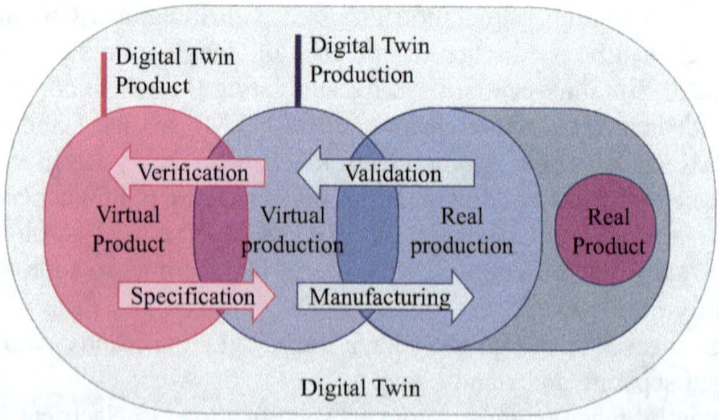

Figure 9.3 Digital twin model used in different stages of product development

applied mainly for health monitoring and online diagnostic of operating machines and converters. In general, a digital twin is a virtual representation of a physical system that virtually shares the same characteristics with its physical counterpart. It enables customers to better understand, optimize, predict, and monitor the performance of an operating systems. The concept of digital twin has been applied in power converters for fault diagnosis recently, which is achieved by comparing the output signals of the digital twin and its physical counterpart in real-time. This concept is also used to estimate the health indicators of the key components in power converter, so as the degradation progress of power converter can be monitored. The digital twin technology includes two parts: the digital presentation of a physical system and an advanced algorithm for the online data analysis.

However, having a digital presentation of an IMD system could enable not only health monitoring in operation but also optimization at design stage. As summarized in Figure 9.3, thanks to an initial modeling effort it is possible to develop a digital twin product that can be used to optimize IMD performance and meet the specification. Once the product is virtually optimized the production and manufacturing steps can be modeled to verify the feasibility of the product. The real production takes to the commissioning of the operating IMD, at this stage all the measurements can be feedback to the production and product digital twin for validation and continuous product improvement.

Due to the large use of emerging technologies, and the combination of heterogeneous techniques, it is essential to build flexible IMD models, which is capable of feeding information from manufacturing and operation to the design stage and the other way around.

9.2.1.2 Design automation and optimization in IMDs

Due to the increasing penetration of electrical devices in modern societies, design automation is of utmost importance to make the required technological advances in short time. Electrical machines and power electronics designer are required to shorten product development cycles, use the most up-to-date devices and produce tailored designs for a demand of growing complexity. These challenges can be addressed just with a systematic and general automatized design approach. Automation can deliver the technologies needed to optimally map the design space into the performance space. For each application, absolute performance limits, labeled as a Pareto front, can be produced.

Optimization strategies can be either stochastic or deterministic. A deterministic algorithm systematically searches the design, based on information about the objective function. Most of them require the calculation of the gradient of the objective function and this is computationally expensive and sometimes impossible. Stochastic algorithms do not need any information about the objective function. Several stochastic optimization strategies have been applied to practical electrical engineering design problems. It has been proved that there is no algorithm which is the best for any class of optimization problem and can avoid the risk of premature convergence to suboptimal solutions. Among the wide variety of available stochastic algorithms, the ones that mimic the natural evolutionary principles, e.g., genetic algorithms (GA), are becoming more and more popular. This class of algorithm makes use of a set of potential solutions (population) that are iteratively modified according to probabilistic rules (inspired by the concepts of natural operators) with the aim of minimizing a set of objective functions for each solution (also called individual or candidate).

As mentioned in previous sections, in power electronics, efforts have been made to develop automated design workflow that consider all the aspects of a converter. This field of power electronic research is known as design automation in power electronics (DAPE) [1]. At present, in power electronics, the main difficulties are combining system-level with component-level designs, the absence of industry interchangeable models, the lack of software standards, and the need for better algorithms and open-source libraries. The key barrier to design automation seems to be the device data sheets, which are not a true representation of the device's performance [2]. For these reasons, current power electronics design follows serial workflows, and it is lacking in automation. However, if a new concept/ technology (e.g., WBG semiconductors) represents a disruptive change, enabling a step forward in converters performance, the gain reduces over time. As the basic concept is given, a significant gain in performance can only be achieved by assigning optimal values to the design variables during the design process. These considerations have led several research groups to look into multi-objective optimization combined to design automation. Figure 9.4(a) shows the optimization workflow for a PFC converter from [3]. In order to perform such an optimization, first a comprehensive mathematical model has to be established as described in previous section. Based on this model, an optimization for multiple objectives, as e.g., efficiency and power density, can be performed. The optimization design process makes best use of all degrees of freedom of a design and also allows to determine the sensitivity of the system performance on base technologies like Figures of Merit of the power semiconductors or properties of the magnetic core materials. Furthermore, different topologies can be easily compared, and inherent performance limits can be identified. In power electronics, this approach has been

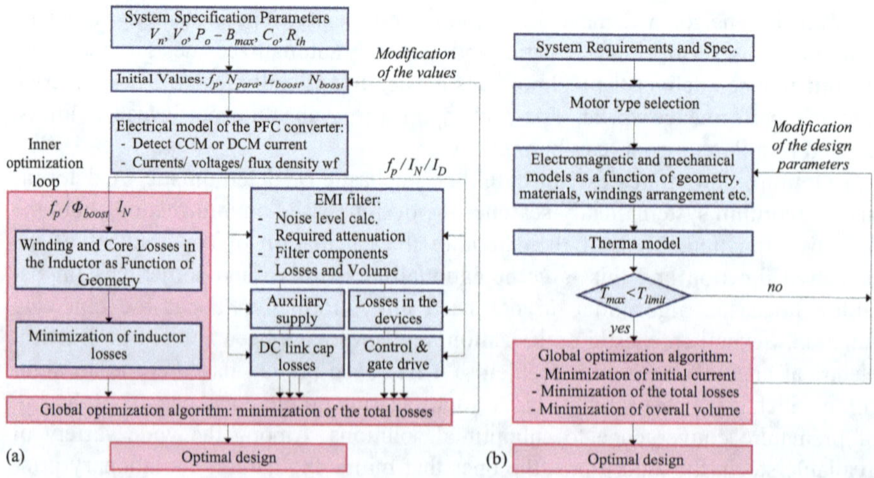

Figure 9.4 Automated optimization workflow for (a) PFC converter and (b) general electrical machine

adopted successfully in many fields of applications from telecommunications power supplies [4–6], to smart transformer application [7], inductive power transfer [8] and integrated power electronics [9], thus allowing to prototype and operate converters with efficiency beyond 99.5% and power densities of 15 kW/L.

Similar to DAPE, optimization methodologies are widely used in electrical machines to improve overall performance. Automated design workflow that optimizes one or multiple machine parameters has been applied to every machine structures. Permanent magnet (PM) volume minimization procedures have been developed for PM assisted synchronous reluctance (PMASR) machines [10]. An automated design methodology for squirrel cage induction machines is presented in, which consists of developing a comprehensive and general model suitable to design the whole machine, with particular focus on the rotor geometry optimization to improve starting torque and current, while maintaining the highest rated efficiency [11,12]. Figure 9.4(b) summarizes the general steps of an automated design workflow for electrical machines. This process begins with the specification requirements and the machine-type selection. A parametric model of the machine is then evaluated varying materials, geometry, windings arrangement, etc. until the optimization algorithm finds optimal combinations of design variables that minimize the objective functions.

While literature presents several examples of PE converter optimization and EM optimization, there are just few papers that address the problem of establishing an automated procedure for the optimization of an entire motor drive at system level. Design automation in IMDs (DAIMD) is, in practice, the only way to guarantee an optimal system performance rather than an assembly of individually optimized components into a system.

The main challenge in optimizing the whole drive at system level is that the design modules (e.g., motor design, converter design and control design) are strongly coupled, and it is not easy to derive the design and optimization flowchart.

Two main approaches to system level optimization can be followed: single level and multilevel optimization [13,14].

The workflow for a single level optimization is shown in Figure 9.5(b). There are five main steps:

- Step 1: Determine the system's requirements, objectives, and constraints, such as cost, weight, torque ripple, and efficiency.
- Step 2: Select motor type, converter topology, and control strategy in terms of system specifications. There are some interactions between them, e.g., some types of controllers fit the given types of motor better than the others, modulation and control strategy can be applied just with converter topology, etc. This step could be included in the optimization process considering the types of motor, converter and controllers as discrete values.
- Step 3: Parametric models for each component of the drive. For motor part, this step mainly includes electromagnetic design, material design, dimension design, mechanical design, thermal design, and so on. For power electronics and control, it mainly includes the design of control circuits and control algorithms.
- Step 4: Construct design optimization models for motor, controller, and the whole system. As drive systems are always high-dimensional and nonlinear design problems, intelligent algorithms, such as GA, DEA, and PSO algorithms, can be good choices in many situations. Therefore, the algorithm parameters should be determined in this step, such as genetic operators in GA. Once the optimization algorithm is chosen, first, the initial population is generated and evaluated. The optimal solutions are found with the following generations building up the Pareto-Front.
- Step 5: The last step consists of a system level validation and evaluation of performance that were not considered in the objective functions. This step consists of two parts. One is the evaluation of steady performance of motor, such as cost and efficiency. The other is the evaluation of dynamic performance of controller or the whole drive system, such as overshoot, settling time, torque ripple, and speed ripple.

While the single level workflow contains a unique optimization step, the multilevel approach presents multiple optimization steps in cascade following a hierarchical approach. Each level corresponds to an optimization step. To give an example, in Figure 9.5, a two-level optimization is presented, where first the motor is optimized and just second the converter and control strategy are optimized. At motor level, the aim is to optimize solely the motor model and evaluate the steady performance of motor, such as cost, weight, output power, and efficiency. Furthermore, motor characteristic parameters should also be calculated in this step, such as resistance, inductance, and magnetic flux of winding for the design optimization of next control level.

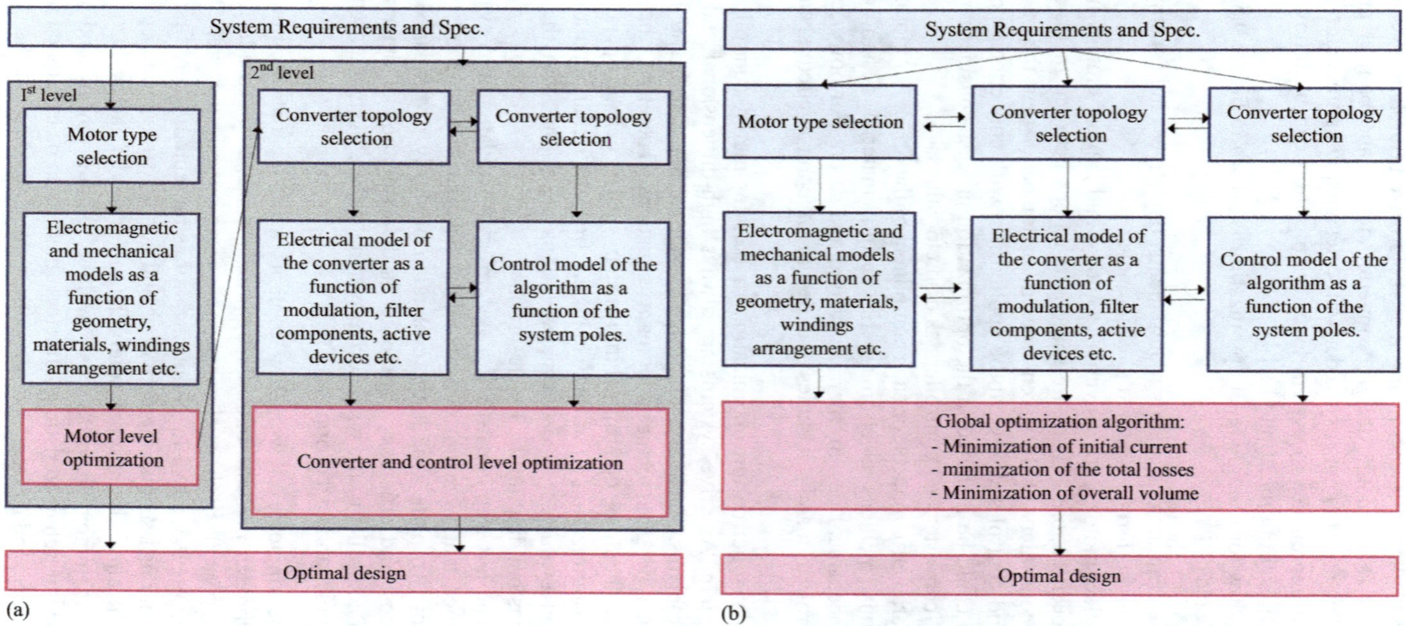

Figure 9.5 System level optimization of motor drive: (a) multiple level optimization and (b) single-level optimization

At control level, the aim is to optimize the control model and evaluate the dynamic performance of the whole system, such as overshoot and setting time. The single-level optimization is the only one that can be really considered a holistic approach; however, its computation cost is always very huge as the design problems of drive systems are generally high-dimensional and nonlinear with strongly coupled multidomain design and analysis. Different domains have different analysis techniques and software, and the computation cost of the whole system is very expensive.

For example, power electronic circuit analysis is needed in the control part, but the needed characteristic parameters of motor are generally calculated from FEM in the motor part; hence, the power electronic circuit design and electromagnetic design are strongly coupled in electrical drive systems. Furthermore, the computation cost of FEM analysis is usually very expensive too in many situations, particularly for some 3D flux complex structure motors. To overcome these problems, a multilevel design optimization method can be followed. In this case, the results of the first optimization are input for the following levels, thus reducing the degree of freedoms of the algorithm. A hierarchical decision is made in establishing the optimizations order.

IMD often represents a disruptive change in drive system design and manufacturing, thus ensuring by default an improvement of certain performance figures, such as power density and overall efficiency. However, as the integration level increases, this gain in performance is likely to reduce. To maintain a significant gain over time, automated and optimal design processes, as the ones described in this section, need to be applied to IMDs. In this way, there can be improvements not only in the traditional performance factors, efficiency and power density but through optimization multiple objectives can be pursued:

- robustness (design centering and tolerancing, and uncertainty)
- resiliency (over environment and compromise)
- reliability.

9.2.2 Modular and scalable design

This section discusses the potential advantage of approaching IMD design following a modular and scalable approach.

In power electronics, the so-called power electronics building block (PEBB) has been largely investigated as it enables modular and scalable system level design. PEBBs are integrated power modules consisting of power semiconductor devices, filters (e.g., decoupling capacitors, inductors), gate drive circuitry, sensors and control [15–17]. The circuit topology is usually a basic one (e.g., half bridge [18,19], H-bridge [20], bidirectional switch [21], three level T-type leg [22], etc.), which allows series or parallel connection to scale up power rating. They can be used in a wide range of power electronics applications, such as inverters for motor drives, converters for power processing equipment, and inverters for grid interface.

The design and fabrication of reliable and high-performance PEBB modules requires high level of packaging integration. Certainly, this poses several technological challenges, mainly because many heterogeneous components and materials have to share the same physical space, thus requiring a shift from planar structures to 2.5D/3D structures. However, since all the circuit elements are closely integrated, it is possible to design the PEBB to minimize both commutation loop inductance and gate driver circuit loop inductance. The reduction of stray inductances and capacitances within the power module results in the reduction of overcurrent and overvoltage stresses induced by high d*v*/d*t* and d*i*/d*t*. This aspect is particularly important when WBG devices are utilized because of the faster switching dynamics compared to Si devices.

Figure 9.6(a) shows an example of integrated H-bridge switching cell as proposed in [23]. In order to reduce manufacturing complexity, commercially available discrete packaged GaN enhancement-mode high-electron-mobility transistor (eHEMT) are used. These devices present all electrical connections on the top surface, while the thermal dissipation pad is on the other side. Performing a single soldering step, it is possible to connect drain, gate and source pads to a pre-manufactured PCB while the thermal pad is soldered on a ceramic DBC substrate. In a second step, the top PCB is populated with gate drive circuitry, decoupling capacitors, and a shunt resistor array for current measurement. The paper experimental results show that this solution is a good compromise between low parasitic inductance, good thermal management, and low capacitive coupling.

A higher level of integration can be achieved if the output filter is also integrated in the PEBB as presented in [24,25]. In this case, the topology is a simple half bridge with LC output low pass filter. In this case, bare SiC dies are used and directly sintered to the DBC and to a flex PCB for all the power connections, thus replacing wire bonds. The filter integration has positive implications on the system operation. First, placing filters close to the switching devices helps to contain HF noise at source thus reducing EMI issues on the sensing and control. Second, since inductors often take a large share of the total converter losses, by integrating them on the power module, see Figure 9.6(b), a better thermal management can be ensured thus giving the possibility of either reduce thermal stress on the filters or increase power density [26].

The PEBB design approach was applied to IMDs in [27,28]. The system comprises a 9-phase salient permanent magnet motor which is configured as 3 separate, 3 phase submachines. The nominal power of the designed motor is 80 kW at 30,000 rpm. At the nominal power, the current level flowing through each of the stator windings is

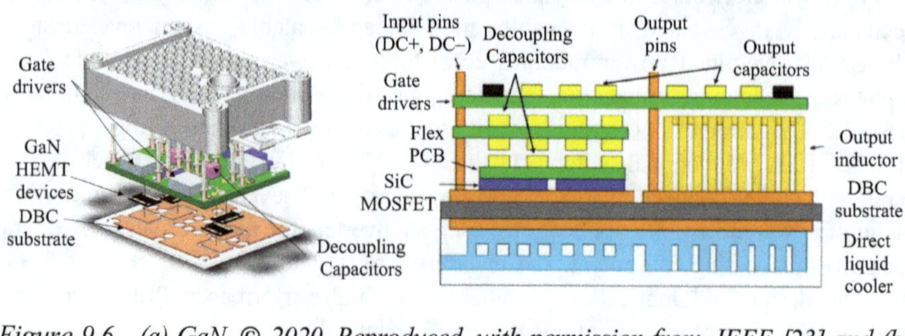

Figure 9.6 (a) GaN, © 2020. Reproduced, with permission from, IEEE [23] and (b) SiC all integrated, © 2020. Reproduced, with permission from, IEEE [24]

105 Arms. In the designed stator, this current level is equivalent to a current density of 28 A/mm^2 within the windings. Consequently, the thermal management of the motor windings is done using a direct liquid cooling system. Since the power electronic parts could be considered to be more susceptible to high temperatures, the same coolant, in this case de-ionized water, flows first through the power module cooling system. As shown in Figure 9.7(a), integrated PEBB design concept is used because each motor winding is connected to one basic custom power module. Each power module is mounted at the end of the stator windings with a coolant guide to ensure effective thermal management of the system. A modular concept is applied also in the control strategy. Each sub-machine has its own control hardware to manage the high-speed, low-level control. The low-level control is responsible for the current control of each sub-machine. Functions include the generation of PWM vectors, fast protection functions, DQ transformations, and current control. A higher-level supervisory system provides the speed and torque control for the whole drive together with other supervisory and communications functions. At hardware level, the control PCBs are implemented for the drive as shown in Figure 9.7(b). From the top, the first PCB in the stack is the master controller board which is connected to three local controller boards (auxiliary board). The second boards from the top are auxiliary controller boards which use the same microcontroller-FPGA structure as the master board. Four integrated ADCs are used to make a synchronous measurement of the 3 local motor phase currents and the local DC bus voltage. In this example, a modular design workflow has been adopted at different stage and levels, thus simplifying the system level design, while focusing at on each independent building block. At power level, each motor phase is supplied by a half bridge which is manufacture as a complete independent unit. Power semiconductors, decoupling capacitors, current sensing, and gate drivers are integrated in a 3D fashion starting from a single DBC tile. At control level, modularity is still present because each submachine of the 9-phase system has a local controller implemented with a separate control board. The clear advantage of this approach its reduction of design time when scaling up phase number or machine

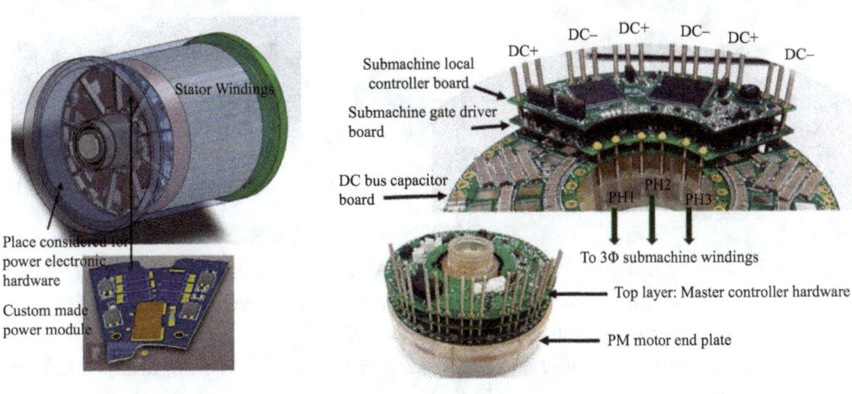

Figure 9.7 NGID: (a) power level building block and (b) control level building block, © 2020. Reproduced, with permission from, IEEE [27]

power because the presented structure could be replicated and just communication between the different boards would require a re-design.

In the previous example, the 9-phase electrical machine presented a modular magnetic structure, but from structural point of view, the three submachines shared the same physical space thus not giving the possibility to manufacture the machine with a modular process. On the other hand, integrated modular motor drives (IMMD) combine the benefits of both worlds of integrated PEBB and modular electrical machine manufacturing. In modular drives, the power electronics converter is divided into modules, each module is supplying a part of the stator windings. The modularity enhances the fault tolerance of the whole drive as each module can be controlled to contribute by certain amount to the whole torque by proper control. Moreover, the system can tolerate the fault in one or more module depending on the number of modules. Having several converter modules sharing the power of the whole drive improves the thermal performance of the power electronics devices as the surface area used by the converter modules increases and requires a smaller voltage rated devices as well. A segmentate winding structure is used in [29], the machine windings are split into several segments, and each segment provides a fraction of the total machine voltage, current, and power, see Figure 9.8(a). The power electronics modules are connected in series at the dc inputs and the motor winding common voltages are biased for different neutral points.

In [30], winding interleaving strategy is adopted for an axial flux permanent magnet machine. The additional degree of freedom of the multiphase control is used to reduce the stator current ripple and its harmonic distortion.

The main goal of a modular and scalable design approach is to be able to first optimize independent switching cells together with their motor segment, and in a second stage to perform a system level optimization. In this way, as shown in

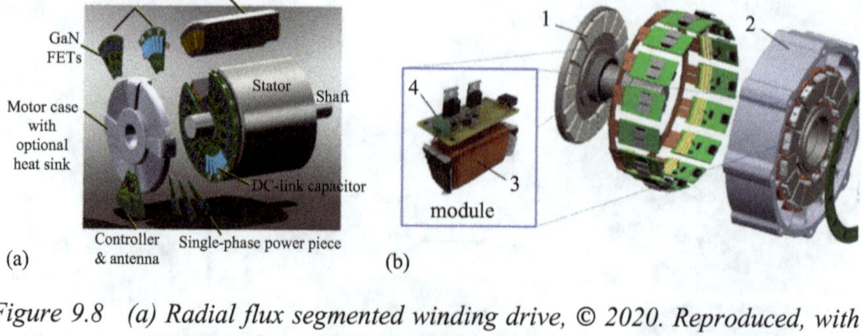

Figure 9.8 (a) Radial flux segmented winding drive, © 2020. Reproduced, with permission from, IEEE [29], (b) axial flux machine, © 2020. Reproduced, with permission from, IEEE [30], (1) two rotor discs with permanent magnets, (2) a motor housing and heatsink, and n *modules. A module consists of (3) a stator core element with a concentrated winding and (4) its dedicated power electronic converter, which is integrated in the machine housing.*

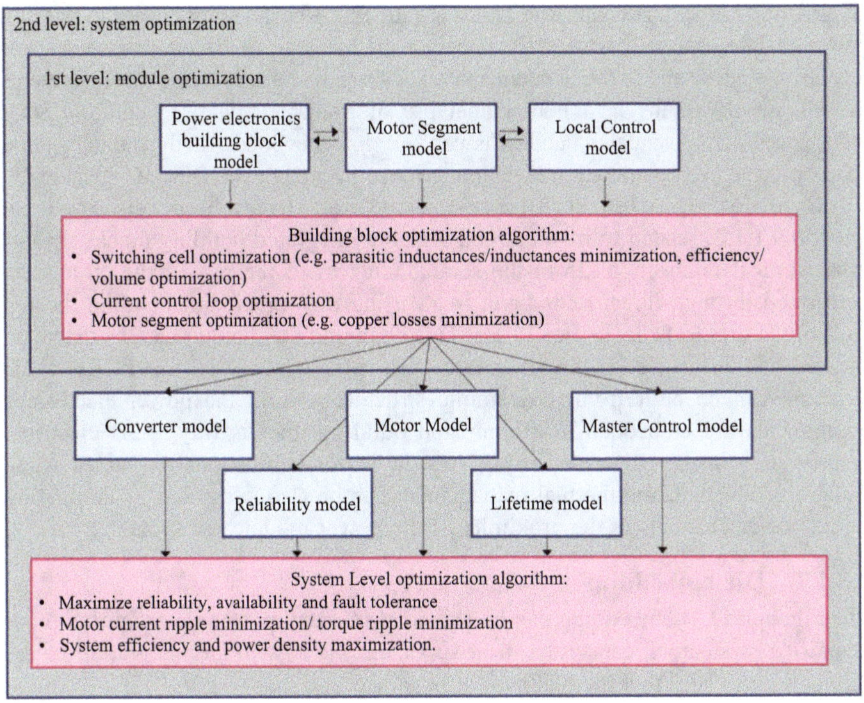

Figure 9.9 System level optimization of motor drive using building block approach

Figure 9.9, the second-level optimization can ignore circuit topology considerations and power electronic switch and associated inductors, capacitors, and other ancillary component selection, focusing just on a systems level related issue. The second level optimization is thus concerned with information and control data flowing in and out of PEBBs and between PEBBs in the larger power system design.

9.3 Emerging assembly technologies

As discussed in Section 9.1.1, IMD manufacturing poses a series of challenges that are often contraindicative, thus there is no technology that is capable to tackle them all. In this section, the authors propose a selection of the most significant and novel assembly technologies for PE converters and electrical machines, with the aim of highlighting in which way they can be employed to solve some of the difficulties encountered in IMD manufacturing, operation, and end-of-life.

9.3.1 PCB embedding

The commercial 2D power electronics module structures are usually based on wire bonding technology, thus posing two main limitations when they are used to package WBG semiconductors. First, the constraint of routing the switching cell on a single plane does not give the chance to minimize and decouple both gate loop and power loop parasitic inductances. Second, wire bonds are the first element that is likely to fail in a power module, especially when the temperature are increased to exploit the advantage of SiC devices compared to the Si counterpart. For these reasons, applying these structures to IMD would results in a non-optimal use of available volume, significant stray inductances and potential reliability issues due to wire bonds. Different integration solutions have been proposed in literature to move from 2D to 2.5D or 3D structures.

Examples of DBC/PCB hybrid structures have been discussed in Section 9.1.1.2, related to modular and scalable integrated building blocks. In this case, vertical connections from the semiconductors to the rest of the circuit are performed through direct or indirect (e.g., using pins or pillars) bonding the top pads of the dies to PCBs or flex PCBs. This integration techniques can be extended further, and instead of having separate parts e.g., DBC, dies, top PCBs, a 3D structure can be built from the ceramic structure and all the power electronics components are embedded in a laminated build-up. In this way, 2.5D electrical connections can be performed using vias, layers of laminations are added when isolation is needed, and thermal management can be also integrated by laminating liquid cooler channels in the proximity of the power module hot spots.

9.3.1.1 Die embedding

Even though PCB embedding can be extended to many power electronics components (e.g., inductors, capacitors, heat sinks, etc.) it is with the integration of the semiconductor dies that the real benefit is gained.

Literature presents just few examples of this emerging technology [31–33], and most of them are performed in-house using prototype-scale equipment, thus leading to slightly different processes depending on the type and accuracy of the equipment. However, the main manufacturing steps, which are described here, are common to all the examples. As shown in Figure 9.10(a), in manufacturing steps 1, the dies are first

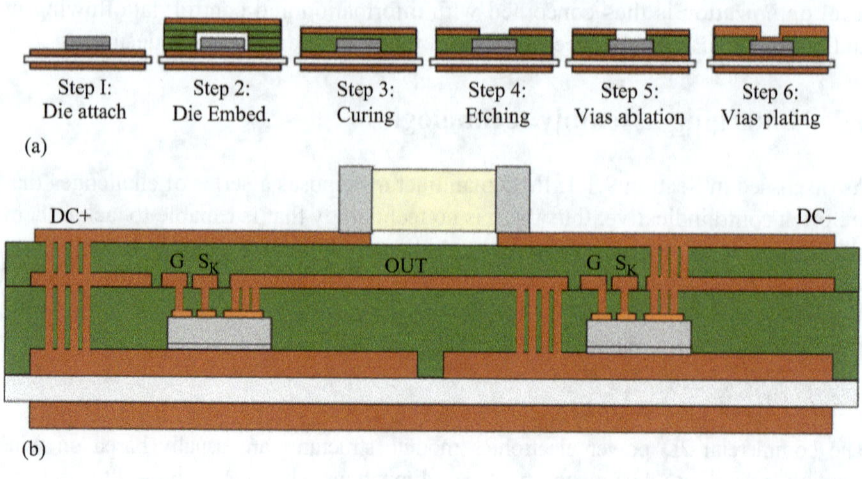

Figure 9.10 PCB die embedding: (a) manufacturing steps and (b) PCB embedded half bridge cell

bonded down onto a substrate, or sometimes copper lead frames. PCB laminate material is then bonded to the substrate embedding the semiconductors (steps 2 and 3). The lamination is topped off with a sheet of copper foil which is patterned by means of photolithography masking and chemical etching (step 4). To make the electrical connections to the devices blind vias are ablated down to the die pads (step 5). The last step (step 6) of the die embedding process consists of electroplating to deposit copper onto the walls of the via, the die pads and the copper foils to form a blind via. Close integration of other circuit components, such as decoupling capacitors or gate driver circuit, is enabled by introducing additional lamination material on top of the first layer. To give an example, in Figure 9.10(b) an integrated half bridge is shown where decoupling capacitor is integrated on the second layer. In this way, by closely integrating capacitors and gate drivers, optimized commutation loop and drive loop can be implemented. In [34], an IGBT-based P2P module (1.2 kV – 25 A) was developed using a laminate embedding technology known as p2 Pack developed by Schweizer Electronic AG. Its dynamic behavior is characterized by double pulse test. Experimental results obtained from this test are compared against the one obtained using a conventionally packaged counterpart (e.g., same device ratings, packaged in a DBC structures with wire bonds). By driving the half bridges using the same circuit, from the turn-OFF and turn-ON di/dt overshoot, it is possible to compare the parasitic inductances, concluding that a reduction of 60 nH can be achieved using die embedding.

By removing wire-bonds, die embedding can result in a more reliable a rugged structure. However, literature lacks reliability studies on these structures, and the presence of different materials (e.g., different CTEs) rigidly encapsulated together poses some challenges. In [35], the authors show the effect of aging on 17 single IGBT embedded demonstrators. It is worth noting that, in this structure, no ceramic substrate is used and the device is bonded on thick copper layer, which effectively constitutes the top layer of insulated metal substrates that are directly liquid cooled on the bottom layer. By using a PCB material made of epoxy resin, glass reinforcement and ceramic fillers, the authors show that after 72,000 active power cycling measurements (temperature range: 80 K, maximum temperature: 120°C, minimum temperature: 40°C), no increase of the collector–emitter–voltage is detected as well as there is no increase of the thermal resistance. However, after thermal cycling (2,500–3,000 temperature shocks from −40°C to 125°C in 30/30 min intervals), optical microscopy on a sample cross-section delamination between glass fibers and epoxy resin. These reliability issues related to the laminate material-led academic research to investigate new materials that can substitute the traditional FR4, in particular, in the higher temperature range were SiC devices can operate. In [36], bismaleimide-triazine (BT) laminate (Mitsubishi Gas Chemical, CCL-HL832NSF double-sided copper-clad laminate) and prepreg (Mitsubishi Gas Chemical, GHPL-830NSF) with high-temperature stability are considered for embedding of SiC MOSFET. These materials proved to have three main advantages compared to epoxy-based FR4: (1) thermal stability under the high temperature of 300°C, (2) breakdown voltage higher than 1.2 kV, and (3) CTE of 5.3 ppm/°C, which match with CTE of 4H-SiC.

9.3.1.2 Cooling system embedding

The PCB lamination process offers the possibility to closely integrate the thermal management to heat sources, such as power semiconductor devices, inductors, and microcontrollers. In literature, two main types of cooling system have been proposed exploiting this concept: direct liquid cooling and mini heat pipes.

In [37], an example of direct liquid cooling is given. Following conventional PCB manufacturing steps, jet-impingement cells are created within the laminates that are directly embedded with a DBC substrate. Figure 9.11(a) shows the topside represented by the ceramic substrate and the inlet/outlet channels for the liquid on the backside of the laminate. After assembling the spigots, the cross-section image of the cooler is shown in Figure 9.11(b). The liquid coolant is directly sprayed on localized areas of the ceramic substrate, thus giving the possibility to direct cool the semiconductor dies, eliminating several thermal interfaces and reducing the risk of hot spots. Experimental results confirm a very low thermal resistance during the cooling transient at different flow rates.

Thanks to the elimination of baseplate and heatsink, direct liquid cooling allows to reduce volume of the thermal management system. However, in such an integrated system, the spigots/connectors, connection pipes and pump constitute the larger part of the volume, reducing the total power density. Heat pipes-based cooling systems, on the other hand, are closed loop system and they result in a significant save of space. The operating principle of heat pipes is depicted in Figure 9.12(a). A closed metal structure contains both the liquid and the vapor phase of a coolant. The region close to the heat source (evaporator) is where the coolant absorbs heat and evaporates, while the region where the coolant gives away heat to the ambient and condensates is kept close to the heatsink (condenser). In between these two regions, the coolant needs to move naturally without exchanging heat with the outside (adiabatic chamber). As indicated by the red arrows, once the hot components make the coolant vaporize, the vapor moves naturally toward the condenser through the adiabatic section by difference in pressure. After condensation, the liquid needs to be forced back to the heat source. To make this

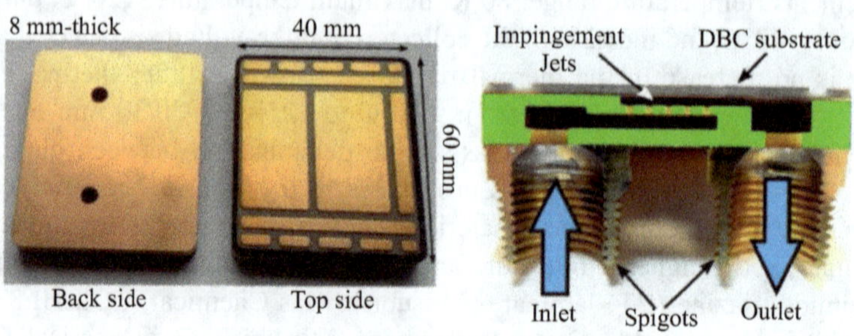

Figure 9.11 Image of the integrated cooler: (a) with a DBC substrate on top and holes that correspond to the inlet and outlet of the coolant at the backside and (b) cross-section image of the integrated cooler showing the impingement jets, © 2020. Reproduced, with permission from, IEEE [37]

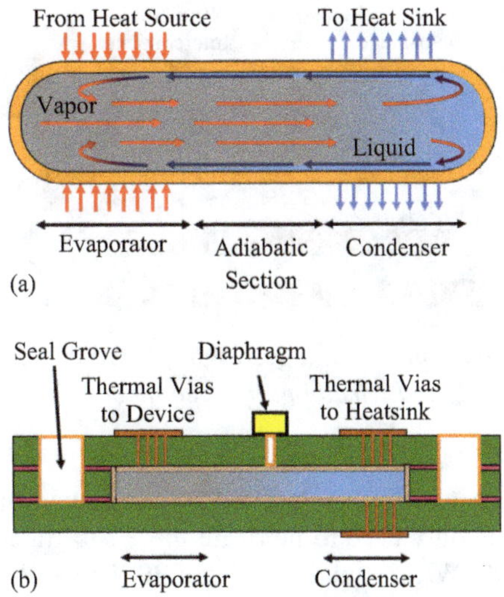

Figure 9.12 (a) Conventional heat pipe structure and (b) PCB embedded cooling system based on laminated heat pipe

happen, the walls of the adiabatic section present a structure, usually a wick, that enhances the capillary pressure caused by the difference in curvature of the liquid menisci. In [38], a heat pipe structure that can be embedded in a PCB is proposed. As shown in Figure 9.12(b), the pipe is created by laminating and plating sequentially FR4 layers, thus resulting in an embedded cavity – fully metalized – inside the PCB. Heat source and condenser are thermally connected to the pipe walls through thermal vias. Instead of a wick structure, the adiabatic chamber walls present axially oriented capillary microgrooves. Dry film lithography selective plating is used to fabricate the microgrooves, thus resulting in a minimum grove width of 50 μm with aspect ratio close to one.

9.3.2 High-temperature packaging techniques

The application of WBG devices in power electronics application enables the movement toward higher operating temperatures. Si-based devices are limited at 150°C of junction temperature, while GaN and SiC devices can withstand up to 250–300°C. However, while the semiconductors characteristics allow them to operate at high-temperature operation without deteriorating their electrical characteristics, conventional power electronics packaging materials cannot operate reliably at such high temperatures. Three aspects of the packaging process need to be furtherly explored, die-attachment, interconnection, and encapsulation.

9.3.2.1 High-temperature die attach

As shown in Figure 9.13, die-attachment is the process in which the semiconductor chips are electrically and thermally connected to the substrates. The most common

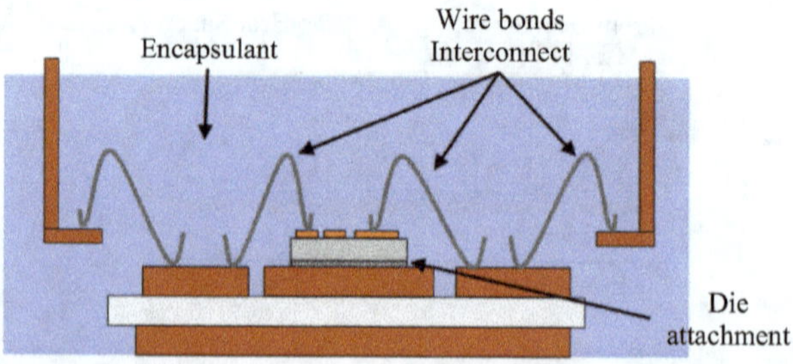

Figure 9.13 Power electronics packages highlighting encapsulant, die attachment, and wire bond interconnect

substrate is ceramic direct bonded copper (DBC) substrates. AlN and Al_2O_3 are the ceramic materials commonly used to fabricate these substrates. AlN shows a thermal conductivity of 180 W/mK, while Al_2O_3 of 30 W/mK. In terms of CTE, with 7.4 ppm/K for Al_2O_3 and 4.6 ppm/K for AlN, they both match well Si (2.6 ppm/K), SiC (4.2 ppm/K), and GaN (5.6 ppm/K), which reduces the thermomechanical stresses produced during assembly and high-temperature operation.

Traditionally, lead-free solders, such as $Sn96.5Ag_3Cu0.5$ and Sn96.5Ag3.5, have been used as alternative to lead solders in Si-based power modules. However, since their melting temperature is around 220°C, these solders are not suitable for high-temperature applications. In order to enable SiC- and GaN-based modules to operate closer to their junction temperature limit, die attachment materials need to possess the following properties [39]:

1. High thermal conductivity to improve heat exchange to the substrate.
2. CTE matching the CTE of substrates and semiconductor devices to reduce the thermomechanical stress induced by temperature cycling in operation and manufacturing.
3. Compatibility with standard die and substrates metallization and surface finishing.
4. High melting point temperature, to prevent melting during operation temperature peaks or during following: manufacturing steps.

Physical properties of the state-of-the-art die attachment materials are summarized in Table 9.1.

As previously discussed, because of their relatively low melting point, lead free solders, such as Sn96.5Ag3.5, cannot be used in high-temperature applications.

On the other hand, gold-based eutectic die-attach systems, such as Au80Sn20 and Au88Ge12, can be used as high-temperature technologies because their melting point is, respectively at 280°C and 356°C. When used on coated surfaces, they do not require flux to avoid oxidation, thus resulting in a more uniform and less porous bond line. However, they the price to pay for such a high melting point is the elevated process temperature and high elastic modulus. The combination of these two properties results in significant thermomechanical stresses during manufacturing.

Table 9.1 State-of-the-art die attachment systems

Materials	Melting *T.* (°C)	Process *T.* (°C)	Thermal cond. (W/mK)	CTE (ppm/K)	Elastic modulus (GPa)
Sn96.5Ag3.5	221	251	78	22	56
Au80Sn20	280	310	57	16	69
Au88Ge12	356	386	44	12	74
AuSn SLID (intermetallics)	650	>232	–	–	76–88
CuSn SLID (intermetallics)	400	>232	–	–	118–134
AgSn SLID (intermetallics)	750	>232	40–60	–	46–50
Ag sintering	961	230–300	100–240	18–23	9–60

Source: © 2020. Reproduced, with permission from, IEEE [39]

An interesting alternative to eutectic die-attach systems are off-eutectic intermetallics bonding techniques. They are known with several names, the two most common are solid liquid phase diffusion (SLID) or transient liquid phase (TLP) systems [40,41]. This technique utilizes two metals, a high-temperature melting metal (e.g., Au, Cu, Ag) and one low-temperature melting metal (e.g., Sn, In). Usually, the high-temperature metals are electroplated on both substrate and dies with layer thickness around 3–5 μm. The low-temperature metal is present as part of a preform foil (e.g., Au80Sn20, AgSn) that is sandwiched between die and substrate, or it is deposited by thin-film deposition techniques such as evaporation and sputtering. The applied processing temperature is higher than the melting point of the low melting point metal, and new intermetallic compounds (IMCs) are formed (e.g., ζ (Au5Sn) and δ (AuSn) phases, η (Cu6Sn5) and ε (Cu3Sn), ε (Ag3Sn) and ζ (Ag85Sn15)). The solidification is isothermal, and the final joint has a higher melting point than the processing temperature. The most interesting solution for power electronics applications is Ag–Sn–Ag because of the mechanical properties of Ag3Sn phase compared to Au5Sn, Cu3Sn, or Cu6Sn5 intermetallic, which is less brittle thus less keen to cracks propagation.

Finally, silver sintering is certainly the most adopted technology for high temperature power electronics application. Sintered silver layers have physical properties that are very close to the bulk silver properties. The sintering process can be carried out with several different procedures, based on the type of equipment and type of Ag paste. Nanoparticle-based Ag paste can be processed at relative low temperature (300 °C) [42] and without the addition of external pressure, while microparticle-based Ag paste are used in pressure assisted [43] process (250°C and variable pressure from 5 MPa to 40 MPa). With pressure-assisted silver sintering, as the pressure increases, the thickness and the porosity of the sintered layer decrease because densification is facilitated, thus resulting in significantly high shear strength up to 65 MPa [44].

9.3.2.2 High-temperature interconnect

Once the die is attached to the substrate, electrical connections need to be carried out from the top pads of the devices (e.g., gate and source pads in vertical power

MOSFETs, gate and emitter pads in vertical power IGBTs, gate source and drain pads in GaN on Si horizontal HEMT) to the substrate and to the external connections. As shown in Figure 9.13, in traditional power modules, planar interconnects are carried out using ribbon or wire bonding. However, since wire bonds are often the most sensitive component during power and thermal cycling, several approaches have been investigated to manufacture wire-bond less structures. Moving towards (3D) integration technologies rather than a planar integration approach is a requirement also for future WBG packages in which parasitic inductances need to be minimized. *Flexible PCB* [24,25,27,28] (see Figure 9.14(a)), *Through metal posts DBC* [19,45,46] (see Figure 9.14(b)), *flip chip with solder balls* [47–49] (see Figure 9.14(c)) are some of the technologies that have been proposed in literature.

- *Flexible PCB* interconnections have been discussed in Section 9.1.1.2 as an example of power electronics build block enabling technology.
- *Through metal posts DBC interconnects* consist of bonding the semiconductor dies between a two DBC sandwich structures. While the bottom pad is directly bonded, to facilitate the connection to the smaller top pads, thin metal posts are interposed between the die and the top interconnect substrate.
- Alternatively, the die top connections can be attached via *flip-chip with solder ball* interconnects the bottom substrate, while the bottom connection is connected to the substrate using a 3-D metal connector.

Regardless the adopted technologies, for high-temperature operation, the interconnects need to exhibit the following properties:

1. Low electrical resistivity to increase current density. Additionally, it is important that the series stray resistance of the interconnect has a temperature dependence that is negligible compared to the internal resistances of the device (e.g., internal gate resistance, ON resistance).
2. Withstand high-temperature getting oxidized. If the oxidation layer compromises the effective cross-section, current capability of each interconnect is reduced.
3. CTE matching the CTE of the substrates to reduce the thermomechanical stress induced at the bonding area to the semiconductor pads.

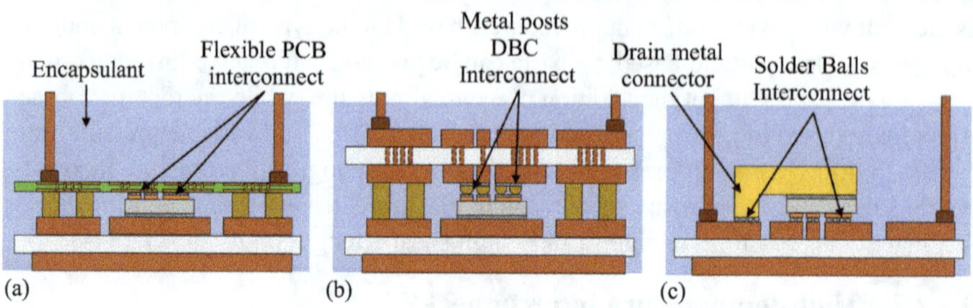

Figure 9.14 Examples of (a) flex interconnects, (b) through metal post DBC interconnect, and (c) flip chip solder balls interconnect

Table 9.2. State-of-the-art wire bonding materials

Material	Wire diameter (μm)	Melting point (°C)	Electrical conductivity (10^{-5} Ω cm^{-1})	Young's modulus (GPa)	CTE (ppm/K)
Al	25–300	660	2.3–2.8	68	24
Au	25–250	1,064	4.5	73	14.2
Ag	25	961	6.3	76	19
Pt	250	1,769	0.94	157	8.8

Wire bonding is an industry standard process, which has the advantage to be flexible and cost effective. While traditional aluminum wire bonds are not suitable to unlock the capability of WBG devices to operate at their maximum junction temperature, in literature several alternative materials have been investigated, such as Au, Ag, and Pt. Table 9.2 gives the details of various interconnection materials.

Gold and platinum wires are interesting because of their high melting temperature, resulting in a reliable operation in a wide range of temperatures. However, their application is limited to low power application because of the practical difficulties in sourcing commercial wires with thick diameters. Additionally, due the material cost Au and Pt, compromise one of the main advantages of the wire bonding process, namely cost effectiveness. Pt has also the disadvantage of a high young modulus resulting which means that high energy is required to deform and bond it.

Thin Ag wires (25 μm) are commercially available and can be used in the packaging of SiC and GaN-based modules, because of the good electrical and thermal properties. As discussed in [50], electro migration of silver causes interconnect property degradation beyond 350°C.

Due to the intrinsic short electrical connections, all the cited systems for 3D integration technologies do not present limitations in current density. However, when using flexible PCB interconnects the copper layer is not thicker than 0.35 μm, thus requiring larger area of copper. As operating temperature increases, if the appropriate die attachment material is selected, thermomechanical stresses can become the limiting factor for a reliable operation. When through metal posts DBC interconnects or flip chip with solder balls are used, CTE matching can be attempted, however due to the high elastic modulus of substrates and metal drain connector, the module results in a rigid structure that cannot accommodate the small deformations. On the other hand, even though by using a flexible PCB a heterogeneous materials is introduced which can cause CTE mismatch, the flex can accommodate significant deformation without causing stress on the device pads.

Although multilayer ceramic substrates, such as LTCC and ceramic PCBs are commercially available, they are not suitable for high-power applications due to limited via sizes and Cu thicknesses.

9.3.2.3 High-temperature encapsulants

The final step in power module assembly is encapsulation. The encapsulant material has the objective of providing protection against mechanical stress, electrical breakdown, and chemical erosions. Five types of encapsulants, including *conformal coatings*, *underfills*, *molding compounds*, *potting compounds*, and *glob tops*, can be identified among the commercial solutions.

- *Conformal coating* is a layer of polymer material sprayed or deposited on an electrical circuit surface for environmental protection and electrical isolation.
- *Underfills and molding compounds* are both rigid encapsulants, which offer in addition to environmental protection and electrical isolation, also reduction and redistribution of mechanical stresses caused by CTE mismatch.
- *Potting compounds and glob tops* are used to protect semiconductor dies from electrical breakdown and chemical erosion. While potting compounds are usually applied into an outer case of metal or rigid plastic, which provides mechanical protection against external forces, glob tops do not require a dam around the circuit components.

Recommended properties for high-temperature encapsulants are:

1. High dielectric strength and small electrical permittivity. These properties need to be stable even at high temperature.
2. Low moisture absorption to prevent water and vapor to compromise encapsulant properties.
3. Low viscosity to facilitate de-airing during potting step.
4. CTE matching those of the components.
5. Low elastic modulus to prevent that CTE mismatch causes thermomechanical stresses.
6. High glass transition temperature (T_g) to maintain nominal CTE and elastic modulus even at higher operating temperatures.

Guidelines for the selection of encapsulant for high-temperature applications can be found in Table 9.3.

Table 9.3 Guidelines for high-temperature encapsulant selection

Properties	**Conformal coating**	**Underfill**	**Molding compound**	**Potting compound and glob top**
Break down field (E_{BD})	20 kV/mm	>10 kV/mm		
Relative permittivity (ε_r)			<5	
Glass transition temperature (T_g)	N/A	>250°C		N/A
Moisture Absorption	<1% weight gain @ 85°C and 85%RH			
Viscosity	N/A	<20 Pa-sec	Low at high temperature	<20 Pa sec
CTE	N/A	20–30 ppm/K	<20 ppm/K	N/A
Elastic modulus	Varies	4–10 GPa		Varies

Source: © 2020. Reproduced, with permission from, IEEE [51]

In epoxy-based encapsulants, to increase T_g beyond 250°C, aromatic molecule structures are introduced in the polymer chain to enhance stiffness of the chain. While benzene is already present in most commercial solutions, with the introduction of naphthalene and fluorine T_g of 260°C can be reached. Addition of ceramic fillers (silica or alumina) in the epoxy matrix improves the thermal conductivity of the encapsulant but does not produce any significant improvement in the glass transition temperature. The drawback of increasing T_g is that also the curing temperature and time increases, resulting in slower processing time or additional thermal stresses during manufacturing.

Polyimide (PI), bismaleimide (BMI), and cyanate ester (CE) show a $T_g > 250$°C; however, their CTE is beyond 50 ppm/K, while to guarantee stability with temperature good values for underfills are between 20 and 30 ppm/K and for molding compounds are usually smaller than 20 ppm/K. To reduce CTE, materials with low CTE such as SiO_2 (silica), Al_2O_3 (alumina), glass, and Si_3N_4 (silicon nitride) have been investigated in literature as fillers. Another candidate is ZrW_2O_8 (zirconium tungstate), which exhibits a negative CTE from 10 to 1,000 K. A three composite system (silica/ZrW_2O_8/CE) with a CTE of 12 ppm/K was tested in [52].

Traditional silicon silicone-based potting compounds are not thermally stable beyond 250°C because chemical processes, such as chain-scission and oxidation of hydrocarbon group embrittle the material. As indicator of degradation, the temperature at which a potting compound has a weight loss of 5% is normally used. Filler addition in the silicon matrix can be used in this case as well to restrain chain mobility by increasing crosslink density. To give an example, by including a 0.7 wt% of montmorillonite (natural rock composed of two tetrahedral sheets of silica sandwiching a central octahedral sheet of alumina) to vinyl silicon rubber the temperature at which there is a 5% of weight loss increase from 230°C to 360°C.

9.3.3 *Additive manufacturing*

As already discussed, an IMD is a complex system, and its conventional manufacturing combines, in the three-dimensional space, a large variety of materials/components each one produced with a different manufacturing technique. Integrating together all these different components/materials requires an additional and complex assembly step which is not trivial.

Additive manufacturing (AM), in principle, offers the capability of creating unconstrained three-dimensional of parts using a wide range of materials (e.g., plastics, ceramics, metal alloys), thus giving the possibility to approach the IMD manufacturing in a more integrated way. Other advantages for investigating AM in future IMD applications are the possibility of producing cost effectively rapid prototypes, and the efficiency in the material utilization which results in almost no waste process. To date, the major limitations to the diffusion of AM techniques is that 3D-printed parts often show physical properties different from their bulk counterpart. If future advances in this technology will overcome this limit, the result would be an improvement in the structural and functional integration of thermal management and power electronics into the electrical machine and its enclosure. Additionally, future development should cover also the transition from prototyping to medium/large scale manufacturing.

During the last decade, many processes have been listed as additive manufacturing technologies. The American Society for Testing and Standards (ASTM) classified these processes under seven types of AM techniques [53]: binder jetting, direct energy deposition, material extrusion, material jetting, powder bed fusion, sheet lamination, and vat photopolymerization. Among those, the processes that are most interesting IMDs applications are [54]:

- *Powder bed fusion* (see Figure 9.15(a)): It uses a laser or an electron beam to selectively transfer energy to the processed raw material. The processed material is in powder form, typically spread on a thin uniform layer. If the energy transfer causes fusion of the powder particles the process takes the name of selective laser melting (SLM). While the terms selective laser sintering (SLS) and direct metal laser sintering (DMLS) are used when particle diffusion take place. SLM can be used to process titanium alloys, stainless steel and aluminum. SLS is used for thermoplastics and ceramics, while DMLS can process any metal alloy.
- *Vat photopolymerization* (see Figure 9.15(b)): Photochemical processes are used to cause chemical monomers and oligomers to cross-link together to form polymers. When ultraviolet light is used to cure a photosensitive resin, the process is called stereolithography (SLA). Polymers (resins) with high thermal stability or enhanced mechanical performance can be used to produce complicated part with high accuracy. When the polymer is loaded with ceramic powder, SLA can be used to produce green state parts that can be sintered later in high temperature ovens.
- *Material extrusion* (see Figure 9.15(c)): A filament of paste or wire is fed into a heated extruder nozzle with a small diameter where it is cured/fused and deposited selectively. Fused deposition modeling (FDM) or fused filament fabrication (FFF) is the techniques that processes thermoplastic filament in the form of solid wires. Paste extrusion, on the other hand, starts from ceramic/binder or metal/binder pastes socked in syringes.

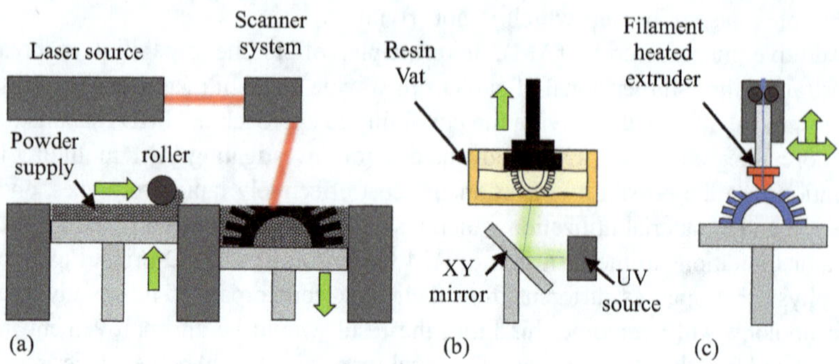

Figure 9.15 (a) Powder bed fusion, (b) stereolithography, and (b) material extrusion

Table 9.4 provides a summary the possible applications for each of the AM techniques listed above. AM technologies have been largely investigated for the realization of *advanced thermal management* solutions alternative to the industry standard. The main advantage of this technology is the capability of producing complex shapes that are difficult or impossible to manufacture using traditional methods in a cost-effective manner. For example, in [55] a Direct Winding Heat eXchanger (3D-DWHX) is proposed, which is installed in direct contact with the concentrated windings of the electrical machines. Material extrusion and FDM are used to process different alternative plastics. The compromise on the choice of the material was between thermal conductivity, glass transition temperature and dielectric strength. Polycarbonate (PC) can withstand up to 15.7 kV/m and has glass temperature of 112°C but has poor thermal conductivity. By loading the PC structure with aluminum both glass transition temperature and thermal conductivity are increased, respectively at 120°C and 1W/mK. AM techniques have been also investigated to enable the application of jet impingement cooling to power electronics application, which would be impractical with traditional manufacturing for the complexity and small size of the components. In [56], an air jet impingement cooler is manufactured using SLA. The air is delivered to the active devices surfaces by curved channels with a continuous reduction of the flow area to compensate the pressure drop. This strategy allowed to the achieve a convective heat transfer coefficients ranging from 5.2 to 17 kW/(m^2·K).

Table 9.4 Possible applications and characteristic of the discussed AM methods

AM process	Materials	Applications	Accuracy
Powder bed fusion	• Copper • Silicone steel • Aluminum • Other metal alloys • Ceramics • Ferrites	• Machine windings, conductive tracks • Machine magnetic cores • Structural and thermal parts • Magnets • Isolators and cooling • Inductor and transformer cores	• SLS: • Layer thickness: 0.1–0.15 mm; • XY res: 0.75–1.25 mm • SLM: • Layer thickness: 0.02–0.038 mm; • XY res: 0.3–0.41 mm
SLA	• High temperature resins • Ceramic loaded resins • Clear resins	• Insulators, thermal parts • Insulators, thermal parts • Structural parts and housing	Layer thickness: 0.025–0.300 mm; XY res: 0.025–0.200 mm
Material Extrusion	• Thermoplastic polymers • Metal/binder and ceramic/binder	• Structural parts and housing • Insulators, thermal parts	Layer thickness: 0.125–0.254 mm; XY res: 0.4–0.6 mm

Magnetic core for stator and rotor of electrical machine usually consists of a lamination stack of FeCo or FeSi sheets, the use of AM processes, on the other hand, results in a solid core. As discussed in [57], where powder bed fusion is applied to produce a switched reluctance motor rotor in FeSi, the solid nature deriving from the AM process lead to higher eddy current losses in this part, compared to commercial counterparts. However, since it is possible to make more geometrically complex, and magnetically optimized rotor the overall performance can be increased.

Powder bed fusion can be also used to manufacture *hard magnetic materials*. In [58], low energy density net shape alnico magnets are manufactured with this technique resulting in similar properties to commercially available solutions, but with the advantage of reducing material waste and higher shape flexibility. Similar process can be used for Nd-F-B magnets, as shown in [59].

While the properties of hard magnets manufactured with AM show similar properties to their bulk counterpart, it is an open challenge to additively manufactured *copper winding* with an electrical conductivity equal to bulk copper. International annealed cooper standard (IACS) establishes that the standard electrical conductivity value for commercially available copper is 58 MS/m. In literature attempts of manufacturing inductor and electrical machine copper windings using DMLS are reported. However, they show that the measured electrical conductivity varies between 50% and 70% of bulk copper [60,61]. For this reason, additively manufactured windings show a higher DC resistance. However, the possibility to improve the filling factor by using preformed winding, and the flexibility of varying cross-section along the winding, have been proved to be an asset to reduce AC resistance losses [62]. Thus, future development in electro-magnetic shape optimization tools for winding and in the additive manufactured process can lead to more efficient windings.

AM processes have been applied also to power electronics. Especially in the fabrication of magnetic components. The technique proposed in [63–65] is based on paste extrusion. Multiple pastes are stocked in different syringes, thus allowing to print at each layer both the magnetic core and the winding. Monomer binders are loaded with ferrite powder or silver powder to be used, respectively, for the core and the winding. After printing the inductor is at green state, and it is kept in shape by the monomer binders. It then needs to be sintered to establish burn the binder and establish a grain structure in the ferrite and in the silver parts. Co sintering of parts with different CTEs poses a problem in terms of stresses generated in the inductor. If the stresses are not balances, warpage, and cracks can occur. In [65], the silver paste is modified, adding a 5% of ferrite powder, to make the winding shrinkage profile closer to the one of the core.

Traditional *power electronics circuits* assembly paradigm, stacking of 2D PCBs, can be restrictive when the power electronics converter needs to be fitted in the IMD housing in proximity with the electrical motor. In Section 9.1.1.3, the use of die embedding has been discussed to move towards a 2.5D/3D integration that can allow to escape the traditional paradigm. However, AM can be an interesting alternative in this direction. Electronic circuit manufacturing requires the ability to process electrical insulations materials as well as conductors, with high accuracy even for small features (e.g., component pads, vias). SLA is the most accurate AM system; however, the use of a tank limits the process to just one material. Additionally, thorough cavities cleaning is required in between steps every time an electronically functional part is inserted. Material extrusion, on the other hand, can process multiple materials with the possibility of placing electronics components while printing. As shown in Table 9.5, this is the most adopted system in latest publications. In particular, in [66], to use thermosetting materials in a material extrusion process, UV-assisted 3D dispensing (3DD) is chosen as the AM technique. Nonconductive and conductive paste can be printed, and UV cured layer by layer creating *x*–*y* and – paths for the current. Final board curing at 80°C is necessary. During the process, cavity can be printed to accommodate integrated chips and electrically connect them to the rest of the circuit.

Table 9.5 Scientific publications regarding additive manufacturing applied to electronic circuit

Paper	**AM system**	**Print resolution – housing**	**Print resolution – conductive tracks**	**Resistivity conductive tracks**
Schmidt *et al.* 2021 [66]	3D dispensing	Min line width: 175 μm Min. layer height: 100 μm	Min track width: 200 μm	16 μΩ cm
Robles *et al.* 2019 [67]	FFF with micro-dispensing	Min line width: 120 μm Min. layer height: 45 μm	Track width: 75 μm	73×10^{-8} Ωcm
Billah *et al.* 2019 [68]	FFF with micro-dispensing	Min line width: N.A. Min. layer height: 254 μm	Bare copper wires, diameter: 405 μm	1.68×10^{-8} Ωcm
MacDonald *et al.* 2018 [69]	Multi3D with subtractive processes	Features: 254 by 175 μm Min. layer height: 254 μm	Copper foil thickness: 36 μm	NA
Bellacicca *et al.* 2018 [70]	FFF with micro-dispensing	Min line width: 100–400 μm Min. layer height: 50 μm	Min track width: 600 μm	1.68×10^{-8} Ωcm
Yu *et al.* 2017 [71]	FFF with liquid metal	Min line width: 500 μm Min. layer height: 200 μm	Cavity diameter: 2 mm	30×10^{-8} Ωcm
Shemelya *et al.* 2015 [72]	FFF with wire embedding	NA	Bare copper wires, diameter: 405 μm	1.68×10^{-8} Ωcm

This raises hope that additive manufacturing technology will eventually provide the key to developing drive electronics modules that can be conveniently designed to fit conformally within the contours of motor enclosures, such as inside the end bell assemblies. There are already some early reports of research efforts seeking to apply 3D printing technology to the fabrication of an inverter power stage. However, the results presented to date suggest that this R&D is still in a very early stage, and there is much to be done to take full advantage of the possibilities made available by additive manufacturing technology. This is intriguing in the context of future embedded IMD designs because it suggests that there might be clever techniques within reach for using these thermally conductive structural materials to enhance the transfer of heat from the power electronics into the motor housing where it can be efficiently dissipated. Much more work is necessary to determine whether such concepts will lead to cost-effective approaches for manufacturing embedded power electronics in the future.

References

[1] J. Biela, J. W. Kolar, A. Stupar, U. Drofenik, and A. M. Muesing, "Towards virtual prototyping and comprehensive multi-objective optimisation in power electronics," in *PCIM Eur. 2010; Int. Exhib. Conf. e Power Electron. Intell. Motion Power Qual.*, no. July 2016, pp. 5–27, 2010.

[2] A. Bindra and A. Mantooth, "Modern tool limitations in design automation: advancing automation in design tools is gathering momentum," *IEEE Power Electronics Magazine*, vol. 6, no. 1, pp. 28–33, 2019, doi: 10.1109/MPEL.2018.2888653.

[3] J. Biela, J. W. Kolar, and G. Deboy, "Optimal design of a compact 99.3% efficient single-phase PFC rectifier," *Conf. Proc. – IEEE Appl. Power Electron. Conf. Expo. – APEC*, pp. 1397–1404, 2010, doi: 10.1109/APEC.2010.5433412.

[4] J. Biela, U. Badstuebner, and J. W. Kolar, "Design of a 5-kW, 1-U, 10-kW/dm3 resonant DC–DC converter for telecom applications," *IEEE Transactions on Power Electronics*, vol. 24, no. 7, pp. 1701–1710, 2009, doi: 10.1109/TPEL.2009.2014377.

[5] J. Biela, U. Badstuebner, and J. W. Kolar, "Impact of power density maximization on efficiency of DC–DC converter systems," *IEEE Transactions on Power Electronics*, vol. 24, no. 1, pp. 288–300, 2009, doi: 10.1109/TPEL.2009.2006355.

[6] U. Badstuebner, J. Biela, D. Christen, and J. W. Kolar, "Optimization of a 5-kW telecom phase-shift dc-dc converter with magnetically integrated current doubler," *IEEE Transactions on Industrial Electronics*, vol. 58, no. 10, pp. 4736–4745, 2011, doi: 10.1109/TIE.2010.2103536.

[7] R. Bosshard and J. W. Kolar, "Multi-objective optimization of 50 kW/85 kHz IPT system for public transport," *IEEE Journal of Emerging and

Selected Topics in Power Electronics, vol. 4, no. 4, pp. 1370–1382, 2016, doi: 10.1109/JESTPE.2016.2598755.

[8] L. F. Costa, G. Buticchi, and M. Liserre, "Optimum design of a multiple-active-bridge DC–DC converter for smart transformer," *IEEE Transactions on Power Electronics*, vol. 33, no. 12, pp. 10112–10121, 2018, doi: 10.1109/TPEL.2018.2799680.

[9] A. Stratta, D. Gottardo, M. Di Nardo, *et al.*, "Optimal integrated design of a magnetically coupled interleaved H-bridge," *IEEE Transactions on Power Electronics*, vol. 37, no. 1, pp. 1–1, 2021, doi: 10.1109/TPEL.2021.3094025.

[10] M. Barcaro, N. Bianchi, and F. Magnussen, "Permanent-magnet optimization in permanent-magnet-assisted synchronous reluctance motor for a wide constant-power speed range," *IEEE Transactions on Industrial Electronics*, vol. 59, no. 6, pp. 2495–2502, 2012, doi: 10.1109/TIE.2011.2167731.

[11] A. Marfoli, M. Di Nardo, M. Degano, C. Gerada, and W. Chen, "Rotor design optimization of squirrel cage induction motor – Part I: results discussion," *IEEE Transactions on Energy Conversion*, vol. 36, no. 2, pp. 1280–1288, 2021, doi: 10.1109/TEC.2020.3020263.

[12] A. Marfoli, M. Di Nardo, M. Degano, C. Gerada, and W. Chen, "Rotor design optimization of squirrel cage induction motor – Part II: problem statement," *IEEE Transactions on Energy Conversion*, vol. 36, no. 2, pp. 1271–1279, 2021, doi: 10.1109/TEC.2020.3019934.

[13] G. Lei, T. Wang, J. Zhu, Y. Guo, and S. Wang, "System-level design optimization method for electrical drive systems – robust approach," *IEEE Transactions on Industrial Electronics*, vol. 62, no. 8, pp. 4702–4713, 2015, doi: 10.1109/TIE.2015.2404305.

[14] G. Lei, T. Wang, J. Zhu, Y. Guo, and S. Wang, "System-level design optimization method for electrical drive systems – a deterministic approach," *IEEE Transactions on Industrial Electronics*, vol. 61, no. 8, pp. 6591–6602, 2014, doi: 10.1109/TIE.2014.2321338.

[15] F. Ponci and H. L. Ginn, "Multi-agent control of PEBB based power electronic systems," *IEEE Power Energy Soc. 2008 Gen. Meet. Convers. Deliv. Electr. Energy 21st Century, PES*, 2008, doi: 10.1109/PES.2008.4596901.

[16] S. Rosado, F. Wang, and D. Boroyevich, "Design of PEBB based power electronics systems," in 2006 IEEE Power Eng. Soc. Gen. Meet. PES, 2006, doi: 10.1109/PES.2006.1709381.

[17] T. Ericsen and A. Tucker, "Power electronics building blocks and potential power modulator applications," in *IEEE Conf. Rec. Power Modul. Symp.*, pp. 12–15, 1998, doi: 10.1109/MODSYM.1998.741179.

[18] S. Haque, K. Xing, G. Q. Lu, D. J. Nelson, D. Borojevic, and F. C. Lee, "Packaging for thermal management of power electronics building blocks using metal posts interconnected parallel plate structure," in *Thermomechanical Phenom. Electron. Syst. -Proceedings Intersoc. Conf.*, pp. 392–398, 1998, doi: 10.1109/ITHERM.1998.689592.

[19] S. Haque, K. Xing, R. L. Lin, C. T. A. Suchicital, and M. Guo-Quan Lu, "An innovative technique for packaging power electronic building blocks using

metal posts interconnected parallel plate structures," *IEEE Transactions on Advanced Packaging*, vol. 22, no. 2, pp. 136–144, 1999, doi: 10.1109/6040.763184.

[20] C. Dimarino, I. Cvetkovic, Z. Shen, R. Burgos, and D. Boroyevich, "10 kV, 120 a SiC MOSFET modules for a power electronics building block (PEBB)," in *2nd IEEE Work. Wide Bandgap Power Devices Appl. WiPDA 2014*, pp. 55–58, Nov. 2014, doi: 10.1109/WIPDA.2014.6964623.

[21] C. Klumpner, F. Blaabjerg, and P. Nielsen, "Speeding-up the maturation process of the matrix converter technology," *PESC Rec. – IEEE Annu. Power Electron. Spec. Conf.*, vol. 2, pp. 1083–1088, 2001, doi: 10.1109/PESC.2001.954263.

[22] A. Deshpande, Y. Chen, B. Narayanasamy, Z. Yuan, C. Chen, and F. Luo, "Design of a high-efficiency, high specific-power three-level T-type power electronics building block for aircraft electric-propulsion drives," *IEEE Journal of Emerging and Selected Topics in Power Electronics*, vol. 8, no. 1, pp. 407–416, 2020, doi: 10.1109/JESTPE.2019.2952367.

[23] A. B. Jorgensen, S. Beczkowski, C. Uhrenfeldt, N. H. Petersen, S. Jorgensen, and S. Munk-Nielsen, "A fast-switching integrated full-bridge power module based on GaN eHEMT devices," *IEEE Transactions on Power Electronics*, vol. 34, no. 3, pp. 2494–2504, 2019, doi: 10.1109/TPEL.2018.2845538.

[24] J. Espina, B. Ahmadi, L. Empringham, L. De Lillo, and M. Johnson, "Highly-integrated power cell for high-power wide band-gap power converters," in *2017 IEEE 3rd Int. Futur. Energy Electron. Conf. ECCE Asia, IFEEC – ECCE Asia 2017*, pp. 146–150, Jul. 2017, doi: 10.1109/IFEEC.2017.7992433.

[25] B. Mouawad, J. Espina, J. Li, L. Empringham, and C. Mark Johnson, "Novel silicon carbide integrated power module for EV application," in *2018 1st Work. Wide Bandgap Power Devices Appl.* Asia, WiPDA Asia 2018, pp. 176–180, May 2018, doi: 10.1109/WIPDAASIA.2018.8734672.

[26] A. Stratta, B. Mouawad, M. Antonini, L. De Lillo, L. Empringham, and M. C. Johnson, "Heterogeneous integration of magnetic component windings on ceramic substrates," *IEEE Journal of Emerging and Selected Topics in Power Electronics*, pp. 1–1, 2020, doi: 10.1109/JESTPE.2020.2989805.

[27] S. Ullah, D. Winterborne, and S. M. Lambert, "Next-generation integrated drive: a high power density permanent magnet synchronous drive with flooded stator cooling," *Journal of Engineering*, vol. 2019, no. 17, pp. 4231–4235, 2019, doi: 10.1049/JOE.2018.8188.

[28] L. De Lillo, B. Ahmadi, L. Empringham, M. Johnson, J. Espina, and R. Abebe, "Next generation integrated drive, NGID: a novel approach to thermal and electrical integration of high power density drives in automotive applications," in *2018 IEEE Energy Convers. Congr. Expo. ECCE 2018*, pp. 1228–1232, Dec. 2018, doi: 10.1109/ECCE.2018.8558001.

[29] J. Wang, Y. Li, and Y. Han, "Integrated modular motor drive design with GaN power FETs," *IEEE Transactions on Industry Applications*, vol. 51, no. 4, pp. 3198–3207, 2015, doi: 10.1109/TIA.2015.2413380.

[30] L. Verkroost, J. Van Damme, D. V. Bozalakov, F. De Belie, P. Sergeant, and H. Vansompel, "Simultaneous DC-link and stator current ripple reduction with interleaved carriers in multiphase controlled integrated modular motor drives," *IEEE Transactions on Industrial Electronics*, vol. 68, no. 7, pp. 5616–5625, 2021, doi: 10.1109/TIE.2020.2992965.

[31] C. Yu, C. Buttay, E. Laboure, V. Bley, C. Combettes, and G. Brillat, "Comparison of topside contact layouts for power dies embedded in PCB," in *2016 6th Electron. Syst. Technol. Conf.*, pp. 1–6, Sep. 2016, doi: 10.1109/ESTC.2016.7764707.

[32] R. Reiner, P. Waltereit, B. Weiss, *et al.*, "Monolithically integrated power circuits in high-voltage GaN-on-Si heterojunction technology," *IET Power Electronics*, vol. 11, no. 4, pp. 681–688, 2018, doi: 10.1049/IET-PEL.2017.0397.

[33] S. Moench, R. Reiner, P. Waltereit, *et al.*, "PCB-embedded GaN-on-Si half-bridge and driver ICs with on-package gate and DC-link capacitors," *IEEE Transactions on Power Electronics*, vol. 36, no. 1, pp. 83–86, 2021, doi: 10.1109/TPEL.2020.3005621.

[34] D. J. Kearney, S. Kicin, E. Bianda, and A. Krivda, "PCB embedded semiconductors for low-voltage power electronic applications," *IEEE Transactions on Components, Packaging and Manufacturing Technology*, vol. 7, no. 3, pp. 387–395, 2017, doi: 10.1109/TCPMT.2017.2651646.

[35] R. Randoll, M. Asef, W. Wondrak, L. Böttcher, and A. Schletz, "Characteristics and aging of PCB embedded power electronics," *Microelectronics Reliability*, vol. 55, no. 9–10, pp. 1634–1639, 2015, doi: 10.1016/J.MICROREL.2015.06.072.

[36] F. Hou, W. Wang, T. Lin, L. Cao, G. Q. Zhang, and J. A. Ferreira, "Characterization of PCB embedded package materials for SiC MOSFETs," *IEEE Transactions on Components, Packaging and Manufacturing Technology*, vol. 9, no. 6, pp. 1054–1061, 2019, doi: 10.1109/TCPMT.2019.2904533.

[37] B. Mouawad, R. Abebe, R. Skuriat, *et al.*, "Cost effective direct-substrate jet impingement cooling concept for power application," in *PCIM Europe Conference Proceedings*, 2018, no. 225809, pp. 189–194.

[38] W. W. Wits and T. H. J. Vaneker, "Integrated design and manufacturing of flat miniature heat pipes using printed circuit board technology," *IEEE Transactions on Components, Packaging and Manufacturing Technology*, vol. 33, no. 2, pp. 398–408, 2010, doi: 10.1109/TCAPT.2010.2041929.

[39] A. A. Bajwa, Y. Qin, R. Reiner, R. Quay, and J. Wilde, "Assembly and packaging technologies for high temperature and high-power GaN devices," *IEEE Transactions on Components, Packaging and Manufacturing Technology*, vol. 5, no. 10, pp. 1402–1416, 2015, doi: 10.1109/TCPMT.2015.2468595.

[40] J. F. Li, P. A. Agyakwa, and C. M. Johnson, "Kinetics of Ag3Sn growth in Ag–Sn–Ag system during transient liquid phase soldering process," *Acta Materialia*, vol. 58, no. 9, pp. 3429–3443, 2010, doi: 10.1016/J.ACTAMAT.2010.02.018.

[41] T. A. Tollefsen, A. Larsson, O. M. Løovvik, and K. E. Aasmundtveit, "High temperature interconnect and die attach technology: Au-Sn SLID bonding," *IEEE Transactions on Components, Packaging and Manufacturing Technology*, vol. 3, no. 6, pp. 904–914, 2013, doi: 10.1109/TCPMT.2013.2253353.

[42] J. G. Bai, J. Yin, Z. Zhang, G. Q. Lu, and J. D. van Wyk, "High-temperature operation of SiC power devices by low-temperature sintered silver die-attachment," *IEEE Transactions on Advanced Packaging*, vol. 30, no. 3, pp. 506–510, 2007, doi: 10.1109/TADVP.2007.898628.

[43] Z. Z. Zhang and G. Q. Lu, "Pressure-assisted low-temperature sintering of silver paste as an alternative die-attach solution to solder reflow," *IEEE Transactions on Electronics Packaging Manufacturing*, vol. 25, no. 4, pp. 279–283, 2002, doi: 10.1109/TEPM.2002.807719.

[44] Y. Liu, H. Zhang, L. Wang, X. Fan, G. Zhang, and F. Sun, "Effect of sintering pressure on the porosity and the shear strength of the pressure-assisted silver sintering bonding," *IEEE Transactions on Device and Materials Reliability*, vol. 18, no. 2, pp. 240–246, 2018, doi: 10.1109/TDMR.2018.2819431.

[45] C. M. Dimarino, B. Mouawad, C. M. Johnson, D. Boroyevich, and R. Burgos, "10-kV SiC MOSFET power module with reduced common-mode noise and electric field," *IEEE Transactions on Power Electronics*, vol. 35, no. 6, pp. 6050–6060, 2020, doi: 10.1109/TPEL.2019.2952633.

[46] C. Dimarino, B. Mouawad, C. M. Johnson, *et al.*, "Design and experimental validation of a wire-bond-less 10-kV SiC MOSFET power module," *IEEE Journal of Emerging and Selected Topics in Power Electronics*, vol. 8, no. 1, pp. 381–394, 2020, doi: 10.1109/JESTPE.2019.2944138.

[47] S. Seal, M. D. Glover, and H. A. Mantooth, "3-D wire bondless switching cell using flip-chip-bonded silicon carbide power devices," *IEEE Transactions on Power Electronics*, vol. 33, no. 10, pp. 8553–8564, 2018, doi: 10.1109/TPEL.2017.2782226.

[48] S. Seal, A. K. Wallace, A. M. Dearien, C. Farnell, and H. A. Mantooth, "A wire bondless SiC switching cell with a vertically integrated gate driver," *IEEE Transactions on Power Electronics*, vol. 35, no. 9, pp. 9692–9701, 2020, doi: 10.1109/TPEL.2020.2974981.

[49] M. Montazeri, D. R. Huitink, A. Wallace, *et al.*, "Vertically stacked, flip-chip wide bandgap MOSFET co-optimized for reliability and switching performance," *IEEE Journal of Emerging and Selected Topics in Power Electronics*, vol. 9, no. 4, pp. 3904–3915, 2020, doi: 10.1109/JESTPE.2020.3032886.

[50] M. Mysliwiec, M. Guziewicz, and R. Kisiel, "Aspects of applying flip-chip technology for SiC power devices assembly," in *Proc. Int. Spring Semin. Electron. Technol.*, pp. 90–93, 2013, doi: 10.1109/ISSE.2013.6648221.

[51] Y. Yao, G.-Q. Lu, D. Boroyevich, and K. D. T. Ngo, "Survey of high-temperature polymeric encapsulants for power electronics packaging," in *IEEE Transactions on Components, Packaging and Manufacturing Technology*, vol. 5, no. 2, pp. 168–181, Feb. 2015, doi: 10.1109/TCPMT.2014.2337300.

[52] W. K. Goertzen and M. R. Kessler, "Three-phase cyanate ester composites with fumed silica and negative-CTE reinforcements," *Journal of Thermal Analysis and Calorimetry*, vol. 93, no. 1, pp. 87–93, 2008, doi: 10.1007/S10973-007-8804-2.

[53] ASTM standard (Designation: F2792 – 12a), "Standard Terminology for Additive Manufacturing Technologies," 2013. doi: 10.1520/F2792-12A.

[54] P. S. Ghahfarokhi, A. Podgornovs, A. Kallaste, *et al.*, "Opportunities and challenges of utilizing additive manufacturing approaches in thermal management of electrical machines," *IEEE Access*, vol. 9, pp. 36368–36381, 2021, doi: 10.1109/ACCESS.2021.3062618.

[55] D. J. Yoon, S. H. Lee, and K. W. Paik, "Effects of polyacrylonitrile anchoring polymer layer solder anisotropic conductive films on the solder ball movement for fine-pitch flex-on-flex (FOF) assembly," *IEEE Transactions on Components, Packaging and Manufacturing Technology*, vol. 9, no. 5, pp. 830–835, 2019, doi: 10.1109/TCPMT.2019.2909095.

[56] B. Kwon, T. Foulkes, T. Yang, N. Miljkovic, and W. P. King, "Air jet impingement cooling of electronic devices using additively manufactured nozzles," *IEEE Transactions on Components, Packaging and Manufacturing Technology*, vol. 10, no. 2, pp. 220–229, 2020, doi: 10.1109/TCPMT.2019.2936852.

[57] L. Gargalis, V. Madonna, P. Giangrande, *et al.*, "Additive manufacturing and testing of a soft magnetic rotor for a switched reluctance motor," *IEEE Access*, vol. 8, pp. 206982–206991, 2020, doi: 10.1109/ACCESS.2020.3037190.

[58] E. M. H. White, A. G. Kassen, E. Simsek, W. Tang, R. T. Ott, and I. E. Anderson, "Net shape processing of alnico magnets by additive manufacturing," *IEEE Transactions on Magnetics*, vol. 53, no. 11, pp. 1–6, 2017, doi: 10.1109/TMAG.2017.2711965.

[59] A. B. Baldissera, P. Pavez, P. A. P. Wendhausen, C. H. Ahrens, and J. M. Mascheroni, "Additive manufacturing of bonded Nd-Fe-B – effect of process parameters on magnetic properties," *IEEE Transactions on Magnetics*, vol. 53, no. 11, pp. 1–4, 2017, doi: 10.1109/TMAG.2017.2715722.

[60] N. Simpson and P. H. Mellor, "Additive manufacturing of shaped profile windings for minimal AC loss in gapped inductors," in *2017 IEEE Int. Electr. Mach. Drives Conf. IEMDC 2017*, Aug. 2017, doi: 10.1109/IEMDC.2017.8002337.

[61] F. Lorenz, J. Rudolph, and R. Wemer, "Design of 3D printed high performance windings for switched reluctance machines," in *Proc. – 2018 23rd Int. Conf. Electr. Mach. ICEM 2018*, pp. 2451–2457, Oct. 2018, doi: 10.1109/ICELMACH.2018.8506845.

[62] N. Simpson, D. J. North, S. M. Collins, and P. H. Mellor, "Additive manufacturing of shaped profile windings for minimal AC loss in electrical machines," *IEEE Transactions on Industry Applications*, vol. 56, no. 3, pp. 2510–2519, 2020, doi: 10.1109/TIA.2020.2975763.

[63] C. Ding, S. Lu, J. Moss, *et al.*, "Additive manufacturing of spiral windings for a pot-core constant-flux inductor," *IEEE Journal of Emerging and

Selected Topics in Power Electronics, vol. 8, no. 1, pp. 618–625, 2020, doi: 10.1109/JESTPE.2019.2934355.
[64] Y. Yan, J. Moss, K. D. T. Ngo, Y. Mei, and G. Q. Lu, "Additive manufacturing of toroid inductor for power electronics applications," *IEEE Transactions on Industry Applications*, vol. 53, no. 6, pp. 5709–5714, 2017, doi: 10.1109/TIA.2017.2729504.
[65] C. Ding, S. Lu, L. Liu, K. D. T. Ngo, and G. Q. Lu, "Additive manufacturing of hetero-magnetic coupled inductors," *IEEE Transactions on Components, Packaging and Manufacturing Technology*, vol. 11, no. 6, pp. 1028–1034, 2021, doi: 10.1109/TCPMT.2021.3083179.
[66] K. Schmidt, B. Polzinger, M. Metry, S. Koppe, and A. Zimmermann, "Hybrid additive manufacturing by embedded electrical circuits using 3-D dispensing," *IEEE Transactions on Components, Packaging and Manufacturing Technology*, vol. 11, no. 3, pp. 510–521, 2021, doi: 10.1109/TCPMT.2021.3054835.
[67] U. Robles, A. Kudzal, and R. C. Rumpf, "Automated hybrid 3-D printing of 3-D meandering interconnects," *IEEE Transactions on Components, Packaging and Manufacturing Technology*, vol. 9, no. 6, pp. 1184–1189, 2019, doi: 10.1109/TCPMT.2019.2909979.
[68] K. M. M. Billah, J. L. Coronel, M. C. Halbig, R. B. Wicker, and D. Espalin, "Electrical and thermal characterization of 3D printed thermoplastic parts with embedded wires for high current-carrying applications," *IEEE Access*, vol. 7, pp. 18799–18810, 2019, doi: 10.1109/ACCESS.2019.2895620.
[69] E. MacDonald, D. Espalin, D. Doyle, *et al.*, "Fabricating patch antennas within complex dielectric structures through multi-process 3D printing," *Journal of Manufacturing Processes*, vol. 34, pp. 197–203, 2018, doi: 10.1016/J.JMAPRO.2018.05.013.
[70] A. Bellacicca, T. Santaniello, and P. Milani, "Embedding electronics in 3D printed structures by combining fused filament fabrication and supersonic cluster beam deposition," *Additive Manufacturing*, vol. 24, pp. 60–66, 2018, doi: 10.1016/J.ADDMA.2018.09.010.
[71] Y. Z. Yu, J. R. Lu, and J. Liu, "3D printing for functional electronics by injection and package of liquid metals into channels of mechanical structures," *Materials and Design*, vol. 122, pp. 80–89, 2017, doi: 10.1016/J.MATDES.2017.03.005.
[72] C. Shemelya, F. Cedillos, E. Aguilera, *et al.*, "Encapsulated copper wire and copper mesh capacitive sensing for 3-D printing applications," *IEEE Sensors Journal*, vol. 15, no. 2, pp. 1280–1286, 2015, doi: 10.1109/JSEN.2014.2356973.

Index

www.ingramcontent.com/pod-product-compliance
Lightning Source LLC
LaVergne TN
LVHW011838200926
848484LV00002B/374
* 9 7 8 1 8 3 9 5 3 1 8 6 6 *